INTRODUCTION TO TISSUE ENGINEERING

IEEE Press
445 Hoes Lane
Piscataway, NJ 08854

IEEE Press Editorial Board
Tariq Samad, *Editor in Chief*

George W. Arnold	Mary Lanzerotti	Linda Shafer
Dmitry Goldgof	Pui-In Mak	MengChu Zhou
Ekram Hossain	Ray Perez	George Zobrist

Kenneth Moore, *Director of IEEE Book and Information Services (BIS)*

INTRODUCTION TO TISSUE ENGINEERING

Applications and Challenges

RAVI BIRLA
*Department of Biomedical Engineering,
Cullen College of Engineering,
University of Houston, Houston, TX*

IEEE Engineering in Medicine
and Biology Society, *Sponsor*

IEEE Press Series in Biomedical Engineering
Metin Akay, *Series Editor*

IEEE PRESS

Copyright © 2014 by The Institute of Electrical and Electronics Engineers, Inc.

Published by John Wiley & Sons, Inc., Hoboken, New Jersey. All rights reserved
Published simultaneously in Canada

No part of this publication may be reproduced, stored in a retrieval system, or transmitted in any form or by any means, electronic, mechanical, photocopying, recording, scanning, or otherwise, except as permitted under Section 107 or 108 of the 1976 United States Copyright Act, without either the prior written permission of the Publisher, or authorization through payment of the appropriate per-copy fee to the Copyright Clearance Center, Inc., 222 Rosewood Drive, Danvers, MA 01923, (978) 750-8400, fax (978) 750-4470, or on the web at www.copyright.com. Requests to the Publisher for permission should be addressed to the Permissions Department, John Wiley & Sons, Inc., 111 River Street, Hoboken, NJ 07030, (201) 748-6011, fax (201) 748-6008, or online at http://www.wiley.com/go/permission.

Limit of Liability/Disclaimer of Warranty: While the publisher and author have used their best efforts in preparing this book, they make no representations or warranties with respect to the accuracy or completeness of the contents of this book and specifically disclaim any implied warranties of merchantability or fitness for a particular purpose. No warranty may be created or extended by sales representatives or written sales materials. The advice and strategies contained herein may not be suitable for your situation. You should consult with a professional where appropriate. Neither the publisher nor author shall be liable for any loss of profit or any other commercial damages, including but not limited to special, incidental, consequential, or other damages.

For general information on our other products and services or for technical support, please contact our Customer Care Department within the United States at (800) 762-2974, outside the United States at (317) 572-3993 or fax (317) 572-4002.

Wiley also publishes its books in a variety of electronic formats. Some content that appears in print may not be available in electronic formats. For more information about Wiley products, visit our web site at www.wiley.com.

Library of Congress Cataloging-in-Publication Data:

Birla, Ravi, author.
  Introduction to tissue engineering : applications and challenges / Ravi Birla.
    p. ; cm. – (IEEE Press series on biomedical engineering)
  ISBN 978-1-118-62864-5 (hardback)
  I. Title. II. Series: IEEE Press series in biomedical engineering.
  [DNLM: 1. Tissue Engineering. QT 37]
  R857.T55
  610.28–dc23

2013035520

Printed in the United States of America

ISBN: 9781118628645

10 9 8 7 6 5 4 3 2 1

*This book is dedicated to:
My parents, Mom and Dad,
My gorgeous and supporting wife, Swati, and
My precious kids Aditya and Pooja*

1

CONTENTS

Preface xiii

Acknowledgments xv

List of Abbreviations xvii

Important Terminology and Concepts xxi

1 Introduction to Tissue Engineering 1

    1.1 Introduction to Tissue Engineering, 2
    1.2 Chronic Shortage of Donor Organs, 3
    1.3 The Tissue Engineering Paradigm, 4
    1.4 Definition of Tissue Engineering, 5
    1.5 Process of Bioengineering 3D Artificial Tissue, 9
    1.6 Design Principles for Tissue Engineering, 12
    1.7 Building Blocks of Tissue Engineering, 14
    1.8 Scientific and Technological Challenges, 15
    1.9 Functional Assessment of Artificial Tissue, 16
    1.10 Seminal Papers in Tissue Engineering, 18
    1.11 Applications of 3D Artificial Tissue, 20
    1.12 Two-Dimensional Versus Three-Dimensional Culture, 22
    1.13 Integration of Core Technologies, 22
    1.14 Growth of Tissue Engineering, 24
    1.15 Disciplines in Tissue Engineering, 26
    1.16 Tissue Engineering and Related Fields, 28
    Summary, 33

Practice Questions, 34
References, 35

2 Cells for Tissue Engineering   40

2.1   Cells and Tissue Engineering, 41
2.2   Cell Structure and Function, 43
2.3   The Dynamic Extracellular Matrix, 47
2.4   Cell Signaling, 48
2.5   Cellular Junctions, 50
2.6   Mammalian Tissue and Artificial Tissue, 52
2.7   Cell Sourcing, 52
2.8   The Cell Transplantation Process, 55
2.9   Cells for Cell Transplantation, 58
2.10 Mode of Action of Cells During Cell Transplantation, 59
2.11 Cell Transplantation and Tissue Engineering, 60
2.12 The Cell Culture Process, 61
2.13 Applications of Monolayer 2D Cell Culture, 64
2.14 Cell Culture Versus Tissue Engineering, 65
2.15 Introduction to Stem Cell Engineering, 66
2.16 Human Embryonic Stem Cells, 70
2.17 Induced Pluripotent Stem Cells, 71
2.18 Adult Stem Cells, 72
      Summary, 72
      Practice Questions, 73
      References, 74

3 Biomaterials for Tissue Engineering   84

3.1   Definition of Biomaterials, 85
3.2   Scheme for Biomaterial Development, 88
3.3   Historical Perspective on Biomaterials, 90
3.4   Tensile Properties, 92
3.5   Modulation of Tensile Properties, 95
3.6   Material Degradation, 97
3.7   Biocompatibility, 100
3.8   Biomimetic Biomaterial, 104
3.9   Classification of Biomaterials, 106
3.10 Biomaterial Platforms, 108
3.11 Smart Materials, 113
3.12 The Dynamic Extracellular Matrix, 114
3.13 Idealized Biomaterial, 116
      Summary, 118
      Practice Questions, 119
      References, 121

4 Tissue Fabrication Technology   130

    4.1 Introduction to Tissue Fabrication Technologies, 131
    4.2 Self-Organization Technology, 133
    4.3 Cell Sheet Engineering, 135
    4.4 Scaffold-Based Tissue Fabrication, 137
    4.5 Cell and Organ Printing, 140
    4.6 Solid Freeform Fabrication, 142
    4.7 Soft Lithography and Microfluidics, 143
    4.8 Cell Patterning, 145
    4.9 Idealized System to Support Tissue Fabrication, 148
    Summary, 149
    Practice Questions, 150
    References, 151

5 Vascularization of Artificial Tissue   156

    5.1 Introduction, 157
    5.2 Seminal Publications in Angiogenesis Research, 159
    5.3 Vascularization Defined, 160
    5.4 Molecular Mechanism of Vasculogenesis, 161
    5.5 Molecular Mechanism of Angiogenesis, 163
    5.6 Molecular Mechanism of Arteriogenesis, 164
    5.7 Therapeutic Angiogenesis, 166
    5.8 Tissue Engineering and Vascularization, 167
    5.9 Conceptual Framework for Vascularization During Artificial Tissue Formation, 169
    5.10 *In Vivo* Models of Vascularization, 172
    5.11 Idealized Vascularization Strategy for Tissue Engineering, 174
    5.12 Flow Chart and Decision Making, 176
    5.13 Biologically Replicated Vascularization Strategies, 179
    5.14 Biologically Mediated Vascularization Strategies, 181
    5.15 Biologically Inspired Vascularization Strategies, 184
    Summary, 186
    Practice Questions, 187
    References, 188

6 Bioreactors for Tissue Engineering   193

    6.1 Introduction to Bioreactors, 194
    6.2 Bioreactors Defined, 195
    6.3 Classification of Bioreactors, 197
    6.4 Design Considerations, 200
    6.5 Idealized Bioreactor System, 202
    6.6 Bioreactors and Tissue Engineering, 205
    6.7 Bioreactors for Mammalian Cell Culture, 207

6.8　Bioreactors for Scaffold Fabrication, 209
6.9　Bioreactors for Scaffold Cellularization, 212
6.10　Perfusion Systems, 215
6.11　Bioreactors for Stretch, 219
6.12　Electrical Stimulation, 221
　　　Summary, 226
　　　Practice Questions, 227
　　　References, 230

7　Tracheal Tissue Engineering　　　　　　　　　　　　　　**237**

7.1　Structure and Function of the Trachea, 238
7.2　Congenital Tracheal Stenosis, 240
7.3　Genetic Regulation of Tracheal Development, 241
7.4　Post Intubation and Post Tracheostomy Tracheal Stenosis, 243
7.5　Treatment Modalities for Tracheal Stenosis, 245
7.6　Design Considerations for Tracheal Tissue Engineering, 247
7.7　Process of Bioengineering Artificial Tracheas, 247
7.8　Tissue Engineering Models for Artificial Tracheas, 250
7.9　Tracheal Tissue Engineering—An Example of a Clinical Study, 253
7.10　Tracheal Tissue Engineering—A Second Example of a Clinical Study, 255
　　　Summary, 258
　　　Practice Questions, 258
　　　References, 260

8　Bladder Tissue Engineering　　　　　　　　　　　　　　**265**

8.1　Bladder Structure and Function, 266
8.2　Neurogenic Bladder Dysfunction, 267
8.3　Surgical Bladder Augmentation, 269
8.4　Development of the Urinary Bladder, 270
8.5　Design Considerations for Bladder Tissue Engineering, 270
8.6　Process of Bioengineering Artificial Bladders, 271
8.7　Cell Sheet Engineering for Bladder Tissue Engineering, 273
8.8　Small Intestinal Submucosa (SIS) for Bladder Tissue Engineering, 275
8.9　Plga as a Biomaterial for Bladder Tissue Engineering, 278
8.10　Acellular Grafts for Bladder Tissue Engineering, 280
8.11　Organ Models for Bladder Tissue Engineering, 283
8.12　Clinical Study for Bladder Tissue Engineering, 284
　　　Summary, 285
　　　Practice Questions, 286
　　　References, 288

9 Liver Tissue Engineering   295

9.1   Structure and Function of the Liver, 296
9.2   Acute Liver Failure, 297
9.3   Liver Transplantation, 299
9.4   Liver Regeneration, 301
9.5   Liver Development, 302
9.6   Design Considerations for Liver Tissue Engineering, 303
9.7   Process of Bioengineering Artificial Liver Tissue, 303
9.8   Stem Cells for Liver Tissue Engineering, 305
9.9   Surface Patterning Technology for Liver Tissue Engineering, 307
9.10 Biomaterial Platforms for Liver Tissue Engineering, 309
9.11 Fabrication of 3D Artificial Liver Tissue, 309
9.12 Vascularization for Liver Tissue Engineering, 311
9.13 Bioreactors for Liver Tissue Engineering, 312
9.14 Spheroid Culture for Liver Tissue Engineering, 313
      Summary, 314
      Practice Questions, 315
      References, 317

Index   323

PREFACE

This book is designed to serve as a textbook for a one-semester tissue engineering class, offered at the senior-undergraduate or first-year graduate level. The first six chapters of the book are focused on covering fundamental principles of tissue engineering and include cell sourcing, biomaterial development, tissue fabrication technology, vascularization strategies, and bioreactors for tissue engineering. These topics are at the heart of tissue engineering. The latter Chapter 3 are focused on applications of tissue engineering, which include development of 3D artificial trachea, 3D artificial bladder, and 3D artificial liver tissue.

The contents of this book are modeled after classes I teach in the Department of Biomedical Engineering at the University of Houston. I teach several classes, one of which is an introductory class in tissue engineering: BIOE 5323—Introduction to Tissue Engineering. BIOE 5323 is designed to serve as an introduction to the field of tissue engineering and is taken by senior undergraduate and first-year graduate students. When I first started teaching BIOE 5323, I put together lecture notes to provide students with a foundation in tissue engineering. Over time, these lecture notes were converted into book chapters and eventually combined into a complete textbook.

The book is designed as a textbook for use in a classroom setting. It is designed as a first text in tissue engineering and as such, does not rely on any other prerequisite classes. The book is self-contained and covers fundamental principles that are necessary to understand tissue engineering. The book is well suited for a one-semester class designed for undergraduate students at the senior level or first-year graduate students.

There is a large question bank that has been included in the book. The questions have been designed to test students' understanding of the principles of tissue engineering and their ability to apply these principles toward the fabrication of 3D artificial tissue. Therefore, all the questions are assay-based questions which require critical thinking; many of the questions are open-ended and can have multiple correct responses. These questions are designed to probe students and test their creativity in designing processes to fabricate 3D artificial tissue.

RAVI BIRLA

ACKNOWLEDGMENTS

I would like to begin by thanking my Department Chair and mentor, Dr. Metin Akay, for his role in the preparation of this manuscript. Dr. Akay has been instrumental in this project, and without his support and encouragement, this book would never have happened. Dr. Akay was involved in this project from conception to completion and has participated in every aspect of the manuscript. Dr. Akay suggested this project to me, identified the need for this book, and shared his vision for the manuscript. Dr. Akay was enthusiastically engaged in every aspect of this project, from reviewing the preliminary proposals to suggesting ways to improve manuscript content. In addition, Dr. Akay also provided the necessary connections with important people at Wiley-IEEE Publishers, which made it all possible. Dr. Akay envisioned this project, believed in my ability to successfully undertake this task, and provided support in every way imaginable; for this, I am deeply indebted.

I would like to thank my wife, Swati, for her support during the preparation of this manuscript. My ability to complete this project in a timely manner required numerous evenings and weekends that were dedicated toward the manuscript, taking time away from personal commitments. Swati was always supportive of this project, encouraged my work throughout, and to the best of my knowledge, did not mind my absence from family commitments—I am still a married man!

I would like to thank my parents for their support and encouragement during the preparation of this book. They have taken a keen interest in this project and have been engaged in the development of the manuscript. They have also been enthusiastically waiting for the publication of this manuscript, and their eagerness to see the completed manuscript served as motivation to complete this project in a timely manner.

I would like to acknowledge the participation of my kids, Aditya and Pooja, in this project. During the writing of this manuscript, Aditya was eight years old and Pooja was six; they were both aware that I was working on this project. Every so often, Aditya and Pooja would come to me and ask *"Dad, what chapter are you on?"* I was encouraged to see the participation of Aditya and Pooja on this project. I was also reminded by my kids that I was behind schedule and needed to spend more time to catch up.

I would like to thank several people for their work in creating the illustrations that have been used in this book. I would like to thank Betsy Salazar and Kristopher Hoffman for creating all the images that have been used throughout the book. Ms. Salazar and Mr. Hoffman have devoted many hours to creating these images and their efforts have enhanced the quality of the book. These illustrations provide a valuable tool for student learning and the work by Ms. Salazar and Mr. Hoffman will go a long way in achieving this objective. I would also like to thank Mohamed A. Mohamed for creating the cover art; the cover image accurately captures the essence of the book.

I would like to thank Ms. Kelley Murfin, with the University of Houston Writing Center, for her assistance in editing and proofreading the manuscript. The time invested by Ms. Murfin in editing the manuscript has ensured accuracy of the material.

LIST OF ABBREVIATIONS

LVAD   Left ventricular assist device
NSF   National Science Foundation
NIH   National Institute of Health
PCR   Polymerase chain reaction
MHC   Myosin heavy chain
MIT   Massachusetts Institute of Technology
2D   Two-dimensional
3D   Three-dimensional
NASA   The National Aeronautics and Space Administration
SERCA   Sarcoplasmic endoreticulum Ca-ATPase
VEGF   Vascular endothelial growth factor
HPCs   Hematopoietic progenitor cells
EPCs   Endothelial progenitor cells
ECM   Extracellular matrix
hES Cells   Human embryonic stem cells
NE   Nuclear envelope
NPC   Nuclear pore complex
ONM   Outer nuclear membrane
INM   Inner nuclear membrane
NUPs   Nucleoporins
RAN   Ras-related nuclear protein

GTPase Guanosine triphosphatase
RAN.GTP Ras-related nuclear protein guanosine triphosphatase
rRNA Ribosomal RNA
mRNA Messenger RNA (mRNA)
tRNA Transfer RNA
ER Endoplasmic reticulum
GAGs Glysoaminoglycans
JAMs Junctional adhesion proteins
ZO Zonula occludens
MSCs Mesenchymal stem cells
iPS induced pluripotent stem cells
HSCS Hematopoietic stem cells
MTS Mechanical testing system
PLA Polylactic acid
HA Hydroxyapatite
MAC Membrane attack complex
PGA Polyglycolic acid
PMMA Polymethyl methacrylate
EGTA Ethylene glycol tetraacetic acid
EDTA Ethylenediaminetetraacetic acid
SDS Sodium dodecyl sulfate
PEO Poly(ethyleneoxide)
PVA Poly(vinyl alcohol)
PAA Poly(acrylicacid)
P(PF-co-EG) Poly(propylene furmarate-co-ethylene glycol)
SCID Severe combined immunodeficient
PPS Poly(propylene sulfide)
MMPs Matrix metalloproteinases
PDMS Polydimethylsiloxane
PIPAAm Poly (N-isopoplyacrylaminde)
CAD Computer-aided design
CAM Computer aided machining
SFF Solid freeform fabrication
RP Rapid prototyping
TAF Tumor angiogenesis factor
EC Endothelial cells
SMCs Smooth muscle cells
MCP-1 Monocyte chemoattractant protein-1

LIST OF ABBREVIATIONS

ICAM-1   Intercellular adhesion molecule-1
VCAM-1   Vascular cell adhesion molecule-1
vWF   von Willibrand factor
ADSCs   Adipose-derived stromal cells
PLAGA   Poly(lactide-co-glycolide)
SAWs   Surface acoustic waves
IDT   Interdigital transducer
Mag-TE   Magnetic force-based tissue engineering
MCLs   Magnetite cationic liposomes
SACs   Stretch-activated channels
VSMCs   Vascular smooth muscle cells
ECs   Endothelial cells
VASP   Vasodilator-stimulated phosphoprotein
ROCK   Rho-associated coiled-coil-containing *protein*
TRPs   Transient receptor potential channels
PECAM-1   Platelet endothelial cell adhesion molecule-1
NO   Nitric oxide
HEPES   4-(2-hydroxyethyl)-1-piperazineethanesulfonic acid
VNS   Vagus nerve stimulation
TENS   Transcutaneous electrical nerve stimulation
NMES   Neuromuscular electrical stimulation
FES   Functional electrical stimulation
PPy   Polypyrrole
PANI   Polyaniline
NSCs   Nerve stem cells
EB   Embryoid bodies
CTS   Congenital tracheal stenosis
LCTS   Long segment CTS
MLB   Microlaryngoscopy and bronchoscopy
IC   Intermittent catheterization
WDs   Wolffian ducts
CND   Common nephric duct
SIS   Small intestinal submucosa
BAMA   Bladder acellular matrix allograft
ACM   Acellular Matrix
BAMGs   Bladder acellular matrix grafts
ALF   Acute liver failure
OLT   Orthotopic liver transplantation

LDLT	Living donor liver transplantation
SLT	Split-liver transplantation
OPTN	Organ Procurement and Transplantation Network
SRTR	Scientific Registry of Transplant Recipients
HCV	Hepatitis C virus
HCC	Hepatocellular carcinoma
HGF	Hepatocyte growth factor
TGF-β1	Transforming growth factor-β1
ADE	Anterior definite endodermal

IMPORTANT TERMINOLOGY AND CONCEPTS

- **TISSUE ENGINEERING**—*Tissue engineering is a multidisciplinary field bringing together experts from engineering, life sciences and medicine, utilizing the building blocks of cells, biomaterials and bioreactors for the development of 3D artificial tissue and organs which can be used to augment, repair and/or replace damaged and/or diseased tissue.*
- **CELL-MATRIX INTERACTIONS**—*When a cell sees any given ECM protein, the cell scans the protein molecule to identify specific binding sites for which it has integrins; for example, the integrin $\alpha 5\beta 1$ binds to the RGD site of the fibronectin molecule. Although the fibronectin molecule is large, there is only a sequence of three amino acids that are recognized by cells having the $\alpha 5\beta 1$ integrin; the binding of the $\alpha 5\beta 1$ integrin to the RGD site on the fibronectin molecule is referred to as a specific cell-matrix interaction.*
- **CELL-CELL INTERACTIONS**—*Cells communicate with other cells via cell-cell interactions, and these are critical in maintaining cell phenotype and tissue function. There are 4 types of cell signaling, known as endocrine, paracrine, autocrine, and contact-dependent signaling. In addition, cellular junctions provide various functions at the cell-cell; gap junctions are one example. The functional coupling of cells with other cells is known as cell-cell interaction.*
- **AUTOLOGOUS CELLS**—*Autologous cells are cells that have been isolated from a tissue biopsy of the person who will also be recipient of these cells; the donor and recipient for autologous cells is the same.*

- **ALLOGENEIC CELLS**—*Allogeneic cells are isolated from a donor and then transplanted into a recipient patient, with the donor and recipient being different people.*
- **CELL TRANSPLANTATION**—*Cell transplantation has been defined as the process by which cells are delivered to the site of injury in order to improve the functional performance of injured tissue. Whole blood transfusions, packed red cell transfusions, platelet transfusions, and bone marrow transplants are examples of cell therapy.*
- **STEM CELL TRANSPLANTATION**—*Stem cell transplantation is a specialized case of cell transplantation, in which the cells being delivered are stem cells. Use of embryonic stem cells, induced pluripotent stem cells, and adult stem cells fall under the classification of stem cell transplantation.*
- **CENTRAL DOGMA OF MOLECULAR BIOLOGY**—*The central dogma of molecular biology states that DNA is transcribed to RNA, which is then translated to proteins.*
- **CHARACTERISTICS OF STEM CELLS**—*Stem cells have three important characteristics that distinguish them from other cell types: self-renewal, unspecialized function, and differentiation potential.*
- **CELL POTENCY**—*Cell potency refers to the differentiation potential of stem cells.*
- **BIOMATERIALS**—*A biomaterial is any substance that simulates the extracellular matrix by functionally interacting with isolated cells to support fabrication and maturation of 3D artificial tissue.*
- **TENSILE PROPERTIES OF BIOMATERIALS**—*The tensile properties of a material are used very frequently in engineering design as an important criterion for material selection. The tensile properties of a material provide information about the strength of the material, its ability to withstand a particular load, and information about elastic properties. All of these properties are extremely important for material selection during tissue fabrication.*
- **BIOMATERIAL DEGRADATION**—*Biomaterial degradation refers to the gradual breakdown of a biomaterial mediated in a controlled manner to support the fabrication of 3D artificial tissue*
- **BIOMATERIAL BIOCOMPATIBILITY**—*The ability of 3D artificial tissue to be accepted by host defense mechanisms upon implantation, while maintaining functional capacity, is known as biocompatibility.*
- **BIOMIMETIC BIOMATERIALS**—*A two-part definition of biomimetic biomaterials has been provided in a recent article: 1) The development of biomaterials for tissue engineering applications has recently focused on the design of biomimetic materials that are able to interact with surrounding tissues by biomolecular recognition, 2) The design of biomimetic materials is an attempt to make the materials such that they are capable of eliciting specific cellular responses and directing new tissue formation mediated by specific*

- **CLASSIFICATION OF BIOMATERIALS**—*Biomaterials are frequently classified based on source (natural and synthetic), based on degradation (biodegradable and non-biodegradable), and based on interatomic bonding forces (metals, polymers, and ceramics).*
- **BIOMATERIAL PLATFORMS**—*There are four platforms that have been widely used for tissue engineering applications: polymeric scaffolds, biodegradable hydrogels, decellular matrices, and self-organization strategies.*
- **DECELLULARIZED MATICES**—*This strategy is based on the utilization of naturally occurring extracellular matrix as the scaffolding material for 3D tissue formation. Tissue specimens are obtained from cadaveric or xenogeneic sources, and cells are completely removed using one of several potential strategies. Removal of cellular components from tissue specimens is known as decellularization, and the material that is obtained after removal of the cells is known as an acellular scaffold.*
- **HYRDOGELS**—*The term hydrogel is composed of "hydro" (water) and "gel," and refers to aqueous (water-containing) gels. To be more precise, it refers to polymer networks that are insoluble in water; they swell to an equilibrium volume but retain their shapes.*
- **POLYMERS**—*Polymers can be viewed as molecules of a high molecular weight that are composed of repeating monomer units.*
- **SELF-ORGANIZATION STRATEGIES**—*Self-organization is prevalent in biological systems; it involves the physical interaction of molecules in a steady-state structure. In a broad sense, self-organization can be viewed as a process that occurs in the absence of any constraining conditions, thereby providing a greater degree of freedom and flexibility.*
- **SMART MATERIALS**—*The most recent generation of biomaterials has been designed to respond to changes in the cellular environment. These materials, known as smart materials, are receptive to changes in the physiological environment and are adaptive to changes in the degree of tissue maturation.*
- **TISSUE FABRICATION TECHNOLOGIES**—*Tissue fabrication technologies can be classified into six categories, which include scaffold-free methods, cell patterning techniques, scaffold-based methods, rapid prototyping technologies, printing technology, and "organ-on-a-chip" models.*
- **SELF-ORGANIZATION TECHNOLOGY**—*Self-organization technology is based on the fabrication of extracellular matrix by cells that then use the newly formed ECM to support artificial tissue fabrication. This technology is an example of a scaffold-free tissue fabrication process and does not require external or synthetic scaffolding; rather, scaffolding is produced by cells.*

- **CELL PRINTING**—*Bioprinting process used for 2D cell patterning by depositing bio-ink on the surface of biopaper.*
- **ORGAN PRINTING**—*Bioprinting process used for fabrication of 3D tissue by depositing bio-ink on the surface of biopaper.*
- **SOLID FREEFORM FABRICATION**—*Solid freeform fabrication refers to a group of technologies that build 3D scaffolds using a layer-by-layer approach. Collectively, these technologies are known as rapid prototyping methods.*
- **SOFT LITHOGRAPHY**—*Soft lithography is a microfabrication technology used to engineer microfluidic devices, particularly microvascular networks.*
- **CELL PATTERNING**—*The process by which the spatial placement of cells is controlled to create an organized pattern of cell monolayers or 3D tissue is known as cell patterning.*
- **VASCULOGENESIS**—*Vasculogenesis refers to initial events in vascular growth in which endothelial cell precursors (angioblasts) migrate to discrete locations, differentiate in situ, and assemble into solid endothelial cords, later forming a plexus with endothelial tubes.*
- **ANGIOGENESIS**—*Angiogenesis refers to the growth, expansion, and remodeling of primitive blood vessels formed during vasculogenesis to form a mature vascular network.*
- **ARTERIOGENESIS**—*Arteriogenesis is the process by which blood vessels increase in diameter to form muscular arteries and incorporate smooth muscle cells and vaso-contraction and vaso-relaxation properties.*
- **THERAPEUTIC ANGIOGENESIS**—*Therapeutic angiogenesis refers to the stimulation of angiogenesis for therapeutic purposes.*
- **BIOLOGICALLY REPLICATED VASCULARIZATION STRATEGIES**—*Biologically replicated processes are influenced by molecular biology, with the objective being the understanding of biological phenomena and defining controlled laboratory conditions to replicate these processes. These strategies are focused on defining in vitro conditions used to drive vasculogenesis, angiogenesis, and arteriogenesis.*
- **BIOLOGICALLY MEDIATED VASCULARIZATION STRATEGIES**—*The term "biologically mediated" refers to the notion that successful implementation of these strategies requires intervention and mediation from recipient of the implanted tissue. Mediation of the vascularization process is a result of implantation of cells or artificial tissue.*
- **BIOLOGICALLY INSPIRED VASCULARIZATION STRATEGIES**—*In this case, inspiration is drawn from biological process with an objective to replicate these processes using innovative in vitro strategies. The goal is not to replicate the biological process, but replicate functionality.*

- **IN VIVO VASCULARIZATION STRATEGIES**—*The concept of in vivo vascularization revolves around culturing bioengineered tissue within specialized chambers that can be implanted to support the formation of new blood vessels within 3D artificial tissue.*
- **BIOREACTORS**—*Bioreactors are devices used extensively in tissue engineering to enable the fabrication of artificial 3D tissue and support the growth, maturation, and development of artificial tissue during controlled in vitro culture.*
- **CLASSIFICATION OF BIOREACTORS**—*Bioreactors are used for cell culture, scaffold fabrication, scaffold cellularization, and bioreactors for stretch, perfusion, and electrical stimulation.*
- **DESIGN CONSIDERATIONS FOR BIOREACTORS**—*The process flow chart for bioreactor design consists of four steps: 1) definition of stimuli, 2) control of processing variables, 3) sensor technology, and 4) stimulation protocol.*
- **BIOREACTORS FOR CELL CULTURE**—*Isolation, culture, and expansion of mammalian cells is a critical prerequisite for tissue fabrication. Automated cell culture bioreactors are designed to undertake all functions of mammalian cell culture using robotic technology.*
- **BIOREACTORS FOR SCAFFOLD FABRICATION**—*Electrospinning is one example of bioreactors that have been used for scaffold fabrication. Electrospinning is a method fabricating individual fibers of a polymer that can be combined in different configurations to promote 3D scaffold fabrication.*
- **BIOREACTORS FOR SCAFFOLD CELLULARIZATION**—*Bioreactors have been developed to aid the cellularization process, and in this section we will discuss six cellularization methods: 1) direct cell injection, 2) cell entrapment using hydrogels, 3) perfusion seeding, 4) surface acoustic waves, 5) centrifugal force, and 6) magnetic nanoparticles.*
- **PERFUSION SYSTEMS**—*In the human body, the circulatory system acts as a distribution network for the delivery of nutrients to cells and tissues while at the same time removing waste products. Perfusion systems are capable of delivering continuous fluid flow to support the metabolic activity of cells and 3D artificial tissue during controlled in vitro culture.*
- **BIOREACTORS FOR STRETCH**—*Cells have biological force sensors, which respond to changes in the force environment, embedded within the cell membrane; these biological force sensors are known as stretch-activated channels (SACs). Bioreactors have been developed to deliver controlled stretch of cells/tissue for the cardiovascular system.*
- **BIOREACTORS FOR ELECTRICAL STIMULATION**—*During normal mammalian function, changes in voltage are used as a mechanism to maintain hemostasis and as a trigger to modulate cell and tissue level*

function. Bioreactors have been developed to deliver controlled electrical stimulation to support the development and maturation of 3D artificial tissue.
- **SMALL INTESTINAL SUBMUCOSA**—*Small intestinal submucosa (SIS) has been extensively used for bladder tissue engineering. SIS is obtained from the submucosal layer of a small intestine segment that has been harvested from porcine donors. During the preparation of SIS, a segment of the small intestine layer is harvested, commonly from pigs, and all layers of the tissue, with the exception of the submucosal layer, are removed mechanically. The submucosal layer is next subjected to a decellularization protocol to remove any cells and cellular components, leaving behind an intact ECM.*
- **POLY (LACTIC-CO-GLYCOLIC ACID) (PLGA)**—*PLGA has been used extensively as a biomaterial for tissue engineering along with many other medical applications. PLGA is a degradable copolymer of lactic acid and glycolic acid; it is often described in terms of the relative percentage of these two monomers. One of the main advantages of PLGA is the nontoxicity of its degradation products; PLGA undergoes hydrolysis, and the degradation products of this reaction are the monomers lactic acid and glycolic acid, both of which are easily metabolized by the body.*

1

INTRODUCTION TO TISSUE ENGINEERING

Learning Objectives:

After completing this chapter, students should be able to:

1. Provide examples of tissue and organ systems being developed using tissue engineering strategies.
2. Describe how tissue engineering can help solve the problem of chronic shortage of donor organs.
3. Discuss the tissue engineering paradigm as it applies to cardiovascular tissue engineering.
4. Define tissue engineering.
5. Describe the process of fabricating artificial tissue.
6. Discuss design principles related to tissue engineering.
7. Identify building blocks for the field of tissue engineering.
8. Describe scientific and technological challenges in the field of tissue engineering.
9. Describe strategies for the functional assessment of 3D artificial tissue.
10. Discuss seminal papers in the field of tissue engineering.
11. Describe potential applications for 3D artificial tissue.
12. Explain the relative advantages of 3D culture over 2D monolayer culture.

Introduction to Tissue Engineering: Applications and Challenges, First Edition. Ravi Birla.
© 2014 The Institute of Electrical and Electronics Engineers, Inc. Published 2014 by John Wiley & Sons, Inc.

13. Describe the collaborative model for tissue engineering research.
14. Discuss the growth in the field of tissue engineering.
15. Discuss the participation rate from different disciplines in tissue engineering.
16. Explain the differences between tissue engineering and other related fields.

CHAPTER OVERVIEW

We begin this chapter by providing a broad overview of tissue engineering research and providing examples of tissue and organ systems that are currently under development using tissue engineering strategies. We next describe the chronic shortage of donor organs and provide a vision of how tissue engineering can help alleviate this problem. In the next section, we describe the tissue engineering paradigm and how it applies to the cardiovascular system. We then provide a formal definition of tissue engineering and describe the process to bioengineer 3-dimensional artificial tissue. In the next section, we describe the design principles related to tissue engineering and identify fundamental building blocks in the field. We then discuss some of the scientific and technological challenges in the field of tissue engineering. Next, we describe strategies for functional assessment of 3D artificial tissue and describe functional, biological and histological metrics. We next discuss seminal publications in the field of tissue engineering and the contribution of these toward the development of the field. We then move on to discuss potential applications of 3D bioengineered artificial tissue. Tissue engineering is a multidisciplinary field, and in the next section, we discuss the multidisciplinary nature of the field and how researchers from many different backgrounds work together. The next section is focused on the growth of tissue engineering as a scientific discipline and some of the drivers of this growth. We end this chapter by providing a description of scientific disciplines that are closely related to tissue engineering.

1.1 INTRODUCTION TO TISSUE ENGINEERING

We begin our discussion of tissue engineering with a broad overview of the field—*what exactly is tissue engineering and why is it important?* While in the next section, we provide a formal definition of tissue engineering, we begin this discussion with a general overview of the field. *Research in the field of tissue engineering is focused on the fabrication of artificial tissue and organs.* The statement of purpose defined for tissue engineering (fabrication of artificial tissue and organs) is very challenging with numerous scientific and technological challenges, many of which we will discuss during the course of this book. However, the important concept to grasp is the simple notion that tissue engineering is equivalent to tissue and organ fabrication, a recurring theme throughout this book.

We have seen that tissue engineering refers to the fabrication of artificial tissue and organ systems; however, this statement requires further clarification. Artificial organ development using mechanical components is a mature field of research with mechanical hearts and left ventricular assist devices being used in patients. The field of tissue engineering should be differentiated from this area of research, as the objective of tissue engineering is to fabricate biological artificial organs that are similar in form and function to mammalian organs. Cells and biomaterials (which simulate mammalian extracellular matrix) are important components of artificial organs fabricated using tissue engineering strategies.

What is the long-term objective of tissue engineering research? The overarching theme in tissue engineering is artificial tissue and organ development. The potential application of artificial organs is obvious: transplantation in patients with damaged or diseased organs. There is a chronic shortage of donor organs, as the number of waitlisted patients is significantly greater than the number of donor organs available. Tissue engineering has the potential to alleviate this problem by fabricating artificial organs that can be used clinically.

Let us continue our discussion on tissue engineering by looking at some areas where active research is being conducted in the fabrication of artificial tissue and organs. Tissue engineering research has expanded significantly in the last decade with active research programs across the country and worldwide encompassing many different tissue and organ systems. There has been significant interest in cardiovascular tissue engineering, with research devoted to the fabrication of artificial heart muscle, blood vessels, valves, cell based cardiac pumps, ventricles, and entire bioartificial hearts. Another active area of research has been in the musculoskeletal system, encompassing fabrication of bone, cartilage, skeletal muscle, and tendons. A significant amount of research has been invested in tissue engineering of the urinary system, which consists of kidneys, urinary bladder, ureters, and urethras. Tissue engineering of the airway system has focused on fabrication of artificial tracheas and artificial lung tissue. The digestive system has been a very active area of tissue engineering research focused on the development of artificial liver tissue, pancreas, intestine tissue, and esophageal tissue. In addition, there is significant interest in the development of artificial skin and tissue engineering strategies for the central nervous system.

1.2   CHRONIC SHORTAGE OF DONOR ORGANS

There is a chronic shortage of donor organs available for transplantation. This can be illustrated by the case of kidney and liver transplantation (Figure 1.1). As can be seen in the figure, the number of patients on the waiting list is significantly greater than the number of donor organs available (1). This chronic shortage of donor organs is evident in other organ systems as well, and highlights the urgency to develop novel strategies to address this problem. The ability to fabricate artificial organs in the laboratory using tissue engineering strategies can alleviate some of the problems associated with chronic shortage of donor organs. Rather than having a patient on a waiting list for a donor organ, the promise of tissue engineering is that artificial organs can be fabricated under controlled conditions in

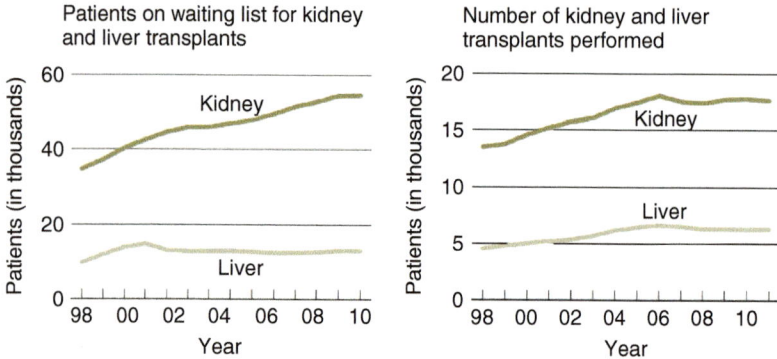

Figure 1.1 Donor Organ Shortage in the US—There is a chronic shortage of donor organs. The number of patients waitlisted for kidney and liver transplants is significantly higher than the number of donor organs available. *Note*–The data and analyses reported in the 2011 Annual Data Report of the Organ Procurement and Transplantation Network and the US Scientific Registry of Transplant Recipients have been supplied by the Minneapolis Medical Research Foundation and UNOS under contract with HHS/HRSA. The authors alone are responsible for reporting and interpreting these data; the views expressed herein are those of the authors and not necessarily those of the US Government.

the laboratory and used for transplantation. This strategy can provide life-saving options for millions of patients around the globe. This is the grand vision of tissue engineering—fabrication of artificial organs that can provide life-saving options for patients around the world.

1.3 THE TISSUE ENGINEERING PARADIGM

In this section, we introduce the tissue engineering paradigm using the cardiovascular system as an example. There are several conditions that can compromise the function of the heart, including acute myocardial failure, atherosclerosis, valve stenosis or hyperplastic left heart syndrome. Several strategies, including pharmacological agents, mechanical devices like pumps, and surgical interventions like heart transplantation, have been developed to help patients with cardiovascular disorders. Undoubtedly, these strategies have helped numerous people and saved many lives. However, heart transplantation is plagued by the chronic shortage of donor hearts, and many of the other treatment strategies also have limitations. The ability to bioengineer artificial hearts and components of the cardiovascular system can provide an alternative treatment modality for many patients; this can lead to an improvement in the quality of life and can also save the lives of many patients.

The field of cardiovascular tissue engineering is focused on the fabrication of artificial heart muscle, blood vessels, tri-leaflet heart valves, cell based cardiac pumps, tissue-engineered ventricles and bioartificial hearts (2). Artificial tissues and organs related to the cardiovascular system can be used in a variety of ways to help patients with cardiovascular disorders. For example, heart muscle can be used to provide functional support to the left ventricle of compromised hearts, thereby assisting in cardiac function. As another example, bioartificial hearts can be used as transplantable organs for patients with end-stage heart failure, thereby providing a life-saving option for many patients across the globe.

The purpose of the discussion presented in this section was to illustrate the tissue engineering paradigm using the cardiovascular system as an example. As we have seen, tissue engineering strategies can be applied toward the fabrication of artificial hearts and components of the cardiovascular system that can be used to repair, replace or augment the functional performance of comprised hearts. This is the fundamental premise of tissue engineering—fabrication of artificial tissue and organs that can be used clinically to help patients by providing functional recovery of diseased or damaged tissue and organs.

1.4 DEFINITION OF TISSUE ENGINEERING

In this section, we define tissue engineering and discuss different terms (regenerative medicine, reparative medicine), that have been used to describe the field. The definition of tissue engineering has been evolving over the last several years. As with any new discipline, the scope of the definition changes with a better understanding of the scope of the field. In addition, due to the diversity of scientific disciplines of researchers participating in the field, the definition changes to accommodate these differences in training. Before presenting our own definition, we would like to consider various definitions provided by renowned researchers in the field as well as those definitions adopted by major scientific governing bodies.

Tissue Engineering—[definition] (National Science Foundation, 1997) (3)

The production of large amounts of functional tissues for research and applications through the elucidation of basic mechanisms of tissue development combined with fundamental engineering production processes.

The NSF's definition of tissue engineering closely reflects the Foundations mission of understanding and promoting science at a very basic level and applying engineering principles for problem solving. These two fundamentals are reflected in the NSF's definition of tissue engineering.

Tissue Engineering—[definition] (Eugene Bell, 1992) (4)

1. *providing cellular prosthesis or replacement parts for the human body;*
2. *providing formed acellular replacement parts capable of inducing regeneration;*
3. *providing tissue or organ-like model systems populated with cells for basic research and for many applied uses such as the study of disease states using aberrant cells;*
4. *providing vehicles for delivering engineered cells to the organism; and*
5. *surfacing non biological devices.*

The definition provided by Dr. Eugene Bell is very much focused on the delivery of end products and is application based. This definition refers to replacement parts for the human body, tissue- or organ-like model systems, and vehicle delivery of engineered cells.

Tissue Engineering—[definition] (Dr. J. P. Vacanti & Dr. R. Langer, 1993) (5)

Tissue Engineering is an interdisciplinary field that applies the principles of engineering and the life sciences toward the development of biological substitutes that restore, maintain, or improve tissue formation.

The definition provided by Dr. Vacanti and Dr. Langer is the one that is most frequently cited in the tissue engineering literature. The publication in which this definition surfaced is one of the seminal papers in the field and is discussed in detail in a later section. The definition provides several governing principles of the field, including the interdisciplinary nature of the research and reference to the end products to improve tissue function.

Tissue Engineering—[definition] (National Institute of Health, 2001) (6)

Reparative medicine, sometimes referred to as regenerative medicine or tissue engineering, is the regeneration and remodeling of tissue in vivo for the purpose of repairing, replacing, maintaining, or enhancing organ function, and the engineering and growing of functional tissue substitutes in vitro for implantation in vivo as a biological substitute for damaged or diseased tissues and organs.

The first point to note about the definition provided by the National Institute of Health (NIH) is the reference to the terms reparative medicine, regenerative medicine and tissue engineering; NIH refers to these three terms as the same scientific discipline. While NIH refers to all three fields asone, there are subtle differences between the fields; these will be discussed in a later section. The second point about the NIH definition is its emphasis on improving human health by the use of the phrase "implantation *in vivo* as a biological substitute for damaged or diseased tissues and organs"; this is consistent with the mission of the NIH, which is to improve and enhance human health.

Tissue Engineering—[definition] (Dr. M. V. Sefton, 2002) (7)

From working with microencapsulated cells and immunoisolation systems for many years, we have learned that successful implementation of a tissue-engineering construct requires (1) an adequate, viable cell mass; (2) the appropriate behavior of the cells; and (3) sufficient durability of the function in vivo. The specific requirements are determined by the application, the nature of the cells, the implantation site, and the biocompatibility of the device.

This definition refers to encapsulation technology as it relates to tissue engineering. This is due to the nature of Dr. Sefton's work, which is focused on the development of encapsulation technology. However, these are two distinct fields and the differences will be provided in a later section. The definition provides a list of requirements for tissue engineered constructs.

Tissue Engineering—[definition] (Dr. A. Atala, 2004)(8)

Tissue engineering, one of the major components of regenerative medicine, follows the principles of cell transplantation, materials science, and engineering toward the development of biological substitutes that can restore and maintain normal function.

Similar to the definition provided by NIH, this definition draws a comparison between tissue engineering and regenerative medicine. While NIH considers tissue engineering and regenerative medicine to be the same, the Atala definition refers to tissue engineering as a component or branch of regenerative medicine.

Tissue Engineering—[definition] (Dr. R. Nerem, 2006) (9)

Whether one uses the term bioengineered tissues, tissue engineering, or regenerative medicine, what one means in general is the replacement, repair, and/or regeneration of tissues and organs.

In this definition, tissue engineering and regenerative medicine are considered to be the same, as mentioned earlier.

Tissue Engineering—[definition] (Dr. C. Mason and Dr. P. Dunnill, 2008) (10)

Regenerative Medicine is an emerging interdisciplinary of research and clinical applications focused on the repair, replacement or regeneration of cells, tissues or organs to restore impaired function resulting from any cause, including congenital defects, disease, trauma, and aging.

This is our final definition and talks exclusively about regenerative medicine, though the definitions refer to many of the guiding principles of tissue engineering, which have been presented in the earlier definitions.

What have we learned by looking at these definitions? There are many definitions of tissue engineering, often based on the principles of the researcher or the scientific organization. The terms tissue engineering, regenerative medicine, and reparative medicine have been used extensively in the literature and often refer to the same field. The term tissue engineering has been extensive in the literature and has been used more often than regenerative medicine and reparative medicine. The use of the terms regenerative medicine and reparative medicine is fairly new, and their exact definitions are still being developed. Tissue engineering, regenerative medicine and reparative medicine are often used interchangeably while, in other instances, tissue engineering is considered to be a sub-group of the two (regenerative medicine and reparative medicine). Often, tissue engineering has been defined as the ability to generate functional 3D tissue constructs *in vitro*, with potential clinical applications to replace, restore and/or augment lost tissue function. Regenerative Medicine has been commonly defined as any strategy directed at stimulating the body's own repair mechanisms, e.g., through the use of gene and/or cell transplantation. Certain authors define regenerative medicine as a broader field, with tissue engineering being one branch.

Although a diversity of definitions has evolved in the recent literature, each one provides a novel insight into the field of tissue engineering. Rather than accepting any one given definition, it is a valuable exercise to study the underlying principles that have evolved in the field of tissue engineering. Based on a survey of the definitions of tissue engineering presented earlier, it was seen that the field has been defined based on participating disciplines (engineering, biology, and surgery), building blocks (cells, biomaterials, and bioreactors), and/or end product applications (tissue repair and/or replacement). Based on this, we have formulated a working definition of tissue engineering, which encompasses these concepts and is illustrated in Figure 1.2.

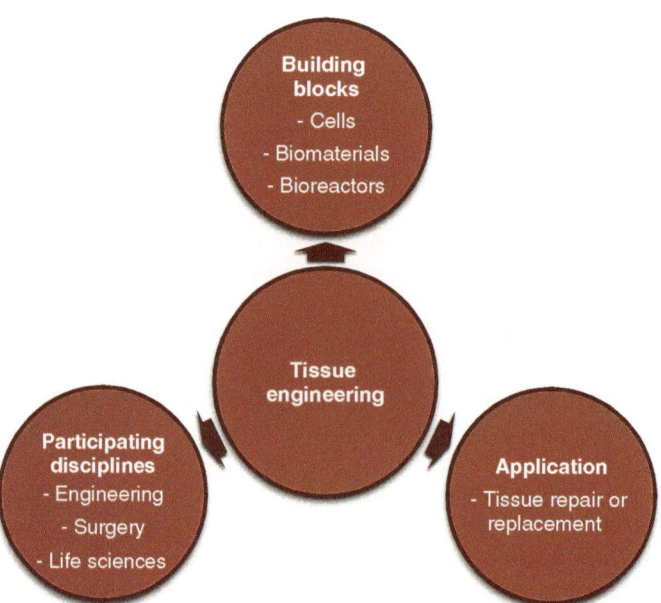

Figure 1.2 Definition of Tissue Engineering—The field of tissue engineering is focused on the development of technologies to support the fabrication of artificial tissue and organs. The building blocks of tissue engineering are cells, biomaterials, and bioreactors. Tissue engineering is a multidisciplinary field with researchers from different backgrounds working together; researchers with training in engineering, medicine, and life sciences have contributed significantly to the development of the field. Potential applications of bioengineered artificial tissue and organs is for repair and/or replacement of damaged or injured tissue.

> *"Tissue engineering is a multidisciplinary field bringing together experts from engineering, life sciences and medicine, utilizing the building blocks of cells, biomaterials and bioreactors for the development of 3-dimensional artificial tissue and organs which can be used to augment, repair and/or replace damaged and/or diseased tissue."*

This definition highlights the multidisciplinary nature of tissue engineering as a scientific discipline, provides insight into the building blocks of the field, and provides information about the potential use of artificial tissue and organs. The terms regenerative medicine and reparative medicine have not been included in this definition. We will adapt this definition throughout the book and will refer to it from time to time. In subsequent chapters, we will also provide an in-depth discussion of many of the underlying principles of tissue engineering.

1.5 PROCESS OF BIOENGINEERING 3D ARTIFICIAL TISSUE

Introduction—The process of tissue fabrication has been aggressively debated over the last couple of years, and different researchers use different processing schemes to fabricate artificial tissue and organs. Nonetheless, several general themes have evolved over the years that provide impetus for the development of a process flow sheet that is required for the fabrication of 3D artificial tissue (Figure 1.3). The steps in the process have evolved from studies at different research institutions and represent a general scheme for the fabrication of artificial tissue and organs. The specific process implemented for any given application will vary, and steps may need to be added or eliminated from the process flow sheet.

Eight Step Process for Tissue Fabrication—In this section, we provide a brief overview of the process of bioengineering 3D artificial tissue (Figure 1.3). The process differs based on the specific tissue system as well as any differences in the tissue engineering strategy adopted; however, several common themes have been identified and can be categorized into 8 stages. Depending on the tissue system and the specific technology, the sequence of steps may also need to be changed. The eight-step process of bioengineering 3D artificial tissue involves: (Figure 1.3)

1. *Cell sourcing*—Cells provide the functional component of artificial tissue. Identification, isolation, purification, expansion, and characterization of a suitable cell source are important steps in cell sourcing. During initial stages of technology development and feasibility studies, cells can also be obtained from animal sources, with cell lines being another option. As the research progresses toward the development of artificial tissue for use in humans, researchers need to determine if the cells will be obtained from autologous or allogeneic sources. The recent expansion in the field of stem cell biology has provided researchers with many different options for cell sourcing, some of which include embryonic stem cells, induced pluripotent stem cells, and adult derived stem cells (hematopoietic stem cells and bone marrow derived mesenchymal stem cells).

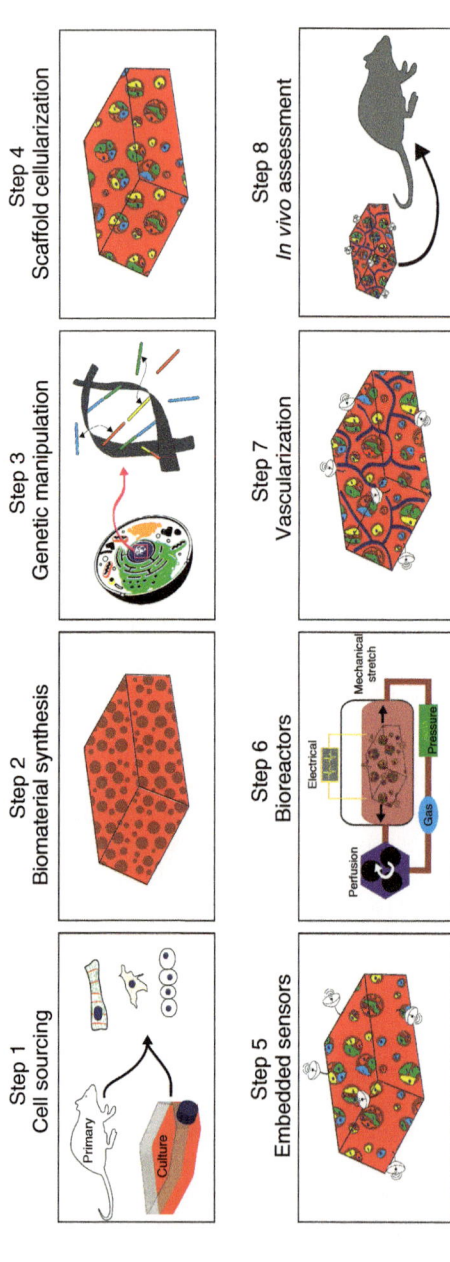

Figure 1.3 Process of Fabricating 3D Artificial Tissue and Organs—(a) **Step 1—Cell Sourcing**—the first step in the process is the isolation and/or expansion of cells, which serve to provide function for artificial tissue and organs. (b) **Step 2—Biomaterial Synthesis**—biomaterials are designed to simulate the properties of the mammalian extracellular matrix and provide structural support during fabrication of artificial tissue and organs. (c) **Step 3—Genetic Manipulation**—genetic properties of cells are modified to improve function and reduce apoptosis and other adverse effects. (d) **Step 4—Scaffold Cellularization**—in this step, cells are coupled with scaffolds. (e) **Step 5—Embedded Sensors**—sensors are embedded to monitor tissue development and maturation. (f) **Step 6—Bioreactors**—bioreactors are used to deliver controlled physiological stimulation to guide development and maturation of artificial tissue and organs. (g) **Step 7—Vascularization**—vascularization is required to support the metabolic activity of 3D artificial tissue and organs. (h) **Step 8—***In vivo* **Assessment**—the final step in the process is to test the functional performance of artificial tissue *in vivo*.

2. *Biomaterial synthesis*—Biomaterials provide structural support during 3D tissue fabrication and serve the role provided by mammalian extracellular matrix. During this stage of the tissue fabrication process, biomaterial synthesis and characterization are important variables that require rigorous optimization. The choice of biomaterial depends on the specific tissue application; there are many different biomaterials to choose from, including polymers, metals and ceramics.
3. *Genetic manipulation*—Prior to scaffold cellularization, the genetic profile of the cells can be modified to increase the likelihood of cell survival or functional integration with the host. Specific genes can be manipulated to reduce apoptosis or increase the expression of specific integrins to increase cell-matrix interactions. In addition, functional genes can be upregulated, like myosin heavy chain for heart muscle, to increase the functional performance of 3D artificial tissue.
4. *Scaffold cellularization*—Scaffold cellularization refers to the process by which isolated cells are seeded within a 3D scaffold. An important variable during the scaffold cellularization process is coupling isolated cells with the scaffold to promote functional integration at the cell-cell and cell-material interface. Successful implementation of the scaffold cellularization process is critical to support 3D tissue formation. The cellularization strategy needs to be optimized to ensure uniformity in cell distribution throughout the scaffold.
5. *Sensor technology*—Sensors are necessary to monitor the overall health of the artificial tissue during the formation, development, and maturation stage of the tissue fabrication process. Embedded sensors are necessary to monitor functional performance of artificial tissue, and data obtained from embedded sensors can be utilized in a feedback loop to regulate processing variables for tissue fabrication. Monitoring of cell behavior, cell-cell interaction, cell-matrix interaction, and tissue formation and function is critical during the tissue fabrication process.
6. *Bioreactors for guidance*—During normal physiological function, mammalian tissue is exposed to a wide array of stimuli, which include electromechanical impulses, fluid stresses, and changes in the chemical environment based on changing concentrations of growth factors, hormones, and cytokines. These signals are important in maintaining tissue function. During the fabrication of 3D artificial tissue, it is critical to develop strategies to deliver these signals. Specialized systems known as bioreactors are designed to deliver physiological signals to 3D artificial tissue, which in turn provides guidance to drive tissue development and maturation.
7. *Vascularization*—Incorporation of blood vessels as an integrated component of the artificial tissue is a critical requirement and is required to support the metabolic activity of 3D artificial tissue.
8. *In vivo assessment*—Once functional 3D artificial tissue has been fabricated, the final step in the process is *in vivo* testing. In this case, the effectiveness

of the tissue graft to repair, replace, and/or augment the function of damaged or diseased tissue is assessed.

Brief Discussion of the Tissue Fabrication Process—Now that we have discussed the process flow sheet of bioengineering 3D artificial tissue, we next present a brief description of how this process comes together for the fabrication of artificial tissue. The identification of a suitable cell source remains a formidable challenge, especially for cardiac applications, as adult derived cardiomyocytes are difficult to obtain and non-proliferative *in vitro*, thereby limiting their applicability. There are several areas of opportunity for cell sourcing, including human embryonic stem cells, adult derived stem cells, and autologous cells derived from patients. The choice of cell source will vary significantly depending on the application; autologous derived skeletal muscle cells can be utilized for cardiac regeneration, while autologous derived cardiac cells may not be the most feasible choice. Selection of suitable scaffolding material depends on the ability of the material to simulate properties of the extracellular matrix (ECM), promote cell viability and proliferation, possess easily controllable degradation kinetics, and have a high degree of immune tolerance when implanted *in vivo*. There are several matrices currently available which meet many of these requirements, while new and improved biomaterials with improved functionality are continuously being developed. The cells are then subjected to strategies for genetic manipulation to increase function and reduce cell apoptosis. The next stage requires successful colonization of the scaffold by the cells; the viability of the cells during culture within the scaffold, the ability of the cells to maintain differentiated phenotype, and the ability of the cells to functionally interact with the biomaterial become important considerations. Sensors are embedded to provide real-time noninvasive monitoring of tissue function and development and are used in a feedback loop to guide processing variables during the fabrication of artificial tissue.

Once cellularization of the scaffold is complete, the next step in the tissue fabrication process involves bioreactors to guide the development and maturation of artificial tissue. It becomes necessary to provide mechanical, electrical, and chemical/hormonal cues to support the functional development of 3D artificial tissue. Bioreactors need to be implemented to induce electro-mechanical stimulation of bioengineered tissue, leading to gene expression that closely resembles the gene expression of *in vivo* tissue. The development of microperfusion systems becomes increasing important to replicate the physiological flow conditions observed *in vivo*. As tissue growth and maturation occurs, vascularization of the bioengineered tissue construct is important. Finally, the ability of the tissue-engineered construct to integrate with the host tissue, without immune rejection and the ability of the construct to both survive and elicit a functional benefit, would need to be demonstrated.

1.6 DESIGN PRINCIPLES FOR TISSUE ENGINEERING

The process for the fabrication of artificial tissue is governed by design principles. In simplest of terms, tissue engineering equals tissue fabrication and, like any

DESIGN PRINCIPLES FOR TISSUE ENGINEERING

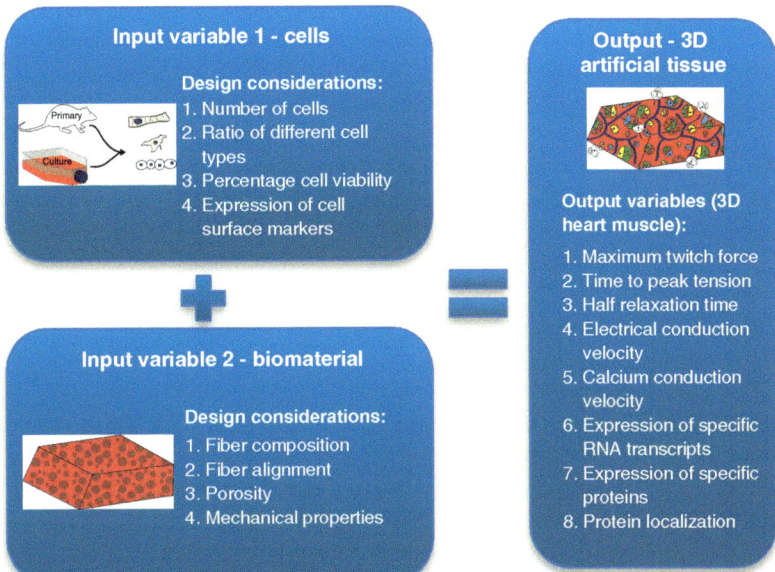

Figure 1.4 Tissue Engineering and Tissue Fabrication—Tissue engineering should be viewed as a process of fabricating artificial tissue; in other words, tissue engineering equates to tissue fabrication. Like any fabrication process, tissue fabrication has inputs (cells and biomaterials) and an output (artificial tissue). At each step of the process, there are critical design variables and design constraints which need to be addressed, some of which are shown in the Figure.

fabrication technology, is based on design principles with critical decision making at every step in the process. Like any other engineering problem, there is an input that feeds into the tissue fabrication process, and there is also an output, 3D artificial tissue. Design considerations for tissue fabrication are shown in Figure 1.4, and, for the sake of simplicity, only two input variables are included in our discussion: cell sourcing and biomaterial synthesis.

For cell sourcing, there are important design considerations that need to be taken into account, some of which include the number and density of cells, relative proportion of different cell types, percentage of viable cells, and expression of specific cell surface markers. All of these variables are under the control of the user and can be changed prior to feeding into the tissue fabrication process. The same argument can be applied toward biomaterial synthesis, and important design considerations include fiber composition and alignment, material porosity, and tensile properties of the materials. Again, all of these variables can be changed by the user prior to feeding into the tissue fabrication process. The input variables just described, cells and biomaterials, are fed into the tissue fabrication process, leading to a specific output—3D artificial tissue. The success of the process is measured by predefined metrics defined by the user, and, depending on the specific application, output criteria will vary.

In the current discussion, we have only provided examples of two steps of the tissue fabrication process (cell souring and biomaterial synthesis), although the same process is valid for all eight steps; for every step in the tissue fabrication process, researchers need to define specific design requirements for artificial tissue fabrication. Process optimization is necessary to achieve predefined values for output variables for any given tissue system. The tissue fabrication process, along with an understanding of input and output variables, is central to the field of tissue engineering.

1.7 BUILDING BLOCKS OF TISSUE ENGINEERING

Earlier in this chapter we described an eight-step process for the fabrication of 3D artificial tissue. All eight steps in the tissue fabrication process are critical and the absence of any one would disrupt the process. However, there are three steps in the tissue fabrication process that are considered to be the building blocks of artificial tissue: cells, biomaterials and bioreactors (Figure 1.5).

Cells are the functional components of artificial tissue; biomaterials are the structural components of artificial tissue while bioreactors provide guidance for tissue development and maturation. In the absence of any one of these three building blocks of tissue engineering, the functional performance of 3D artificial tissue will be significantly compromised. At the start of any tissue fabrication process, the

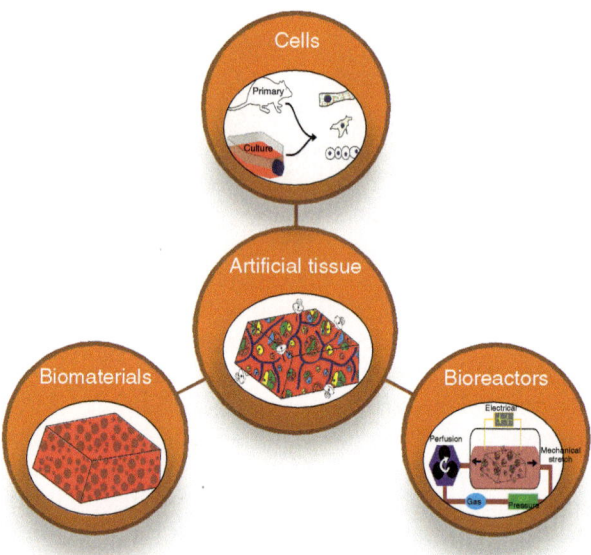

Figure 1.5 Building Blocks of Tissue Engineering—The building blocks of tissue engineering are cells, biomaterials and bioreactors.

researcher must identify the source of cells, the biomaterial to be utilized, and the guidance stimuli to be used; collectively, these three provide the platform to initiate any tissue fabrication study. Using these three building blocks, researchers can build an initial prototype for first-generation artificial tissue. This can be viewed as an entry point into the tissue fabrication process, similar to laying the foundation for building a house.

1.8 SCIENTIFIC AND TECHNOLOGICAL CHALLENGES

The process of tissue fabrication is very convoluted and complex, and at each step of the process, there are numerous scientific and technological hurdles that need to be overcome (Figure 1.6).

1. *Cell sourcing*—Where will the cells come from? Human embryonic stem cells, induced pluripotent stem cells or mesenchymal stem cells? If using stem

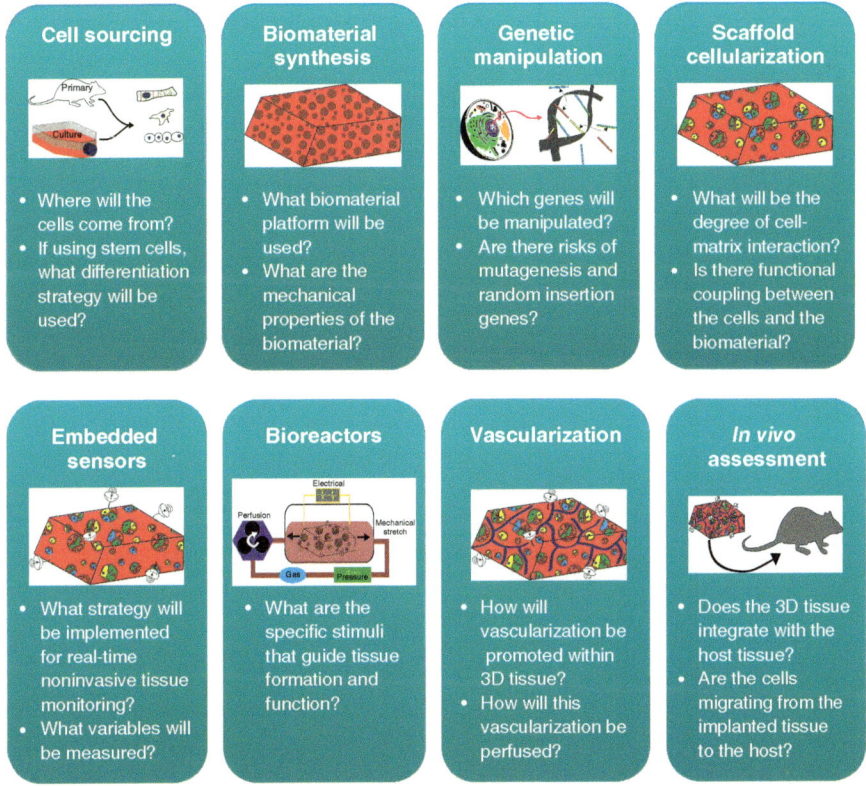

Figure 1.6 Scientific Challenges in Tissue Engineering—At every step of the tissue fabrication process, there are numerous scientific challenges, some of which are illustrated in the Figure.

cells, what differentiation strategy will be implemented? What is the differentiation efficiency? How will the differentiated cells be separated from the non-differentiated cells?

2. *Biomaterial synthesis*—What biomaterial platform will be used? Polymeric scaffolds, biodegradable hydrogels, decellularized scaffolds, or scaffold-free technologies? What are the mechanical properties of the biomaterial? What is the porosity, pore diameter, and pore orientation? Does the biomaterial have attachment sites for cell surface receptors? Is the material degradable and, if so, what are the degradation kinetics? Can the degradation kinetics be modulated?

3. *Genetic manipulation*—Which genes will be manipulated? What delivery mechanism will be implemented? Are there risks of mutagenesis and random insertion of genes?

4. *Scaffold cellularization*—What will be the degree of cell matrix interaction? Will there be functional integration at the cell material interface? Is there functional coupling between the cells and the biomaterial, or are the cells passively resting on the biomaterial? Does the presence of the cells change the properties of the biomaterial? What is the viability of the cells after cellularization, and how does this viability change with time?

5. *Embedded sensors*—What strategy will be implemented for real-time noninvasive monitoring of tissue function? What variables will be measured? How will the data be used in a positive feedback control loop?

6. *Bioreactors*—What are the specific stimuli that guide tissue formation and function? How important is stretch, electrical stimulation, and perfusion? What should be the spatial and temporal variations in physiological stimuli? Will different stimuli be used at different points during the tissue fabrication process?

7. *Vascularization*—How will vascularization be promoted within 3D tissue? Will *in vivo* methods of vascularization be used? Will *in vitro* methods of vascularization be used? How will this vascularization be perfused?

8. *In vivo assessment*—Does the 3D tissue construct integrate with the host tissue? Is there mechanical coupling? Is there electrical coupling? Are the cells migrating from the implanted tissue to the host? Does implantation of the 3D artificial tissue lead to functional improvement of the host tissue or organ?

1.9 FUNCTIONAL ASSESSMENT OF ARTIFICIAL TISSUE

The objective of tissue engineering is to fabricate 3D artificial tissue—after completing the tissue fabrication process, how do we measure performance of the bioengineered tissue? The ability to define performance metrics, which accurately reflect critical functional variables, is extremely important for the tissue fabrication process. Performance metrics need to be carefully defined and must accurately assess the function of artificial tissue.

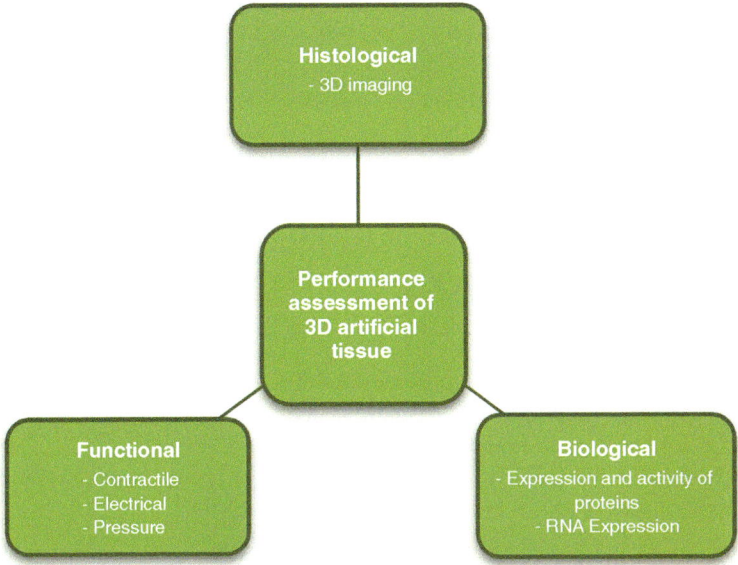

Figure 1.7 Performance Metrics for 3D Artificial Tissue—The success of artificial tissue can be assessed based on functional performance metrics, biological performance metrics, and, finally, based on histological performance metrics.

How exactly do we define performance metrics for 3D artificial tissue? The objective of the tissue engineering process is to bioengineer artificial tissue that is similar in form and function to mammalian tissue equivalents. Therefore, the best way to measure functional performance of 3D artificial tissue is by direct comparison with functional performance of mammalian tissue; the closer the two, the better.

There are three categories of metrics designed to assess the performance of 3D artificial tissue: functional, biological, and histological (Figure 1.7).

Functional performance metrics are designed to assess function of artificial tissue. Some examples include contractile force, intraluminal pressure and electrical properties. Depending on the function of the tissue, different performance metrics are used; the primary function of heart muscle is force generation, therefore, the contractile properties of artificial heart muscle are important functional metrics. Biological metrics refer to the expression and activity of specific proteins using western blotting or the expression of mRNA transcripts using rt-PCR. The cells of all mammalian tissue perform specialized functions, and in order to perform specialized functions, have a characteristic gene/protein expression pattern. For example, in order for heart muscle to generate force, specific proteins like myosin heavy chain (MHC) are expressed; the greater the expression of MHC, the higher the twitch force. Therefore, measurement of MHC expression for 3D artificial heart muscle proves to be an important assessment tool. Histological metrics refer to the localization of specific proteins, either in the extracellular matrix or intracellular proteins. Histological tools allow visualization of the cells relative to the extracellular matrix; this visualization in turn provides information about cellular organization and tissue level architecture.

None of the three performance metrics (functional, biological, and histological) are more important than the others. The collective information gathered from all three performance metrics provides an accurate assessment of the success of the tissue fabrication process and the quality of bioengineered artificial tissue.

1.10 SEMINAL PAPERS IN TISSUE ENGINEERING

Seminal work in the area of tissue engineering was conducted by Dr. Robert S. Langer, who is widely recognized as the founder and father of the field. The early work in tissue engineering, when the field was unknown to the general public and to other researchers in the field, was conducted in the laboratory of Dr. Langer at Massachusetts Institute of Technology (MIT). Several publications by Dr. Langer are seminal in the field and have provided the foundation for many researchers to build upon. Even today, Dr. Langer leads one of the largest and most respected academic research laboratories in the country. In addition to his scientific endeavors, Dr. Langer has trained numerous scientists who have gone on to hold prominent positions in both academia and industry. In recognition of his vast contribution to the field, this section begins with a brief biography of Dr. Langer, followed by a discussion of his seminal papers in the field. All information presented in Dr. Langer's biography has been obtained from public sources and is cited appropriately. This biography has not been reviewed, validated, or endorsed by Dr. Langer.

Dr. Robert S. Langer was born on August 29th, 1948 in Albany, New York (11). Dr. R. Langer received his undergraduate degree in Chemical Engineering from Cornell University in 1970 and his Ph.D, also in Chemical Engineering, from MIT in 1974, with Dr. Clark Colton serving as his doctoral advisor[1] (12). Dr. Langer did post-doctoral training with Dr. Judah Folkman at the Children's Hospital of Boston from 1974–77[1] (13). Dr. Langer is currently the Germeshausen Professor of Chemical and Biomedical Engineering at MIT (14) and was honored with an Institutional Professorship (the highest honor award to a faculty member) in 2005 (15).

During his professional career, Dr. R. Langer has won more than 150 major scientific awards. Dr. R. Langer was awarded the National Medal of Science, the nation's highest scientific honor, on July 27th 2006; the award was presented by President George W. Bush (16). In 2002, Dr. R. Langer was awarded the Charles Stark Draper award (equivalent of the Nobel Prize for Engineers) by the National Academy of Engineering (17). Also in 2002, Dr. R. Langer was awarded the Lemelson-MIT award, the highest recognition for inventorship (18). In addition, Dr. R. Langer was named as one of the 15 most influential innovators worldwide by Forbes Magazine in 2002 (19) while CNN and Time Magazine (2001) named Dr. R Langer among America's Best in Science and Medicine (20).

Dr. R. Langer is an author of more than 950 scientific papers, many of which describe seminal work in the field of tissue engineering. His work has provided the foundation for several areas of tissue engineering, including controlled drug delivery, heart valves, heart muscle, lungs, livers, bioreactors and microperfusion. Dr. R. Langer's research has been protected by over 600 patents, with more than 100 of these being licensed to companies and at least 35 products, at the time of publishing, either in the market or in clinical trials. Dr. R. Langer has founded several companies and was instrumental in the commercialization of the first tissue engineering product, skin.

Three papers published by Dr. R. Langer are considered seminal in the field of tissue engineering (5,21–22).

In 1976, Dr. R. Langer provided evidence for the controlled release of large molecules from 3D polymeric matrices, providing the foundation for the field of controlled drug release [21]. In this study, several materials were screened based on inflammatory response, and based on the results, two materials, hrydron-S and ethylene-vinyl acetate copolymer, were tested as controlled release vehicles. Several proteins were embedded into the material architecture, including soybean trypsin inhibitor, alkaline phosphate, and catalase. The release kinetics were monitored for up to 100 days. The release kinetics was shown to approach zero-order kinetics and the activity of proteins after being released from the polymer was confirmed. This study is seminal since it was the first demonstration of the release of large macromolecules from polymer substrates and laid the foundation for the field of controlled release technology (21).

In 1988, Dr. R. Langer published an article providing evidence to support the culture of primary hepatocytes derived from rodent livers within 3D matrices fabricated from different polymers (polyglactin, polyorthoesters, polyanhydride) (22). The cells were cultured within the 3D matrices *in vitro*. Scaffold fabrication was conducted using several different technologies, including solvent casting, compression molding, and filament drawing. Cells were plated onto the polymer scaffold at a concentration of 1×10^5 or 1×10^6 cells/ml, and cellularized scaffolds were maintained in culture for 3 to 4 days in a 10% CO_2 environment. The cellularized scaffolds were then transplanted onto the omentum of recipient animals after a partial hepatectomy. Histological evidence showed engraftment of cells within 3D matrices during *in vitro* culture. Cell survival was also demonstrated *in vivo*, during the implantation period. This paper is seminal as it was the first time primary cells were cultured within a 3D polymeric scaffold and shown to maintain viability during *in vitro* culture (22). The strategy used in this study, one consisting of cellularization of custom fabricated scaffolds followed by 3D culture, has now become a hallmark of the field of tissue engineering, with numerous publications using this strategy. Clearly, this publication had a significant impact on the field and is responsible for significant growth that has occurred within the last two decades.

In 1993, Dr. R. Langer published a review article on tissue engineering (5). In this article, Dr. R. Langer provided a broad overview of the field as well as his visionary goals for the potential impact, in terms of clinical applications, for tissue-engineered products (5). Equally important, a now widely adopted definition of tissue engineering was also proposed. According to Dr. R. Langer (5), *"tissue engineering is an interdisciplinary field that applies the principles of engineering and* the life sciences toward the development of biological substitutes that restore, maintain, or improve tissue formation." In addition to providing a definition of the field, several examples of potential applications of tissue engineering technologies in different systems were explained, including skin, cartilage, and bone. At the time of his article's publication, the field of tissue engineering was unknown to the general public and to researchers across the country. However, this publication appeared in a journal, *Science*, which is both broad based and very influential, and therefore served in creating awareness for the field. Researchers recognized the importance of the field and the potential impact successful tissue engineering therapies could have on human health. This served to encourage researchers to enter the field and resulted in an increase in the number of scientific publications describing tissue engineering research.

The three papers published by Dr. R. Langer are seminal due to the awareness and recognition provided to the field of tissue engineering and their establishment of scaffold cellularization as a tissue engineering strategy. Collectively, these three publications provided the foundation for the field of tissue engineering as we know it today.

1.11 APPLICATIONS OF 3D ARTIFICIAL TISSUE

During the course of this chapter, we have studied the process of bioengineering 3D artificial tissue. Prior to moving on, it is a good exercise to discuss potential applications of artificial tissue—*how exactly is bioengineered artificial tissue going to be used?* Based on our discussion thus far, it may be fairly obvious that the most significant application of artificial tissue and organs is clinical, with the objective to develop novel treatment modalities that can have an impact on human health. This is indeed the ultimate goal and long term vision for the field of tissue engineering: to fabricate artificial tissue and organs that can be implanted in humans to support, repair, augment and/or replace damaged or diseased tissue and organs.

However, in addition to the potential clinical applications, there are several other areas in which 3D artificial tissue can be used. Some potential applications for artificial tissue are as models for basic research, tools to study the effect of space radiation on human health, tools for high throughput screening assays, and as grafts for the attachment of mechanical devices to host tissue.

Artificial tissue can be used as a model for basic research to gain an understanding of the processes related to tissue formation and development and as an insight into cell-cell and cell-matrix interaction. There are several models that are currently used, with 2D monolayer culture of isolated cells being a commonly used model. Monolayer 2D culture systems are based on the isolation and culture of cells in a 2D environment (23); these cells are then be subjected to different interventions, some of which include controlled exposure to pharmacological agents and environmental toxins. The response of the cells during 2D monolayer culture can be studied in a controlled *in vitro* environment. Monolayer cell culture is a standard technique and is used extensively around the globe for numerous applications. One of the primary advantages of monolayer cell culture is the ability to study physiological effects of regulated stimuli in the absence of confounding systematic variations from *in vivo* systems and therefore lead to an understanding of cause and effect relationships. While monolayer cell culture techniques have tremendously enhanced our understanding of basic cell biology, monolayer culture systems are conducted in 2D and lack 3D tissue level architecture and therefore do not provide a true representation of mammalian tissue. Artificial tissue, which has been engineered in the laboratory, overcomes this limitation by replicating many of the features found in mammalian tissue; 3-dimensionality is a significant advantage of artificial tissue when compared with cells maintained in monolayer culture. Just as 2D monolayer culture is a standard technique across research laboratories, 3D artificial tissue has the potential to replace these models and become the staple for basic research across the country.

Artificial tissue can prove to be a powerful tool to study the effect of space radiation on human health. When astronauts travel to space, they are exposed to harsh environments consisting of space radiation, microgravity, and oxidative stress, all of which have adverse effects on human health (24–27). The specific effects of space radiation and other stimuli on human tissue are not known due to the lack of models to undertake systematic studies. It is critical to gain an understanding of the dose–response behavior of specific stimuli observed in space to develop countermeasures necessary to ensure safety of astronauts. Agencies like The National Aeronautics and Space Administration (NASA) can benefit from tissue engineering models to gain an understanding of the effects of space radiation, microgravity, and oxidative stress on human health and use this information to develop countermeasures to eliminate harmful effects.

Artificial tissue can be used to develop high throughput screening assays for pharmacological agents (28–31) or environmental toxins (32–35). Many of these agents are tested in 2D monolayer cultures followed by *in vivo* assessment with no intermediate steps in between. Artificial tissue can serve as an intermediate step between 2D monolayer and *in vivo* assessment. In the case of new drug development, candidate compounds are screened using *in vitro* monolayer cell culture models, and potential candidates are tested using small animal models. At each step of the process, the number of compounds is reduced from a few thousand to a few hundred. The size and scale of these studies is very large and can add significantly to the total cost of the development of a new drug. If tissue engineering models are used prior to *in vivo* assessment, the total number of compounds that need to be tested using animal models can be reduced; this can lead to a reduction in development time and associated costs.

Mechanical devices are routinely used during surgery and, while they have proven to be effective in several conditions, the attachment of the device to mammalian tissue often proves to be challenging. As an example, left ventricular assist devices (LVADs), which are used in cases of chronic heart failure, serve to pump blood directly from the left ventricle to the aorta (36–38). The LVAD is attached to the apex of the heart using a very invasive procedure, which requires the use of numerous sutures; the interface between the LVAD and mammalian tissue is nonfunctional. This can be significantly enhanced by the use of artificial tissue at the interface between LVAD and mammalian tissue; which will lead to an increase in adhesion strength and will provide functional coupling between the mechanical device and mammalian tissue. In the future, bioengineered artificial tissue can be developed to replicate many functions of mechanical devices like LVADs; however, one short-term objective would be to serve as anchoring points between mechanical devices and mammalian tissue.

The long-term application of 3D artificial tissue is clinical, designed to replace and restore tissue function for damaged or diseased tissue. However, before achieving the clinical objective, there are numerous applications for artificial tissue, some of which have been discussed here and many of which can be developed in the near future.

1.12 TWO-DIMENSIONAL VERSUS THREE-DIMENSIONAL CULTURE

Cell culture techniques have been developed and optimized to maintain and expand isolated cells on the surface of 2D tissue culture plates (23,39–41) These techniques have become standard across academic research laboratories. We will provide a detailed discussion of the topic in Chapter 2. The culture of isolated cells is critical for tissue engineering studies as well, due to the large number of cells required to support tissue fabrication. During the last several decades, monolayer culturing of cells has added tremendously to our understanding of concepts related to cell biology and pharmacology. The importance of monolayer cell culture techniques cannot be overstated, and these techniques continue to be a mainstay for many areas of investigation. However, there are some limitations associated with monolayer cell culture. During normal mammalian function, cells are not maintained under 2D conditions, but rather under 3D conditions; the cells are in constant communication with other cells and with components of the extracellular matrix. Cell-cell interactions and cell-matrix interactions are important in maintaining cell phenotype and tissue function. These physiologically important cues are not fully reproduced during 2D monolayer culture; this limitation can be overcome by tissue engineering models. Rather than maintaining isolated cells in 2D culture, researchers now have the ability of culturing 3D artificial tissue and utilizing these models to answer many questions in cell biology and physiology that cannot be addressed by 2D culture systems. Tissue engineering models offer several advantages over 2D culture systems, the most important of which is the ability of these models to replicate complex 3D architecture of mammalian tissue; this in turn supports cell-cell and cell-matrix interactions and can increase our understanding of cell biology and cell physiology.

1.13 INTEGRATION OF CORE TECHNOLOGIES

Development of tissue engineering technologies requires collaborative efforts from diverse scientific disciplines. This model of scientific collaboration is fairly well-established in many scientific endeavors; however, the novelty of tissue engineering places additional challenges in implementing successful collaborations. Development of core technologies for tissue engineering requires expertise from engineering, medical sciences, and life sciences disciplines. At every stage of technology development, experts from engineering, medical sciences, and life sciences participate in different aspects of the process. We will study the technology development process as it applies to the fabrication of 3D artificial heart muscle and assess the relative contributions of experts from different disciplines.

The technology development process is divided into three phases: fabrication of first-generation heart muscle, development of mature heart muscle, and, finally, fabrication of 3D artificial heart muscle similar in form and function to mammalian heart muscle. This can be viewed as early-stage, mid-level, and later stages of technology development. During each of these three stages, researchers from different disciplines (engineering, medical sciences, life sciences) have a very specific role to play during the technology development process.

Let us begin this discussion by assessing the fabrication of early-stage heart muscle tissue. The first step in the technology development process is identification of the need—*why do we need to bioengineer artificial heart muscle?* The need for artificial heart muscle will generally begin in the surgical suite during the treatment of patients with myocardial infarction. From the point of view of the surgeon, while there are several therapeutic options available to treat acute myocardial infarction, each one has severe limitations. Therefore, there is an urgent need to develop novel treatment modalities for patients with myocardial infarction; the ability to fabricate 3D artificial heart muscle is one such treatment modality. The two most essential components of 3D artificial heart muscle are cells and biomaterials; these are necessary pre-requisites to initiate the technology development process. Researchers in life sciences, particularly cell biology, are best trained to assume the responsibilities of cell sourcing while researchers in engineering, particularly material science, are best trained to assume responsibilities for biomaterial fabrication.

Cell sourcing requires identification, isolation, purification, and characterization of a suitable cell source, typically carried out by cell biologists. Development of biomaterials, particularly biomimetic biomaterials, can be spearheaded by the engineering team and would require expertise in biomaterial synthesis, characterization, and induction of bioactivity thereby allowing functional interaction with cells. The ability of cells to functionally interact with biomaterials and promote the formation of 3D heart muscle depends on many factors like attachment of cells to fibers of the biomaterials via integrin mediated binding and ability of cells to maintain differentiated phenotype during scaffold colonization. Understanding and manipulating cell-material interactions necessitates scientific input from engineering as well as life sciences experts.

During the next stage of technology development, the objective is to progress from early stage heart muscle to mature heart muscle. During this stage, there are three important areas of research. First, there needs to be an effort directed toward the development of small animal models to test the effectiveness and safety of bioengineered heart muscle. This work is best spearheaded by surgeons, who have the necessary skill set to undertake studies of this nature. Surgeons are also best capable of assessing success of the artificial heart muscle as it relates to recovery of myocardial function. Second, the development of bioreactor technology is essential and required to simulate physiological conditions and modulate fluid stress environments. Bioreactors are required to deliver controlled mechanical stretch, electrical stimulation, and controlled fluid flow to guide development and maturation of artificial heart muscle; engineers are best trained to undertake these studies. Third, an accurate assessment of tissue function is required, which involves gene and protein expression along with ultra-structural analysis. These performance metrics are essential to guide the success artificial heart muscle, and this work should be spearheaded by experts in the life sciences.

Let us move on to the final stage of the technology development process, which is focused on the development of 3D artificial heart muscle similar in form and function to mammalian heart muscle. Again, each of the three groups of experts has a specific role in the technology development progress. Large animal models are required to test the effectiveness and safety of tissue grafts; advanced bioreactors are needed with real-time functional assessment of processing variables and noninvasive methods to measure tissue function. As we have seen before, researchers in medical sciences, life sciences, and engineering have specific roles to play in this late stage of technology development; this again demonstrates the need for true collaboration by scientists from each discipline.

A true collaborative effort between various disciplines is imperative to the success of each phase, and it is crucial to promote the exchange of technology between each phase, revisiting the problem definition during every stage of the process. This simple example serves to demonstrate the degree of complex interactions and exchange of information required at the very early stages of scientific development between scientists from medical sciences, engineering, and life sciences. Development of a successful model to accomplish this degree of scientific and technological collaboration will be a significant challenge for the field of tissue engineering, yet is an essential ingredient for success.

1.14 GROWTH OF TISSUE ENGINEERING

The field of tissue engineering has seen significant growth during the last decade (Figure 1.8).

As has been the case with the development of any new technology, either within academia or industry, initial work starts due to the vision of a single person and is localized to the surrounding environment of this visionary person. During the early stages of technology development, preliminary work is always focused on establishing the feasibility of the work. Expansion of the technology is associated with successful development and evolution beyond initial feasibility studies. A very famous example is the story of Apple, which is now one of the most valuable companies on earth and has touched the lives of hundreds of millions of people around the globe. With such phenomenal success, it is hard to imagine the modest beginning of Apple, founded as Apple Computer Inc., in a small suburban garage by Steve Jobs, widely regarded as one of the greatest innovators of our time. As we will see in the current discussion, the field of tissue engineering has followed a similar trajectory, starting in the laboratory of a single investigator, Dr. Robert Langer, and expanding to its current status with national and international acceptance and recognition.

Early work in the field began in the Boston area, particularly in the laboratory of Dr. R. Langer. His early work focused on controlled drug delivery, with his first publication in this area appearing in 1976. However, the first publication describing tissue engineering appeared much later in 1988. It showed the survival of primary hepatocytes within 3D scaffolds. During these early years, there were few publications about either controlled drug delivery or tissue engineering. As with most new technologies, the initial work was restricted to a few research centers, which were knowledgeable in the scientific field. However, over time, there was an expansion in the field, as can be seen by the increase in the number of publications from 1990–2011 (Figure 1.8).

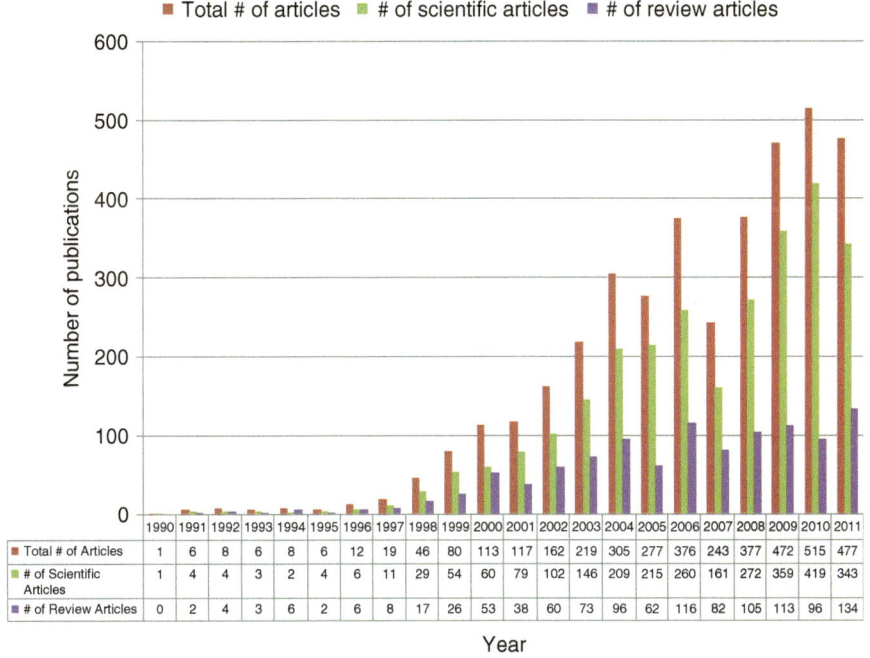

Figure 1.8 Growth of Tissue Engineering (1990–2011)—the growth of tissue engineering as a scientific discipline has seen significant growth over the period spanning from 1990 to 2011. The growth in tissue engineering is evident in terms of the number of scientific articles and review articles that have been published during this time period.

We evaluated the growth in tissue engineering based on the number of scientific publications. We conducted a title search, using tissue engineering as the keyword in Medline. Our search was limited to articles published in English. We distinguished between the total number of articles published and the number of review articles; the difference between these two was considered to be the total number of scientific articles. The results of this search are presented in Figure 1.8. As can be seen from the figure, there was a significant growth in the number of publications in the field of tissue engineering during the time period from 1990–2011. In 1990, there was one scientific publication with the word tissue engineering in the title; this number increased to 113 by the year 2000 and 515 by the year 2010. The growth of tissue engineering is not fully reflected in our data, as we limited our search to articles published with the phrase "tissue engineering" in the title. There are numerous articles which were published in the field and do not have the phrase tissue engineering in the title but focus on diverse research areas like bioreactors, biomaterials, and vascularization.

The growth of tissue engineering is attributed a review article published by Dr. R. Langer and Dr. V.C. Vacanti in the journal Science in 1993. This article was simple and did not provide a great deal of scientific and/or technological insight into the field. However, this article did provide a framework for tissue engineering, defining the scope, challenges, and future potential of success in the field. Publication of this article in a broad based journal like *Science* created a certain awareness of the field. The concept was particularly well received and endorsed by the scientific community. It encouraged researchers to participate in a shared vision, perhaps motivated by the potential impact of the development of tissue engineering technologies. Another driving factor leading to growth in tissue engineering was the availability of existing technologies that could translate into this field. The two major components defined in the paper were cell biology and material science; expertise in these areas was readily available in many research centers across the globe.

The field of tissue engineering has expanded significantly over the last decade, and there are several large research laboratories and tissue engineering centers across the country and across the globe. While the field was initially limited to certain research laboratories in the Northeast region of theUnited States, it has grown nationally to many regions in the United States and globally, particularly in Europe and Asia. This growth is seen in the number of publications in the field, in the number of new research laboratories being established and in the increase in annual research expenditure associated with the field. While the initial recognition of the field took time, tissue engineering as a scientific discipline has been well grounded for future expansion.

1.15 DISCIPLINES IN TISSUE ENGINEERING

The field of tissue engineering has traditionally been dominated by engineers, particularly chemical engineers. This is due to the fact that Dr. Langer, the founding father of the field, is a chemical engineer. Traditionally, this research has been well supported by Biomedical Engineering Departments, with almost all BME Departments having faculty members whose primary research area is in the tissue engineering or regenerative medicine space. In recent years, there has been a significant expansion in the number of BME Departments in the United States with significant number of faculty hires, many of whom list tissue engineering as an area of interest. This is also reflected in the data, as engineering disciplines accounted for up to two-thirds of publications in the field in 1995.

During later years of tissue engineering, researchers from Medical Schools, particularly from surgical disciplines, entered into the field. This was due to the nature of tissue engineering research, which is focused on the development of artificial tissue/organs for repair and replacement; surgeons are the end users of this technology. There was and continues to be a mutual interest in the development of this technology, and engineers and surgeons have become partners in this work (Figure 1.9). This is reflected in the data, where the participation rate of surgeons in scientific publications has gradually increased over the years, from about 25% in 1995 to about 35–40% in recent years. This trend has been fairly consistent over the years and continues to move forward in a positive manner, with a good partnership between both parties.

Biologists are the third major contributing partner in the field of tissue engineering and represent about 10–15% of scientific papers in the field. The contribution of cell biologists in the development of tissue engineering technology cannot be

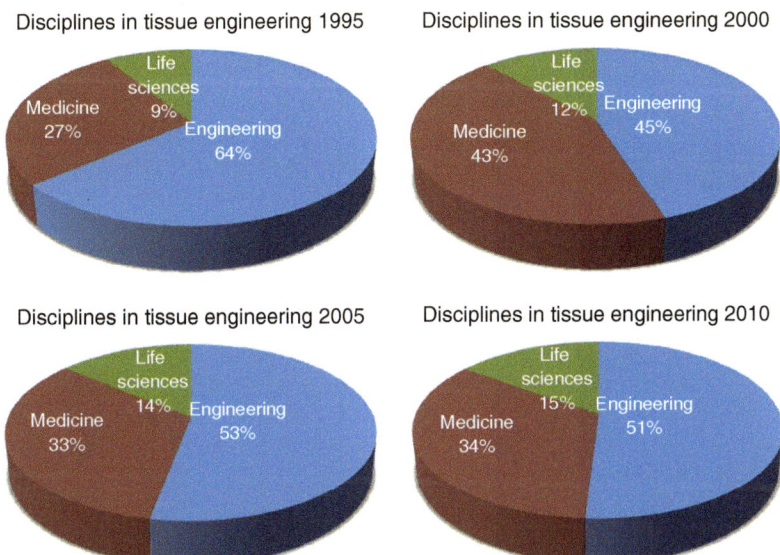

Figure 1.9 Scientific Disciplines in Tissue Engineering—An approximation of the participation rate of researchers from different backgrounds is shown for different time points during the period from 1995 to 2010.

overstated due to the seminal nature of cells in the tissue fabrication process. While the actual percentage may seem lower for researchers from life sciences, this is due to the fact that the research conducted in these fields is not directly geared toward tissue engineering, but rather to gain an understanding into molecular mechanisms of cell behavior; this work then indirectly feeds into the tissue fabrication process. For example, there are numerous publications in the field of stem cell engineering that are focused on developing novel technologies to drive the differentiation fate of stem cells toward a specific lineage. This work is not conducted in the context of tissue engineering, but is important for tissue/organ fabrication. As such, the numbers for life sciences are artificially skewed.

Tissue engineering continues to enjoy strong participation from researchers in engineering, medical sciences, and life sciences, and continued participation of all will remain a central theme for the field, which is important for successful tissue fabrication.

1.16 TISSUE ENGINEERING AND RELATED FIELDS

There are a few fields that are closely related to tissue engineering and will be discussed in this section. There has often been some confusion, disagreement, and debate over the exact definitions for each of these scientific disciplines. The field of tissue engineering is still very fluid and is being defined and redefined on a regular basis. This problem is compounded by the fact that researchers use different terminology to refer to different things; this is expected due to the novelty of the field and due to personal preferences and unique interpretations of the field of tissue engineering. In this section, we present the current understanding of tissue engineering and several related fields, which include gene/protein therapy, controlled release, cell transplantation, cell encapsulation, tissue/organ engineering, and regenerative/reparative medicine. We discuss relative advantages and disadvantages of these fields and scientific and technological challenges for the respective areas of research. We also discuss interrelationships between these scientific disciplines.

The field of tissue engineering is closely related to the field of controlled release and can be viewed as an extension of gene/protein/cell therapy. While it is often difficult to draw comparisons between scientific disciplines, identifying common trends may provide a broader perspective. Lessons learned from the more mature fields like gene and cell therapy will invariably be valuable to fields like tissue engineering, which are still in their infancy. Although research in these fields continues in parallel, the ability to umbrella this body of work under one scientific discipline may be desirable.

Gene Therapy—Gene therapy is defined as the process by which genes, small DNA, or RNA molecules are delivered to human cells, tissues, or organs to correct a genetic defect, or to provide new therapeutic functions for the ultimate purpose of preventing or treating diseases[42]. The primary aim of gene therapy is to either increase or decrease the level of a specific protein within target tissue in order to modify cellular function of the targeted cell or to effect changes to surrounding tissues by altering secreted proteins (42).

The candidate genes for cardiovascular disorders can easily be categorized based on target tissue, which includes the myocardium, vasculature, and cardiac conducting system (43). As one example, the calcium handling proteins play a critical role in maintaining cardiac contractility via excitation contraction coupling. Calcium release from the sarcoplasmic reticulum, in response to depolarization, is regulated by the ryanodine receptors while calcium uptake is controlled by sarcoplasmic endoreticulum Ca-ATPase (SERCA): SERCA activity in turn is regulated by phospholamban. Studies have shown that congestive heart failure in humans is associated with a decrease in SERCA2 (44), measured in terms of gene expression (45), protein level, and activity (46). Several studies have shown that adenovirus-mediated gene transfer of SERCA2 has resulted in improvement of calcium transients in cardiac myocytes after heart failure (47–49).

There are several critical challenges in the field of cardiac gene therapy which need to be addressed (42): 1) mode of delivery, 2) viral myocardial tropism 3) timing of expression, 4) insertional mutagenesis. The mode of delivery of genes has been via transduction, which involves the utilization of viral vectors, or via transfection, which involves non-viral vectors. Transduction is utilized more frequently due to increased efficiency in gene transfer while non-viral vectors are considered safer. The challenge remains to increase the safety of transduction and the efficiency of transfection. Viral myocardial tropism refers to the ability of the virus to specifically target the myocardium, as transfection of non-target tissue can have detrimental effects due to nonfunctional expression of proteins and/or immunogenic effects (42). The timing of expression is important, and it may be desirable to shut off the expression after therapeutic benefit has been observed, thereby limiting the risk of tumorigenesis (42). Insertional mutagenesis refers to the integration of DNA within coding regions of genes and/or regulatory elements; this must be avoided because it can disrupt normal function of the gene[4] (42).

Protein Therapy—Protein therapy is defined as the delivery of proteins to cells, tissues or organs to provide a therapeutic function. Therapeutic angiogenesis, a specialized case of protein therapy, involves delivering angiogenic growth factors to support the formation of new blood vessels, thereby increasing perfusion to infarct tissue (50). In this section, we will discuss therapeutic angiogenesis as it relates to protein therapy in order to gain an understanding of the field and some of its associated challenges. Angiogenesis is defined as the formation of new blood vessels from preexisting vasculature via activation of endothelial cells that proliferate and migrate to construct new capillaries (51).

Vascular endothelial growth factor-A (VEGF-A) has been evaluated extensively to support angiogenesis, and several mechanisms, perhaps acting concurrently, have been proposed for potency of VEGF-A (50). VEGF-A binding to VEGFR-2 promotes survival, proliferation, and migration of endothelial cells, while binding to VEGFR-1 results in vascular permeability and has been shown to promote migration of circulating monocytes and recruitment of hematopoietic progenitor cells (HPCs) from bone marrow to ischemic sites (52–54).

The major challenges that need to be addressed in the field of therapeutic angiogenesis are the mode of delivery of growth factors, concentration of growth factors, and time of delivery. The options for delivery of proteins are the same as for delivery of cells and include intravenous, intracoronary, or intramyocardial delivery (50). The concentration and timing of delivery is also important, as evidence from clinical studies have shown that administration of a single high dose of angiogenic growth factors leads to unstable vessels, while exposure at lower concentrations for extended time periods promotes stable vessel formation (50).

Cell Therapy—Cell therapy involves use of isolated cells that are delivered to damaged or diseased tissue for therapeutic purposes. When applied to the heart, cell therapy is referred to as cardiac cell therapy (55–57); we will discuss cardiac cell therapy to illustrate the challenges associated with cell therapy as a scientific discipline. The field of cardiac cell therapy is centered on the premise that localized delivery of cells to the site of an infarct will result in an increased contractile performance. Initial work involved use of satellite cells and progenitor cells for skeletal myoblasts. The motivating factors to utilize these cells were the availability of an autologous source and contractile activity of mature myoblasts.

Since the initial work involving satellite cells, several cell types have been evaluated as potential candidates for cardiac cell therapy: 1) mesenchymal cells derived from bone marrow and/or adipose tissue, 2) circulating progenitor cells, 3) endothelial progenitor cells (EPCs), 4) resident myocardial progenitor cells, and 5) human embryonic stem cells (55).

The exact mode of action has not been fully elucidated for every cell type, with multiple pathways being postulated (55): 1) differentiation of uncommitted stems to cardiomyocytes resulting in direct increase of contractility, 2) differentiation of EPCs to endothelial cells thereby promoting vascularization, 3) differentiation of stem cells to smooth muscle cells thereby increasing neovascularization, 4) eliciting a paracrine effect thereby activating endogenous stem cells, and/or 5) stimulating survival of border zone cells.

While the field of cardiac cell therapy has tremendous opportunities, several challenges need to be overcome prior to realizing the full potential of the field. These include: 1) mode of delivery for the cells (intravenous, intracoronary or intramyocardial), 2) cell retention at the infarct site, and 3) long term engraftment and survival.

Comparison Between Gene Therapy, Protein Therapy, Cell Therapy and Tissue Engineering—The field of tissue engineering can be viewed as an extension to work in protein, gene and cell therapy. While the goal of these latter fields is to deliver specifically targeted cellular components at the site of infarct tissue, the goal of tissue engineering is to deliver 3D tissue at the site of injury. In the case of gene and protein therapy, the therapeutic agent is the gene or protein of interest, respectively. These strategies are focused at the molecular level and are selective in terms of targeting a specific function. Insertion of a gene or delivery of a protein will replace a very specific function of the cell that has been lost due to injury (as an example, VEGF can be used to increase neovascularization). While gene/protein therapy is designed to mediate changes at the molecular level, cell therapy works at a larger scale and is designed to mediate changes at the cellular level. While gene/protein therapy aims to selectively replace one function of the cell, the goal of cell therapy is to completely replace the cellular component of damaged or diseased tissue. The major limitation of cell therapy is low retention at the site of delivery, and this problem is addressed by the field of tissue engineering. In the case of cell therapy, isolated cells are transplanted at the site of injury, and a small percentage of these cells are retained at the site of delivery. Tissue engineering strategies are focused on bioengineering 3D artificial tissue using isolated cells and then transplanting bioengineered tissue to the site of injury; the therapeutic agent is 3D tissue, and the problem of low retention has been solved. Cells within 3D artificial tissue are retained at the site of delivery as they are integrated as part of a system.

Controlled Release—The field of controlled release is very closely related to the field of tissue engineering, partly due to research in both fields being started by the same person, Dr. Robert Langer. Controlled release strategies involve the release of a specific drug over time. Rather than providing a single dose at the time of administration, the goal is to provide sustained release over time, preferably with zero order kinetics, which means that the release rate is consistent with time (21,58,59). The main advantage of controlled release strategies involves the ability to maintain therapeutic plasma drug levels without reaching extremely high or dangerously low concentrations. Polymeric scaffolds have been utilized extensively to bind therapeutic drugs; controlled degradation of the polymeric results in the release of drugs. While the field of controlled release is promising, polymer design with controllable degradation kinetics and safe degradation by-products is important. In addition, surgical implantation of the polymer may be required, thereby necessitating biocompatible biomaterials.

Controlled Release and Tissue Engineering—There are clear differences and similarities between the two fields. The most important difference between the two fields is that controlled release is focused on delivery of a therapeutic agent to the injury site and does not involve cells. Tissue engineering, as we have seen, revolves around cells and the development of strategies to support tissue fabrication. This is a very important distinction between the two, as controlled release technologies are focused on delivery of drugs while tissue engineering strategies are focused on development of cell based technologies. The commonality between the two fields is the use of scaffolds. In the case of controlled release technology, properties of the scaffold regulate release kinetics of drugs, which in turn dictate effectiveness of the therapy. Controlled release of the therapeutic agent depends on degradation kinetics of the material—as the material degrades, the therapeutic agent is released within the tissue, and the rate of drug release correlates with the rate of material degradation. In the case of tissue engineering, properties of the biomaterial are important to support tissue fabrication. In one strategy, scaffold degradation is replaced by extracellular matrix (ECM) produced by cells, which in turn support formation of 3D artificial tissue. In the case of tissue engineering, the rate of material degradation correlates with the rate of ECM production by cells, which in turn supports artificial tissue fabrication.

Cell Encapsulation—Cell encapsulation is focused on the culturing of cells within a scaffold, which regulates the release of a therapeutic agent produced by cells into the culture environment[7]. Cells are encapsulated within a 3D scaffold to protect from the host immune system; immune cells like neutrophils and macrophages cannot penetrate the barrier created by the scaffold. The scaffold acts as a semi-permeable membrane and blocks host immune cells; however, nutrients like oxygen and glucose can pass through the scaffold and reach cells. In addition, therapeutic agents, like insulin, produced by the cells can leave the scaffold. The properties of the scaffold are designed to protect cells from the host immune cells, while at the same time supporting diffusion of nutrients to support cell viability and supporting release of therapeutic agents (7). As can be appreciated from the forgoing discussion, cell encapsulation is closely related to controlled release, as both cases require the use of a scaffold for encapsulation of drugs or cells. In the case of controlled release, polymer degradation is used to regulate the release kinetics of the drug while in the case of cell encapsulation, material degradation is not a prerequisite for success of the therapy; rather, the scaffold acts as a semipermeable membrane to support the release of therapeutic agents by encapsulated cells.

Cell Encapsulation and Tissue Engineering—The relationship between the two fields is the utilization of cells for therapeutic purposes. In the case of cell encapsulation, cells function to release specific proteins in the host environment, which serves a therapeutic purpose. In the case of tissue engineering, cells are used to support artificial tissue fabrication, which then acts to replace or restore function in damaged or diseased tissue.

Organ Engineering and Tissue Engineering—The term organ engineering refers to the design and fabrication of entire bioartificial organs and can be considered an extension of the field of tissue engineering (60–62). The holy grail of tissue engineering is indeed the fabrication of bioartificial organs. There needs to be one distinction—the success of tissue engineering technologies should not be judged by the ability to bioengineer bioartificial organs. Artificial tissue itself is a successful endpoint, and in many clinical applications, bioengineered artificial tissue can provide lifesaving options for patients. There are also cases when entire organ transplantation will be necessary and artificial organs will be needed for treatment. As an example, in the case of cardiovascular tissue engineering, technology is being developed to bioengineer artificial heart muscle, tri-leaflet heart valves, blood vessels, heart pumps, ventricles and bioartificial hearts. Depending on the clinical condition, individual tissue constructs may be required and can prove to be beneficial in restoring lost functionality. However, in the case of end-stage heart failure, heart transplantation may be the only viable treatment option, and a complete bioartificial heart will be required. Therefore, both tissue engineering and organ engineering are important and need to be pursued.

Regenerative Medicine and Reparative Medicine—The terms regenerative medicine (63–66) and reparative medicine (67–70) are broad terms used to define therapeutic strategies aimed at regenerating or repairing damaged or diseased tissue, irrespective of the mechanism or therapeutic agent involved. The therapeutic agent could be a drug, protein, gene, cell (encapsulated or not), or artificial tissue. The fields that we have discussed thus far—gene/protein therapy, cell transplantation, controlled release, cell encapsulation and tissue engineering—are all subcategories of regenerative or reparative medicine. Regenerative and reparative medicine can be viewed as broad overarching themes that refer to any strategy aimed to regenerate or repair damaged or diseased tissue. The specific fields that we have discussed should be viewed as specific therapeutic strategies to achieve this end objective. The term regenerative medicine has been used extensively in the literature while reparative medicine has not been very dominant. We do not distinguish between the two and consider both to be the same. However, due to the dominance of the term regenerative medicine relative to reparative medicine, we will use the term regenerative medicine to refer to any therapeutic strategy with the potential to regenerate mammalian tissue. The term reparative medicine is not used in the remainder of this book and has been included in our discussion for the sake of completion.

Regenerative Medicine and Tissue Engineering—We end this section with a brief discussion distinguishing the fields of regenerative medicine and tissue engineering. Based on our prior discussion, the reader will already have an understanding about the differences between the two fields. However, due to the importance of these two fields in the presentation of the material in this book, we include this discussion. We consider tissue engineering to be a specific therapeutic strategy aimed and repairing, replacing, and/or restoring lost tissue function, which is a subcategory of the broader field of regenerative medicine.

SUMMARY

In this chapter, we have provided a framework for the field of tissue engineering. We started the chapter explaining the chronic shortage of donor organs around the globe and the ability of tissue engineering to provide a viable clinical strategy for artificial tissue and organ development. We provided a formal definition of tissue engineering and outlined an eight-step process to fabricate 3D artificial tissue and organs. We discussed the building blocks of tissue engineering (cell, biomaterials, and bioreactors) and looked at some of the scientific and technological challenges in the field of tissue engineering. We discussed seminal work by Dr. Robert Langer and his contribution to the development of tissue engineering; we also described seminal publications in the field by Dr. Robert Langer. We looked at several applications of the 3D artificial tissue and compared the relative advantages and disadvantages of 2D versus 3D culture. We discussed an integrative model for tissue engineering including participation from different disciplines and the relative contribution of researchers from different disciplines toward development of tissue engineering models. We also presented some data demonstrating the significant growth in the field of tissue engineering and looked at drivers of growth in the field. We concluded this chapter by presenting a comparison of tissue engineering with related fields including cell transplantation, controlled research, regenerative medicine, gene/protein transplantation, and encapsulation technology. Many of the concepts that were introduced in this chapter will be important in later chapters. In addition, many of the concepts that were introduced in this chapter will be expanded upon in subsequent chapters including cells, biomaterials, bioreactors and vascularization. In conclusion, during the course of this chapter, we have looked at an eight-step process to bioengineer artificial tissue and have identified cells, biomaterials, and bioreactors as the building blocks for tissue engineering.

PRACTICE QUESTIONS

1. Provide a general description of tissue engineering. Without using information from the chapter and using any technical terms, based on your understanding of the field, talk about the field of tissue engineering. Based on your current understanding, what exactly is tissue engineering? Why is it important? What are some potential outcomes of successful tissue engineering technologies?
2. Describe how the development of artificial organs using tissue engineering strategies could alleviate problems associated with the shortage of donor organs.
3. Define tissue engineering.
4. In this chapter, we provided an eight-step process flow sheet for the fabrication of artificial tissue and/or organs. Describe the process of bioengineering artificial tissue and/or organs.
5. In the previous question, you were asked to describe an eight-step process of fabricating artificial tissue and/or organs. If you were to add two additional steps to the process, what would those be and why?
6. During our discussion of the tissue fabrication process, we introduced sensor technology to monitor 3D artificial tissue fabrication. Describe why sensors are needed. What are some important variables that should be monitored? Explain what is meant by real-time noninvasive monitoring of tissue function and explain why this is important.
7. What are the building blocks of tissue engineering? Explain your answer.
8. Describe and discuss five scientific challenges associated with artificial tissue fabrication.
9. What are some potential applications of artificial tissue and organs? Provide four examples, only two of which can be from the chapter.
10. How can 3D artificial tissue be used to support the drug development process?

11. Tissue engineering is a multidisciplinary research field. Describe some of the participating disciplines and explain the relative contribution of each to the process of tissue fabrication.
12. The functional performance of 3D artificial tissue can be assessed by measuring functional, biological, and histological metrics. What do each of these three terms mean?
13. How would you measure the functional performance of 3D artificial heart muscle? Explain in terms of functional, biological, and histological metrics.
14. In this chapter, we discussed three seminal publications in the field of tissue engineering. Explain why these papers are considered to be seminal. What was the contribution of each of these to the field of tissue engineering?
15. What are some of the differences between 2D and 3D culture? Discuss the relative advantages and disadvantages of 2D culture and 3D culture.
16. Cell culture techniques using 2D monolayer systems have been used for decades. The technology for 2D cell culture is well-established. Tissue engineering offers the potential to develop 3D culture systems for isolated cells. The technology for 3D cell culture is not well-established, as the field is very young. What needs to be done to standardize 3D culture techniques?
17. Compare the fields of cell transplantation and tissue engineering. Start by describing these fields. Compare the relative advantages and disadvantages of each of the two fields.
18. What does the term regenerative medicine refer to? What is the relationship between regenerative medicine and tissue engineering?
19. Describe the terms tissue engineering and organ engineering. Select any organ system, excluding the cardiovascular system, and explain how tissue engineering and organ engineering can be used to develop therapeutic strategies.
20. If you had an opportunity to bioengineer any artificial tissue or organ, which one would it be and why?

REFERENCES

1. Organ Procurement and Transplantation Network and Scientific Registry of Transplant Recipients 2010 Data Report. Am. J. Transplant. 2012;12:1–154.
2. Khait L, Hecker L, Blan NR, et al. Getting to the heart of tissue engineering. J Cardiovasc Transl Res 2008;1(1):71–84.
3. Lal B, Viola J, Hicks D, Grad L. Emergence of Tissue Engineering as a Research Field. 2003 Oct 13.

4. Bell E. Tissue Engineering, an Overview. In: Bell E, editor. Tissue Engineering: Current Perspectives. Boston, MA: Birkhäuser; 1993. p 3–15.
5. Langer R, Vacanti JP. Tissue Engineering. Science 1993;260(5110):920–926.
6. Reparative Medicine: Growing Tissue and Organs. 2001.
7. Sefton MV. Functional considerations in tissue-engineering whole organs. Reparative Medicine: Grow. Tissues Organs 2002;961:198–200.
8. Atala A. Tissue engineering and regenerative medicine: Concepts for clinical application. Rejuvenat. Res. 2004;7(1):15–31.
9. Nerem RM. Tissue engineering: The hope, the hype, and the future. Tissue Engng. 2006;12(5):1143–1150.
10. Mason C, Dunnill P. A brief definition of regenerative medicine. Regenerative Medicine 2008;3(1):1–5.
11. Langer R. Biomaterials 1990;11(9):613.
12. Robert Langer. Nat. Rev. 2005;Drug(5):366.
13. Langer R. Robert Langer, ScD--Engineering medicine. Interview by M. J. Friedrich. JAMA 2005;294(13):1609–1610.
14. Robert Langer MIT Faculty Page. MIT Department of Chemical Engineering 8 A.D. January 7;Available at: URL: http://web.mit.edu/cheme/people/faculty/langer.html.
15. Elizabeth A. Thomson. MIT Newsoffice: Bob Langer named an Institute Professor; 2005 Mar 2.
16. President Bush Presents Awards to 2005 and 2006 National Medal of Science and Technology Recipients. The White House Press Room 2006 Jul 27.
17. 2002 Recipient of the Charles Stark Draper Prize. National Academy of Engineering 2002.
18. Langer wins $500,000 Lemelson-MIT award. MIT Newsoffice 1998 Apr 15.
19. A Celebration Of Business Innovators And Ideas. Forbes. 12-5-0002. Ref Type: Magazine Article.
20. CNN/Time Magazine's America's Best Science and Medicine. CNN and Time Magazine. 2001. Ref Type: Magazine Article.
21. Langer R, Folkman J. Polymers for the sustained release of proteins and other macromolecules. Nature 1976;263(5580):797–800.
22. Vacanti JP, Morse MA, Saltzman WM, Domb AJ, Perez-Atayde A, Langer R. Selective cell transplantation using bioabsorbable artificial polymers as matrices. J. Pediatric Surg. 1988;23(1:Pt 2):t-9.
23. Helmrich A, Barnes D. Animal cell culture equipment and techniques. Methods in Cell Biology, 57 1998;57:3–17.
24. Fry RJM, Nachtwey DS. Health-Effects of the Radiation Environment in Space. Radiat. Res. 1983;94(3):541.
25. Schimmerling W, Sulzman FM. The Nasa Space Radiation Health-Program. Life Sci. Space Res. Xxv(2) 1994;14(10):133–137.
26. Cirio R, Cucinotta FA, Durante M. Proceedings of the "1(st) International Workshop on Space Radiation Research and 11(th) Annual NASA Space Radiation Health Investigators' Workshop", Arona (Italy), May 27–31, 2000—Preface. Physica Medica 2001;17:III-IIV.

27. Frey MA. Protecting Astronaut Health by Limiting Career Exposure to Space Radiation. Aviat. Space Environ. Med. 2009;80(8):741–742.
28. Goel N, Chaturvedi S, Goel A. Impact of technologies on the drug development process. Indian J. Pharmacol. 2008;40:203.
29. Munoz SG, Oksanen CA. Process modeling and control in drug development and manufacturing. Comput. Chemi. Eng. 2010;34(7):1007–1008.
30. Kaitin KI. Deconstructing the Drug Development Process: The New Face of Innovation (Vol 87, pg 356, 2010). Clinic. Pharmacol. Therap. 2011;89(1):148.
31. Jorgensen JT. A challenging drug development process in the era of personalized medicine. Drug Discov. Today 2011;16(19–20):891–897.
32. Jayapal M, Bhattacharjee RN, Melendez AJ, Hande MP. Environmental toxicogenomics: A post-genomic approach to analysing biological responses to environmental toxins. Int. J. Biochem. & Cell Biol. 2010;42(2):230–240.
33. Hyman MA. Environmental Toxins, Obesity, and Diabetes: An Emerging Risk Factor. Alternat. Therap. Health Med. 2010;16(2):56–58.
34. Brown P, Morello-Frosch R, Brody JG et al. Institutional review board challenges related to community-based participatory research on human exposure to environmental toxins: A case study. Environ. Health 2010;9.
35. Chin NP. Environmental Toxins: Physical, Social, and Emotional. Breastfeeding Med. 2010;5(5):223–224.
36. Nguyen E, Stein J. Functional Outcomes of Adults with Left Ventricular Assist Devices Receiving Inpatient Rehabilitation. Pm&R 2013;5(2):99–103.
37. Landis ZC, Soleimani B, Stephenson ER, El-Banayosy A, Pae WE. Severity of End Organ Damage as a Predictor of Outcomes after Implantation of Continuous Firm Left Ventricular Assist Devices (LVAD). J. Heart Lung Transplant. 2013;32(4):S222.
38. Jacobs S, Geens J, Rega F, Burkhoff D, Meyns B. Continuous-flow left ventricular assist devices induce Left ventricular reverse remodeling. J. Heart Lung Transplant. 2013;32(4):466–468.
39. Wilson G. Cell-Culture Techniques for the Study of Drug Transport. Euro. J. Drug Metab.Pharmacokinet. 1990;15(2):159–163.
40. Ahern H. Cell and Tissue-Culture Techniques A Combination of Science and Art. Scientist 1995;9(24):18–19.
41. Mozdziak PE, Petitte JN, Carson SD. An introductory undergraduate course covering animal cell culture techniques. Biochem. Mol. Biol. Edu. 2004;32(5):319–322.
42. Lyon AR, Sato M, Hajjar RJ, Samulski RJ, Harding SE. Gene therapy: targeting the myocardium. [Review] [94 refs]. Heart 2008;94(1):89–99.
43. Ly H, Kawase Y, Yoneyama R, Hajjar RJ. Gene therapy in the treatment of heart failure. [Review] [86 refs]. Physiology 2007;22:81–96.
44. Wankerl M, Schwartz K. Calcium transport proteins in the nonfailing and failing heart: gene expression and function. [Review] [92 refs]. J. Mol. Med. 1995;73(10):487–496.
45. Arai M, Alpert NR, MacLennan DH, Barton P, Periasamy M. Alterations in sarcoplasmic reticulum gene expression in human heart failure. A possible mechanism for alterations in systolic and diastolic properties of the failing myocardium. Circ. Res. 1993;72(2):463–469.

46. Hasenfuss G, Reinecke H, Studer R, et al. Relation between myocardial function and expression of sarcoplasmic reticulum Ca(2+)-ATPase in failing and nonfailing human myocardium. Circ. Res. 1994;75(3):434–442.
47. del MF, Williams E, Lebeche D et al. Improvement in survival and cardiac metabolism after gene transfer of sarcoplasmic reticulum Ca(2+)-ATPase in a rat model of heart failure. Circulation 2001;104(12):1424–1429.
48. Logeart D, Vinet L, Ragot T, et al. Percutaneous intracoronary delivery of SERCA gene increases myocardial function: a tissue Doppler imaging echocardiographic study. Am. J. Physiol.—Heart Circulat. Physiol. 2006;291(4):H1773–H1779.
49. Davia K, Bernobich E, Ranu HK, et al. SERCA2A overexpression decreases the incidence of aftercontractions in adult rabbit ventricular myocytes. J. Mol. Cell. Cardiol. 2001;33(5):1005–1015.
50. Molin D, Post MJ. Therapeutic angiogenesis in the heart: protect and serve. [Review] [44 refs]. Curr. Opini. Pharmacol. 2007;7(2):158–163.
51. Carmeliet P. Manipulating angiogenesis in medicine. [Review] [294 refs]. J. Int. Med. 2004;255(5):538–561.
52. Pipp F, Heil M, Issbrucker K, et al. VEGFR-1-selective VEGF homologue PlGF is arteriogenic: evidence for a monocyte-mediated mechanism. Circ. Res. 2003;92(4):378–385.
53. Olsson AK, Dimberg A, Kreuger J, Claesson-Welsh L. VEGF receptor signalling—in control of vascular function. [Review] [139 refs]. Nat. Rev. Mol. Cell Biol. 2006;7(5):359–371.
54. Heissig B, Rafii S, Akiyama H, et al. Low-dose irradiation promotes tissue revascularization through VEGF release from mast cells and MMP-9-mediated progenitor cell mobilization. J. Exp. Med. 2005;202(6):739–750.
55. Laflamme MA, Murry CE. Regenerating the heart. [Review] [135 refs]. Nat. Biot. 2005;23(7):845–856.
56. Segers VF, Lee RT. Stem-cell therapy for cardiac disease. [Review] [77 refs]. Nature 2008;451(7181):937–942.
57. Lyon A, Harding S. The potential of cardiac stem cell therapy for heart failure. [Review] [56 refs]. Cur. Opin. Pharmacol. 2007;7(2):164–170.
58. Folkman J. How the field of controlled-release technology began, and its central role in the development of angiogenesis research. [Review] [35 refs]. Biomaterials 1990;11(9):615–618.
59. Laurencin CT, Langer R. Polymeric controlled release systems: new methods for drug delivery. [Review] [78 refs]. Clinics Lab. Med. 1987;7(2):301–323.
60. Van Dyke M, Oberpenning F, Meng J, Soker S, Yoo JJ, Atala A. Total organ replacement using tissue engineering. Faseb J. 2007;21(5):A140.
61. Gaujoux S, Larghero J, Cattan P. Tissue engineering: A solution for organ replacement? J. Chirurgie 2009;146(2):109–111.
62. Rustad KC, Sorkin M, Levi B, Longaker MT, Gurtner GC. Strategies for organ level tissue engineering. Organogenesis 2010;6(3):151–157.
63. Couto DS, Perez-Breva L, Cooney CL. Regenerative Medicine: Learning from Past Examples. Tissue Eng. Part A 2012;18(21–22):2386–2393.
64. Mhashilkar AM, Atala A. Advent and Maturation of Regenerative Medicine. Cur. Stem Cell Res. Therapy 2012;7(6):430–445.

REFERENCES

65. Fisher MB, Mauck RL. Tissue Engineering and Regenerative Medicine: Recent Innovations and the Transition to Translation. Tissue Engineering Part B-Reviews 2013;19(1):1–13.
66. Slingerland AS, Smits AIPM, Bouten CVC. Then and now: hypes and hopes of regenerative medicine. Trends Biotechnol 2013;31(3):121–123.
67. Chaikof EL, Matthew H, Kohn J, Mikos AG, Prestwich GD, Yip CM. Biomaterials and scaffolds in reparative medicine. Reparative Medicine: Growing Tissues Organs 2002;961:96–105.
68. Yip C. Biomaterials in reparative medicine—Biorelevant structure–property analysis. Reparative Medicine: Grow. Tissues Organs 2002;961:109–111.
69. Parenteau NL, Young JH. The use of cells in reparative medicine. Reparative Medicine: Grow. Tissues Organs 2002;961:27–39.
70. Sipe JD. Tissue engineering and reparative medicine. Reparative Medicine: Grow. Tissues Organs 2002;961:1–9.

2

CELLS FOR TISSUE ENGINEERING

Learning Objectives:

After completing this chapter, students should be able to:

1. Explain why cells are important for tissue engineering.
2. Describe the structure and function of organelles in eukaryotic cells.
3. Describe the structure and function of the mammalian extracellular matrix.
4. Discuss different mechanisms for cell signaling.
5. Describe the structure and function of cellular junctions.
6. Discuss cell-cell and cell-matrix interactions in terms of tissue engineering.
7. Discuss potential cell sources for tissue engineering and tissue fabrication.
8. Explain the principles of cell transplantation, and discuss how this field relates to tissue engineering and the tissue fabrication process.
9. Provide examples of cells that have been used for cell transplantation.
10. Compare the fields of cell transplantation and tissue engineering.
11. Describe the process of 2D monolayer cell culture, and explain how it relates to tissue engineering and the tissue fabrication process.
12. Provide examples of applications of monolayer cell culture.
13. Compare the fields of 2D cell culture and 3D tissue engineering.

Introduction to Tissue Engineering: Applications and Challenges, First Edition. Ravi Birla.
© 2014 The Institute of Electrical and Electronics Engineers, Inc. Published 2014 by John Wiley & Sons, Inc.

14. Describe characteristics of stem cells, and explain the concepts of cell differentiation and cell potency.
15. Discuss the use of human embryonic stem cells, induced pluripotent stem cells, and adult stem cells for fabrication of 3D artificial tissue.

CHAPTER OVERVIEW

We begin this chapter by explaining the importance of cells during the tissue fabrication process. We proceed with a discussion of cell structure and function and then discuss the dynamic extracellular matrix. We then discuss cell signaling and cellular junctions. We next discuss the role of cell biology in tissue engineering and follow this up with discussion on cell sourcing for tissue fabrication. The last three sections of this chapter are focused on mammalian cell culture, cell transplantation, and stem cell engineering. We describe the process of cell transplantation, including steps that are involved in the development of cell transplantation therapies. During our discussion on mammalian cell culture, we describe the process of cell culture and the relative importance of cell culture during tissue fabrication. We also provide a discussion comparing the relative advantages and disadvantages of cell transplantation and tissue engineering. We next move on to describe mammalian cell culture. We describe several steps in the cell culture process, including isolation, culture, and expansion of primary cells. We discuss applications of cell culture and provide a comparison of cell culture versus tissue engineering. We then move on to discuss stem cells and the role of stem cell engineering in tissue engineering. We describe the properties of stem cells, along with several stem cell sources that have been used in tissue engineering, including human embryonic stem cells, induced pluripotent stem cells, and mesenchymal stem cells.

2.1 CELLS AND TISSUE ENGINEERING

An understanding of cells is central to the field of tissue engineering and regenerative medicine. As we have seen in Chapter 1, cells are the functional components of 3D artificial tissue and are one of the building blocks of the field. Isolation, culture, and expansion of cells are routinely conducted as a part of the tissue fabrication process. In addition, stem cell engineering has evolved to play a critical role in tissue engineering (1–4). Genetic programming of stem cells can drive the differentiation fate of these cells toward a given lineage, and these cells can be used to support fabrication of artificial tissue (5–7). An understanding of cell biology, cell culture, and stem cell engineering is critical for success in tissue engineering and will be the focus of this chapter.

Cells are the fundamental units of life and are one of the building blocks for tissue engineering. There are several aspects of cell biology that are important for the tissue fabrication process, and, as illustrated in Figure 2.1, cell biology, cell culture, cell transplantation, and stem cell engineering are important during the

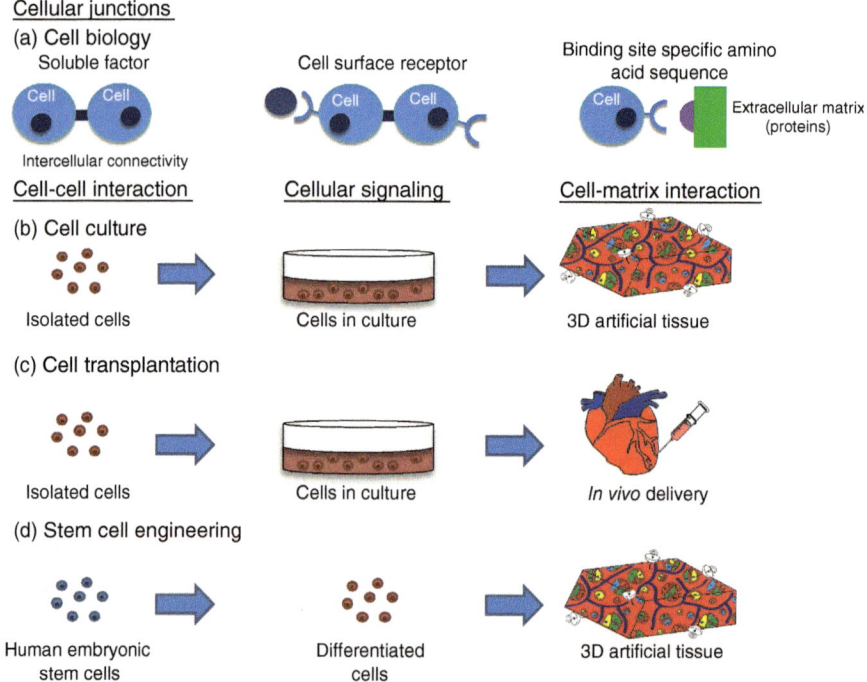

Figure 2.1 Cells and Tissue Engineering—Cells are important for tissue engineering, and four important areas are (a) Cell Biology, (b) Cell Culture, (c) Cell Transplantation, and (d) Stem Cell Engineering.

tissue fabrication process. These four topics are the subject of this chapter and are briefly introduced in this section.

Let us begin this discussion with cell biology as it relates to tissue engineering. Some of the concepts are illustrated in Figure 2.1a and will be discussed throughout this chapter, including cell-cell interactions, cell-matrix interactions, intercellular connectivity, and cellular signaling. Individual cells form the basic unit of mammalian tissue and are a critical function component of 3D artificial tissue. Mammalian tissue is composed of different cell types, each of which has a unique role to support functional performance of tissue. In addition to cells, mammalian tissue is composed of an extracellular matrix (ECM), which provides structural support during tissue function. There are functional interactions between adjacent cells and between cells and mammalian ECM; these are referred to as cell-cell interactions and cell-matrix interactions respectively, both of which are critical determinants of tissue function. Cells do not function in isolation; they are part of a large biological system. As such, cells have a sophisticated mechanism to sense and respond to changes in the local environment through a process known as cell signaling.

In addition to understanding cell behavior as it relates to the tissue fabrication process, another important area related to tissue engineering is cell culture

(8–11) (Figure 2.1b). In order to undertake any tissue engineering experiment, it is important to isolate, maintain, and expand cells in culture. In order to achieve this objective, a technique known as cell culture is used. Cell culture techniques are important for the isolation of primary cells, along with the culture and expansion of these cells prior to utilization for fabrication of 3D artificial tissue. There are several fundamental principles related to mammalian cell culture, which are discussed later in this chapter.

Cell transplantation was introduced in Chapter 1 and involves the delivery of isolated cells *in vivo* in order to improve functional performance of damaged or diseased tissue (12–14) (Figure 2.1c). Prior to *in vivo* transplantation, isolated cells are cultured using cell culture techniques to increase the number of cells, as large cell numbers are required for cell transplantation studies.

The most important question relating to cells during the tissue fabrication process is: *where exactly will the cells come from?* There are many options available for cell sourcing. Stem cell engineering has received a lot of attention in recent literature (Figure 2.1d), with human embryonic stem (hES) cells being one example (15–17). The primary advantage of hES cells is the totipotent potential of these cells, which means that they can be differentiated to form any cell type in the human body. Therefore, starting with a single cell type (hES cells), researchers have the ability to obtain many different cell types, which can then be used to fabricate artificial tissue. During the course of this chapter, we will study many scientific challenges associated with achieving this objective and will also study different sources for stem cells.

2.2 CELL STRUCTURE AND FUNCTION

Cells are the fundamental units of life and building blocks of all mammalian tissue, which is made up of multiple cell types interconnected with extracellular matrix components. Mammalian cells are extremely complex and can undertake hundreds of functions at any given time. Cells consist of many different structures known as organelles, each of which has a specialized function. Cells are categorized as prokaryotes if they do not contain a nucleus; they are categorized as eukaryotes if the cells contain a nucleus. Both prokaryotic and eukaryotic cells have DNA. In the case of prokaryotic cells, DNA is not separated in membrane bound structures from the cytoplasm; rather, it is concentrated within a region of the cell known as the nucleoid. One the other hand, the nucleus of eukaryotic cells is contained in a specialized membrane-bound nucleus which separates it from the cytoplasm.

Nucleus—The nucleus serves as a storage site for genetic material in eukaryotic cells and consists of several components: nuclear envelope (NE), nuclear pore complex (NPC), chromosomes, chromatin, and nucleolus (18–26).

The NE consists of two lipid bilayer membranes which surround the nucleus and separate it from the cytoplasm (27–30). The membrane that is closer to the cytoplasm is known as the outer nuclear membrane (ONM) and the membrane that is on the nuclear side is known as the inner nuclear membrane (INM). The NE also consists of nuclear lamina and the nuclear pore complex (NPC). As many as 80 proteins that make up the ONM and INM have been identified, many of which participate in specific interactions with cytoplasmic proteins and proteins in the nuclear lamina and are important in maintaining the proper assembly and structure of the NE. Proteins in the nuclear lamina, including the lamin proteins LA, LC, LB1, and LB2, are known to interact with proteins in the INM, including emerin, SUN1/2, LAP2β, and LBR (31–35). There are also several proteins in the ONM that interact with proteins in the cytoplasm. Nesprin 1, 2, 3 and 4 are known to bind to microtubules and actin molecules as well as participating in plectin-mediated binding to intermediate filaments. These proteins serve to anchor and stabilize the nucleus relative to the cytoplasm by specific protein-protein interactions.

NPCs are porous channels that connect the nucleus and the cytoplasm and serve to regulate flow of molecules in both directions. NPCs consist of proteins known as nucleoporins (NUPs), and have a central pore that is connected to the nuclear envelope, cytoplasmic rings with eight associated filaments, and a nuclear ring, which also has eight associated filaments. Small molecules with a diameter of 4–5 nm or less can passively diffuse through the NPCs, to and from the cytoplasm and nucleus, without any selective membrane transport mechanisms in place. Large molecules are transported through the NPCs by selective membrane transport via specialized transport proteins known as importins and exportins (21). The transport of molecules from the cytoplasm through the NPC to the nucleus is known as nuclear import and makes use of transport proteins known as importins (21). Similarly, the transport of molecules from the nucleus through the NPC to the cytoplasm is known as nuclear export and makes use of transport proteins known as exportins. In the case of nuclear import, the molecule of interest binds to the importin on the cytoplasmic side and is transported through NPCs to the nucleus; once inside the nucleus, the importin undergoes a conformational change after binding Ras-related nuclear protein (RAN) guanosine triphosphatase GTPase, (RAN.GTP) which allows release of the molecule of interest inside the nucleus (21). The reverse is true for nuclear export, where the molecule of interest binds to exportin proteins and RAN.GTP complex in the nucleus; this allows transport of the molecule to the cytoplasm, where conformation changes in the exportin proteins cause release of the transported molecule (21).

The genetic material within the nucleus is organized into chromosomes (36–39). Chromosomes are organized into primary, secondary, and tertiary structures using proteins to form these complex structures; this results in the formation of chromatin. The organization of chromosomes to form chromatin requires the use of an equal mass of proteins. Nucleosomes are repeating units of chromatin and consist of 145–147 base pairs of DNA around a histone octamer core. Histones are proteins that are used to compact and package DNA to form nucleosomes. Histones serve in compaction of DNA and serve to stabilize nucleosome structure by protein-protein interactions, which stabilize the histone core. In addition, electrostatic interactions and hydrogen bonds between DNA and the histone protein help stabilize the entire nucleosome structure. Individual nucleosomes are held together by linker DNA and this leads to the formation of the primary chromatin structure, resembling beads on a string. Compaction and organization of primary chromatin structures leads to the formation of secondary and tertiary chromatin structures.

The nucleolus is the largest compartment within the nucleus of eukaryotic cells; the core function of the nucleolus is synthesis and assembly of ribosomal RNA (rRNA) (40–45). The nucleolus has three major compartments known as the fibrillar center, dense fibrillar region, and granular component, each with a specialized function. The nucleolus also contains genes for pre-rRNA, which are transcribed into ribosomal proteins. The inner fibrillar center of the nucleolus is where transcription of pre-rRNA takes place while rRNA processing occurs in the dense fibrillar region, and early steps of ribosome assembly occur in the granular region of the nucleolus.

Ribosomes—Ribosomes are the sites of protein synthesis in eukaryotic cells and are made in the nucleolus of cell nucleus, as we have seen before (46–52). Ribosomes can either be suspended in the cytosol or bound to the endoplasmic reticulum. Free ribosomes translate proteins that are used in the cytoplasm, where membrane-bound ribosomes translate proteins that are used in the membrane or exported from the cell. The number of ribosomes present in a cell relates to the protein synthesis activity of the cell, and certain cells can have up to 10 million ribosomes. Ribosomes are characterized based on their rate of sedimentation, and are designated as 80S for eukaryotic cells. Eukaryotic ribosomes consist of two subunits, the larger of which is known as the 60S subunit and contains about 45 proteins; the smaller one is known as the 40S subunit and contains about 30 proteins. Consistent with their role in protein synthesis, eukaryotic ribosomes have a binding site for messenger RNA (mRNA) and three binding sites for transfer RNA (tRNA), known as the E site, P site and the A site.

Mitochondria—Mitochondria are the primary organelles for energy generation within eukaryotic cells (53–57). Mitochondria are responsible for the production of ATP, which is the primary currency of energy in cells, from breakdown of carbohydrates and fatty acids via oxidative phosphorylation. The number of mitochondria present in eukaryotic cells is related to the metabolic activity of the specific cell type; cells like cardiac myocytes, which have a high metabolic demand, are known to have a high concentration of mitochondria. The mitochondria of eukaryotic cells consist of two membranes, referred to as the outer membrane and the inner membrane; the space between the outer and inner membranes is known as the intermembrane space. The inner region of mitochondria is known as the matrix, and the inner membrane folds into matrix-forming indentations known as cristae. The inner region of the mitochondria contains enzymes responsible for the citric acid cycle, which is the primary mechanism for production of ATP in eukaryotic cells. The citric acid cycle in the mitochondria interfaces with glycolysis, which takes place in the cytoplasm. Glycolysis is responsible for the breakdown of glucose to pyruvate, which then feeds into the citric acid cycle for ATP production. Through this process, mitochondria are able to meet the metabolic demands of the cells.

Lysosomes—Lysosomes are membrane-bound structures, consisting of a single membrane which contains digestive enzymes (58–61). Lysosomes can be viewed as membrane-bound sacs of digestive enzymes with a primary catabolic function in eukaryotic cells. The primary function of eukaryotic lysosomes is digestion and breakdown of biological molecules, including proteins, nucleic acids, carbohydrates, and lipids. Lysosomes contain about 50 different digestive enzymes and are involved in phagocytosis and autophagy (62,63). Phagocytosis is the process by which external agents like bacteria are internalized within the cell and then destroyed by digestive enzymes of the lysosome. The external agent is first engulfed by the cell membrane of a host cell and then internalized to form a membrane-bound vesicle known as a phagosome. The phagosome fuses with a host lysosome to form phagolysosomes. The digestive enzymes of the lysosome destroy this external agent within the phagolysosome. Lysosomes are also responsible for a process known as autophagy, which involves digestion of internal organelles. The organelle to be digested is first enclosed within a membrane that is derived from the endoplasmic reticulum. The newly formed vesicle is known as an autophagosome. As in the case for phagocytosis, the autophagosome fuses with lysosomes, and digestive enzymes in the lysosome degrade this organelle.

Endoplasmic Reticulum (ER) and Golgi Apparatus—The ER is an extensive network of membranes found in the cytoplasm of eukaryotic cells that functions in protein synthesis and distribution (64,65). The membrane network of the ER accounts for more than half of the total membrane structures in eukaryotic cells, and the luminal space of the ER can account for up to ten percent of total cell volume. There are two types of ER, referred to as rough ER and smooth ER, each with distinct functions. The rough ER is covered with ribosomes on the outer surface and is involved in protein synthesis. The smooth ER does not have ribosomes on the surface and is involved in synthesis of lipids and in carbohydrate metabolism. While the ER is involved in protein synthesis, the Golgi apparatus is involved in storage and modification of these proteins and then further distribution of the modified proteins to other organelles in eukaryotic cells. The ER and the Golgi apparatus define the secretory pathway of proteins within the cells. Proteins are synthesized in the rough ER and then transferred to the Golgi apparatus for protein modification, and from the Golgi apparatus, the proteins are transported in vesicles to other organelles within the cell and can also be transported outside of the cell.

Cytoskeleton—The cytoskeleton is a network of fibers that provides structural and mechanical support for eukaryotic cells (66). The composition, distribution, and alignment of the cytoskeleton's fibers varies based on cell type and physiological state of cells, but the cytoskeleton consists of three types of fibers: microtubules, microfilaments, which are intertwined strands of actin, and intermediate filaments. Microtubules have a diameter of about 25 nm, intermediate filaments have a diameter of about 10 nm, and actin filaments have a diameter of about 7 nm. Microtubules are involved in cell motility and movement of chromosomes and movement of cell organelles and function to maintain cell shape. Some examples of microtubules include α-tubulin and β-tubulin. Actin filaments are one of the most abundant cytoskeletal proteins and, together with the contractile protein myosin, are involved in muscle contraction. Intermediate filaments play a structural role by providing mechanical support to eukaryotic cells and play an important role in maintenance of cell shape. Some examples of intermediate filaments include desmin in muscle cells, nuclear lamins in the nuclear lamina of all cell types, and nestin found in stem cells of the central nervous system.

Plasma Membrane—The plasma membrane provides a barrier around eukaryotic cells and provides a selective barrier to regulate the flow of molecules in and out of the cell. The plasma membrane consists of several phospholipids that are organized asymmetrically into two regions, with the inner leaflet toward the cytosol of the cell and the outer leaflet toward the outside of the cell (67–70). The plasma membrane contains several phospholipids, glycolipids, cholesterol, and protein molecules. Most eukaryotic plasma membranes contain an equal amount of lipid and protein molecules on a weight basis. The four most abundant phospholipids in the plasma membrane are phosphatidylcholine and sphingomyelin in the outer leaflet and phosphatidylethanolamine and phosphatidylserine in the inner leaflet. The lipid molecules provide structural support to the plasma membrane while the protein molecules play a functional role. The proteins in the plasma membrane are categorized as either peripheral proteins or integral membrane proteins. Peripheral proteins are not embedded in the phospholipid bilayer, but are associated with the membrane through protein-protein interactions. Integral membrane proteins, on the other hand, are embedded within the phospholipid layer of the plasma membrane. Integral membrane proteins, which span the lipid bilayer, are known as transmembrane proteins. Many transmembrane proteins serve as transport channels that selectively allow specific molecules to enter and leave the cell. This transport function of these proteins and selective permeability of plasma membranes to specific molecules is a critical function for eukaryotic cells.

2.3 THE DYNAMIC EXTRACELLULAR MATRIX

The ECM provides structural support for cells during normal tissue formation and function (71–74). The ECM in mammalian tissue is a complex mixture of proteins and other molecules and has specific binding sites for cells. During 3D tissue formation, cells form intercellular connections with neighboring cells and attach in a specific way to the ECM. These cell-cell interactions and cell-matrix interactions modulate tissue function, thereby allowing changes in the physiological state of the tissue. Cell-matrix interactions refer to the binding of cell surface receptors known as integrins to specific domains on ECM molecules. Binding of cell surface integrins to specific regions on ECM molecules triggers a cascade of intracellular signaling events, leading to changes in cell behavior and/or tissue organization. There are specialized cell types within any given tissue responsible for ECM production. Some examples of ECM-producing cells are fibroblasts, chondrocytes, and osteoblasts.

Structure and Function of ECM—The ECM can be viewed as the structural component of mammalian tissue and functions to stabilize 3D tissue by providing mechanical support. The ECM also functions to modulate tissue properties through specific cell-matrix interactions. The ECM consists of two major components, which are protein molecules and glysoaminoglycans (GAGs) (71–74). Some examples of ECM proteins include collagen, elastin, fibronectin, and laminin, while some examples of GAGs include hyaluronan, chondroitin sulfate, and heparin. The specific composition, distribution, and alignment of ECM proteins and GAGs varies from one tissue to another and changes in response to the physiological state of the tissue, including changes in stress environment or tissue remodeling in response to specific pathological conditions. While the ECM has many different functions, two primary functions of the ECM are to provide mechanical support/stability and to regulate cellular properties through specific cell-matrix interactions, which elucidate intracellular signaling events leading to changes in cell behavior/phenotype.

Cell-Matrix Interactions and Integrins—We have seen that cells and the ECM are major components of mammalian tissue. We have also seen that ECM components are large macromolecules like collagen and fibronectin and consist of hundreds of amino acids. Cells and ECM talk to each other in a very specific manner known as cell-matrix interaction. It is very intriguing to think about the way in which cells and the ECM communicate and/or interact. ECM proteins and other components are very large macromolecules with hundreds of amino acids. *How exactly does the cell "see" this large ECM protein and how exactly does the cell interact or communicate with the ECM protein?* Cells have cell surface receptors known as integrins, which recognize specific amino acid sequences on ECM proteins (75–80). Integrins consists of two subunits known as α and β subunits. Eighteen α and eight β subunits that form 24 distinct integrins have been identified, each with a specific ECM binding site. When a cell sees any given ECM protein, the cell scans the protein molecule to identify specific binding sites for which it has integrins; for example, the integrin α5β1 binds to the RGD site of the fibronectin molecule (81,82). Although the fibronectin molecule is large, there is only a sequence of three amino acids that are recognized by cells having the α5β1 integrin (81,82); the binding of the α5β1 integrin to the RGD site on the fibronectin molecule is referred to as a specific cell-matrix interaction.

Why is cell-matrix interaction important? Cell-matrix interaction is the only way for cells to functionally interact with ECM proteins. In the absence of specific cell-matrix interactions, cells will passively interface with ECM proteins without any functional benefit to cells. When cell surface integrins interact with specific ECM protein binding sites, this interaction triggers a series of intracellular signaling events that impact cell survival and proliferation, regulation of transcription and protein synthesis, and cytoskeletal organization and remodeling. Cell-matrix interaction is therefore important for cell function, behavior, and phenotype, and the absence of cell-matrix interaction leads to abnormal cell behavior and phenotype.

2.4 CELL SIGNALING

In the previous section, we studied the ECM and ways in which cells communicate with components of the ECM via cell-matrix interactions. Cells also communicate with other cells via cell-cell interactions, and these are critical in maintaining cell phenotype and tissue function (83). There are four types of cell signaling, known as endocrine, paracrine, autocrine, and contact-dependent signaling (83). The general principle of cell signaling in all four cases is the same—the signaling molecule is secreted by a specific cell and acts on a second cell (or the same cell in the case of autocrine signaling) by binding to a cell surface receptor specific to the signaling molecule. The binding of the signaling molecule to the cell surface receptor leads to a cascade of signaling events that regulate cell behavior and phenotype.

Endocrine Signaling—In the case of endocrine signaling, hormones are secreted by specialized cells in endocrine glands, and secreted hormones are distributed throughout the body by the circulatory system (84,85). Some examples of molecules that participate in endocrine signaling are thyroid hormone and adrenaline. Thyroid hormone is secreted by thyroid glands and released into the circulatory system; it acts on several cell types to regulate the rate of metabolism. Adrenaline is secreted by the adrenal gland and is responsible for the well-known fight and flight response, which causes an increase in blood pressure and heart rate during time of stress and anxiety.

Paracrine Signaling—In the case of paracrine signaling, the molecule is secreted into the local environment and rather than being transported through the circulatory system, the molecule acts on cells that are in close proximity (86–89). Neurotransmitters like acetylcholine use paracrine signaling; these molecules are released by nerve endings and travel across a synapse to act on the target cell.

Autocrine Signaling—In the case of autocrine signaling, the molecule is released by a cell and binds to cell surface receptors on the same cell from which it is released. Using this mechanism, cells can modulate their own behavior and phenotype in response to any given external cues (90–92). An example of autocrine signaling is in the case of T lymphocytes in response to antigenic stimulation. When T lymphocytes are exposed to specific antigens, they release a growth factor into the local environment; this growth factor then binds to cell surface receptors on T lymphocytes. Binding of the growth factor to cell surface receptors triggers an intracellular signaling cascade; this in turn results in T lymphocyte proliferation (93–95). This process leads to an increase in the number of T lymphocytes and therefore provides a mechanism to increase the intensity of the immune response by host cells.

Direct Cell-to-Cell Signaling—In this case, the signaling molecule is not released into the local environment or into the circulatory system; rather, the signaling molecule remains attached to the secretory cell. An adjacent cell that has a cell surface receptor for this molecule binds to the molecule while remaining attached to the secretory cell. Direct cell-to-cell signaling plays an important role in human embryology and development. During human development, cells start off as unspecialized stem cells and eventually give rise to specialized cells. Initially, there is a large pool of unspecialized stem cells, and only a certain proportion of these cells differentiate toward a specific lineage, while additional cells remain undifferentiated or differentiate to a different lineage. Direct cell-to-cell signaling plays an important role in regulating the differentiation fate of stem cells toward a specific lineage. In the case of neural differentiation, direct contact of developing nerve cells with neighboring cells inhibits differentiation of the neighboring cells to a neural lineage (96,97).

Intracellular Signaling—In the previous section, we looked at cell-matrix interaction and subsequent intracellular signaling events that modulate cell behavior and phenotype. The same thing occurs with cell-cell interactions—when signaling molecules bind to specific cell surface receptors, this binding triggers a series of intracellular signaling events that can lead to changes in cell function, including regulation of metabolic activity, regulation of gene expression, and changes in cytoskeletal organization and distribution.

2.5 CELLULAR JUNCTIONS

Introduction—In the previous section, we looked at several mechanisms used by cells for signaling, which depend on binding of a specific molecule to a cell surface receptor. We examined autocrine, paracrine, endocrine, and direct cell signaling. In this section, we continue our discussion on cellular signaling and look at cellular junctions, which provide various functions at the cell-cell or cell-matrix interface. There are five categories of cellular junctions: tight junctions, adherens junctions, desmosome junctions, gap junctions, and hemidesmosome junctions. Cellular junctions can be classified based on function: occluding junctions, anchoring junctions, or communication junctions. Tight junctions are an example of occluding junctions, which prevent the movement of molecules between adjacent cells. Anchoring junctions connect one cell to another cell or to components of the ECM and include adherens junctions, desmosome junctions, and hemidesmosome junctions. Communicating junctions allow the flow of molecules between adjacent cells and include gap junctions.

Tight Junctions—Tight junctions are found between adjacent cells, are prevalent in epithelial tissue, and serve to provide adhesion and barrier functions, hold cells together, and provide a semipermeable barrier (98–101). Tight junctions provide cells with a semipermeable size-specific and ion-specific barrier and restrict the diffusion of apical and basolateral membrane components. Tight junctions also function as landmarks by spatially confining signaling molecules and polarity cues and serving as docking site for vesicles. The structural organization of proteins has been identified, and at least three major components come together to at the site of tight junctions: claudins, occludins, and the family of junctional adhesion proteins (JAMs). The claudin family of proteins spans the intracellular space between epithelial cells and consists of at least 24 members, which vary significantly between tissue systems and organs. These variations in the claudin family of proteins between different tissue systems serve to induce selectivity of tight junctions. Occludin also spans the extracellular space, and although a specific function for this protein has not been elucidated, occludins have been implicated in the organization of tight junctions and as a regulatory protein. JAMs are a group of transmembrane structural proteins that anchor the adjacent cells together at the tight junction. In addition to claudin, occulin, and the JAM family of proteins, tight junctions also contain scaffolding proteins known as the zonula occludens (ZO) group of proteins. The ZO group of proteins directly interacts with claudins, occulins, and JAMs and provides a link with intracellular actin; ZO proteins provide a scaffolding function by stabilizing tight junction proteins and anchoring these proteins to intracellular components.

Adherens Junctions—The primary function of adherens junctions is to connect adjacent cells together (102). In addition to bridging adjacent cells, adherens junctions connect the intracellular actin bundles of adjacent cells, thereby providing increased structural stability and a mechanism to translate changes in the extracellular environment to intracellular remodeling of cytoskeletal proteins. There are two major protein complexes at adherens junctions: cadherin-catenin complexes and nectin-afadin complexes. In the case of cadherin-catenin complexes, the cadherin molecule is a transmembrane protein with five repeating subunits in the extracellular space that are stabilized by calcium ions. The intracellular portion of cadherin interacts with p120-catenin, β-catenin, or plakpglabin, which in turn interacts with intracellular cytoskeletal proteins. The nectin-afadin complex acts similarly to the cadherin-catenin complex, with nectin providing the primary scaffolding function and associating cytoskeletal proteins by the intracellular protein afadin.

Desmosome Junctions—Desmosome junctions anchor adjacent cells together and perform a function similar to tight junctions and adherens junctions (103,104). Desmosome junctions contain components of three protein families: cadherins, armadillo proteins, and desmoplakin. The cadherins span the extracellular space between two adjacent cells and provide physical anchoring of cells. The cadherin proteins are connected to plakins on the intracellular side and the plakin proteins are connected to keratin filaments; this mechanism provides continuity in protein interactions, stabilizes participating cells, and provides a direct link between extracellular and intracellular components. The armadillo family of proteins facilitates tethering of desmoplakin and keratin filaments to desmosomes; this serves to regulate clustering of desmosomal components and mediate signal transduction pathways.

Gap Junctions—The first three cellular junctions (tight junctions, adherens junctions, and desmosome junctions) are examples of anchoring junctions, which couple adjacent cells and provide structural and mechanical stability. On the other hand, gap junctions are functional coupling points between adjacent cells and regulate the flow of molecules between coupled cells (105–108). Gap junctions are composed of repeating units of a protein family known as the connexin family of proteins. Gap junctions consist of porous channels formed between adjacent cells and are made of six protein connexin subunits; the actual channel is referred to as a connexon. Gap junctions act as physical channels that connect adjacent cells and allow the flow of ions or other molecules through these cells. This is particularly important in excitable tissues like heart muscle, where the flow of ions through gap junctions is critical for depolarization of the heart.

Hemidemosome Junctions—Hemidemosomes junctions anchor cells to the underlying basement membrane by connecting intermediate filaments within the cells to specific components of the ECM via integrin mediated binding (109). Hemidemosome junctions contain many different proteins on the intracellular side, including the plectin and plakin family of proteins. The primary function of the hemidesmosome junction is to anchor cells to the underlying ECM.

2.6 MAMMALIAN TISSUE AND ARTIFICIAL TISSUE

During the course of this chapter, we have looked at cell architecture and organelle function, ECM structure and function, and cell-cell and cell-matrix interactions. In this section, we provide a brief description of how many of the concepts introduced in this chapter are relevant in the formation and function of mammalian tissue; we can also apply these concepts to the fabrication of artificial tissue.

Let us begin our discussion with a review of some of the concepts as they relate to mammalian tissue. As we have seen, mammalian tissue is composed of different cell types and ECM components, which provide functional and structural support, respectively. Individual cells connect with other cells using cellular junctions that include tight junctions, adherens junctions, desmosome junctions, and gap junctions. In addition, cells communicate with other cells using a variety of cellular signaling methods including autocrine, paracrine, endocrine, and direct cell-cell signaling. *The functional coupling of cells with other cells is known as cell-cell interaction.* In addition to cells, mammalian tissue consists of the ECM, which serves to provide mechanical support; components of the ECM functionally couple with cells by binding to cell surface receptors known as integrins. *The functional coupling of cells with the ECM is known as cell-matrix interaction.* Cell-cell and cell-matrix interactions trigger a cascade of intracellular signaling events, which modulate cell behavior and phenotypes and, as a result, have a direct impact on 3D tissue architecture and function.

Now that we have seen how many of the concepts presented in this chapter relate to mammalian tissue, the next critical question is—*how do these concepts apply to tissue engineering and fabrication of 3D artificial tissue?* As we have seen in chapter one, tissue engineering is focused on the development of technologies to support the fabrication of artificial tissue. As we have seen during our discussion of the tissue fabrication process, isolated cells are cultured within 3D scaffolds to support the formation of 3D artificial tissue. During the tissue fabrication process, it is critical to support the formation of functional cell-cell and cell-matrix interactions between isolated cells and biomaterials. Successful implementation of strategies to support cell-cell and cell-matrix interactions is critical to the development of 3D artificial tissue.

2.7 CELL SOURCING

Requirements of an Ideal Cell Source—During the tissue fabrication process, isolated cells are coupled with biomaterials to fabricate 3D artificial tissue.

CELL SOURCING

During this process, one critical decision needs to be made regarding the source of cells—*where exactly will the cells come from?* In order to help researchers determine the most suitable cell source for any given tissue engineering application, the following criteria have been established for an ideal cell source (110): 1) cells need to be safe and not trigger tumor creation, 2) improve functional performance of host tissue, 3) functionally integrate with host tissue and host vasculature, 4) cells need to be tolerant of noninvasive delivery methods, which are increasingly common in the operating room, 5) not trigger the host immune response, 6) in the case of stem cells, be sensitive to social and ethical issues, 7) when working with stem cells, the cells should have a clearly demonstrated potential to be differentiated with high efficiency to the cell type of interest, 8) any cell type should tolerate the processing conditioning required to develop off-the-shelf therapies. While there are several choices for cells in tissue engineering, none satisfy all of the stated design requirements. There are four cell sources which have been used extensively in tissue engineering: cell lines, animal derived cells, stem cells (adult-derived mesenchymal stem cells, embryonic stem cells, or induced pluripotent stem cells), and, finally, human derived cells. We will discuss these in detail in subsequent sections and start here by presenting some general considerations for cell sourcing, as shown in Figure 2.2.

Autologous versus Allogeneic—Cells can either be autologous or allogeneic. Autologous cells are cells that have been isolated from a tissue biopsy of the person who will also be recipient of these cells; the donor and recipient for autologous cells is the same. This strategy has clear advantages, as it circumvents issues related

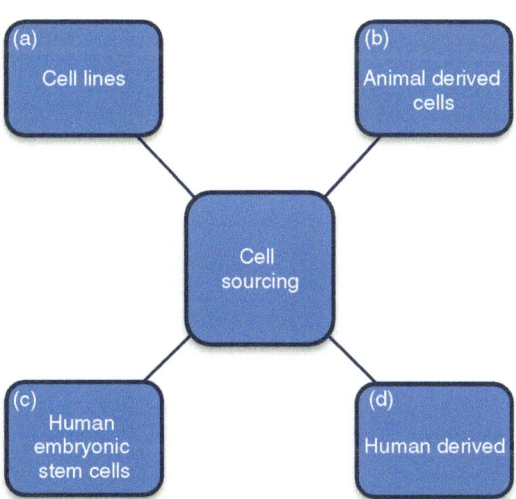

Figure 2.2 Cell Sourcing for Tissue Engineering—There are four sources of cells that are commonly used during the fabrication of 3D artificial tissue: (a) Cell lines, (b) Animal Derived Cells, (c) Human Embryonic Stem Cells, and (d) Human Donor Cells.

to host immune rejection and eliminates any issues related to identifying potential donors. Autologous cells are aggressively being evaluated in stem cell therapies involving the use of bone marrow-derived mesenchymal stem cells (MSCs). MSCs can be purified from a patient's bone marrow aspirate and processed *in vitro* and expanded in culture, followed by differentiation to a given cell lineage. The differentiated cells can later be implanted into the same patients at the site of injury. Allogeneic cells are isolated from a donor and then transplanted into a recipient patient, with the donor and recipient being different people. Blood transfusions are the most popular examples of allogeneic cell transplantation. The main limitations of allogeneic-based cells are the challenges in identifying potential donors and the potential for immune rejection by the host.

Animal-Derived or Human-Derived Cells—Cells can either be animal-derived or isolated from human tissue specimens. Clearly, if the cells are being used for clinical applications, cells isolated from humans would be the only choice. Animal-derived cells are used extensively during early stages of research, particularly in proof-of-concept studies, model development and optimization studies, and safety/efficacy studies. The primary advantage of using animal-derived cells is availability, as tissue specimens can be obtained based on need.

Cell Lines—Cell lines are used extensively for tissue engineering studies and can easily be purchased form commercial vendors. Primary cells are maintained in culture for extended periods of time and at regular intervals, can be sub-passaged to increase the cell yield (discussed in a subsequent section of this chapter). After several sub-passages, the cells in most cases tend to decrease in proliferative capacity; however, in some cases, these cells undergo genetic alterations that allow the cells to proliferative for extended periods of time. When this occurs, the cells are known as a cell line (111). This property of cell lines, the ability to proliferate over extended periods of time, proves to be advantageous for tissue engineering studies. During the tissue fabrication process, a large number of cells are required to support 3D tissue formation. The proliferative capacity of cell lines allows researchers to obtain these cell numbers during routine cell culture.

Stem Cell Engineering—We will discuss stem cell engineering in detail in subsequent sections, but will use this opportunity to introduce the general concept. Cell sourcing for tissue engineering requires cells to have tissue specific functionality—fabrication of artificial heart muscle requires cardiac myocytes, fabrication of artificial cartilage tissue requires chondrocytes, and so forth. It is well-known that during early stages of human embryogenesis, all cells are the same, and during the course of development, they differentiate to become different cell types (112,113). These early stem cells are known as embryonic stem cells. The interest in stem cells, particularly embryonic cells, arises from the differentiate potential of these cells. For example, if a researcher wants to bioengineer vascularized heart muscle, several different cell types would be required, including cardiac myocytes, fibroblasts, endothelial cells, and smooth muscle cells. Since embryonic cells have the potential to differentiate into any cell type, a single cell source can be used to obtain multiple cell types, which in turn can be used to bioengineer multicellular tissue, like vascularized heart muscle.

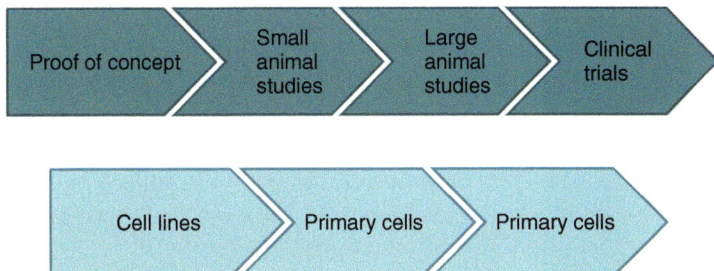

Figure 2.3 Cell Sourcing During the Progression of Tissue Engineering Studies—Different cell sources are suited for different parts of the technology development process for fabrication of 3D artificial tissue.

Cell Sourcing Strategy in Tissue Engineering Research—In tissue engineering, there is a common trajectory used for cell sourcing: progressing from proof-of-concept studies to clinical trials (Figure 2.3).

During the initial stages of project development, cell lines are used to conduct proof of concept studies. Cell lines are considerably easier to work with, reasonably inexpensive, and have reproducible phenotypes during cell culture. Once proof-of-concept studies have been successful, the work is translated to small animal studies, which are designed to validate the results obtained with cell lines. At this stage of the study, primary cells that are isolated from animal tissue are more suitable. The next step in the technology progression is from small animals to large animals; in this case, animal-derived cells would be more suitable. The final stage of the tissue engineering study is clinical trials, which require testing of the artificial tissue in human patients. The cells used in clinical trials have to be autologous, with stem cells being commonly used, particularly bone marrow-derived MSCs.

2.8 THE CELL TRANSPLANTATION PROCESS

Introduction to Cell Transplantation—In the first chapter of this book, we introduced the concept of cell transplantation or cell therapy, both of which refer to the same thing. The concept of cell transplantation is based on delivery of isolated cells to the site of injury; these therapies are based on the hypothesis that transplanted cells will support functional recovery of damaged or diseased tissue (110). From a conceptual standpoint, this is a very intriguing premise and is based on the hypothesis that isolated cells have the capacity to perform tissue-level functions. This is in sharp contrast to the field of tissue engineering, which is based on the premise that lost tissue function can be replaced by transplanted 3D tissue. The primary difference between the field of cell transplantation and tissue engineering is the therapeutic agent, with isolated cells being used in the former and 3D artificial tissue in the latter. Prior to describing the details of the process, we introduce a few fundamental concepts related to the field and provide some definitions.

- Cell transplantation has been defined as the process by which cells are delivered to the site of injury in order to improve the functional performance of injured tissue (114,115). Whole blood transfusions, packed red cell transfusions, platelet transfusions, and bone marrow transplants are examples of cell therapy.
- Stem cell transplantation is a specialized case of cell transplantation, in which the cells being delivered are stem cells (116). Use of embryonic stem cells, induced pluripotent stem cells, and adult stem cells fall under the classification of stem cell transplantation. Cell transplantation and cell therapy are broad terms and apply when any cell type is used; stem cell engineering is a specialized case of cell transplantation that utilizes stem cells as the therapeutic agent.

Steps in the Process of Cell Transplantation—The process of cell transplantation is shown in Figure 2.4.

While there are several choices for stem cells, the general scheme that we present is based on hES cells (117). In this case, hES cells are isolated from the inner cell mass of a blastocyst at the 64 to 200 cell mass stage. The cells are isolated and

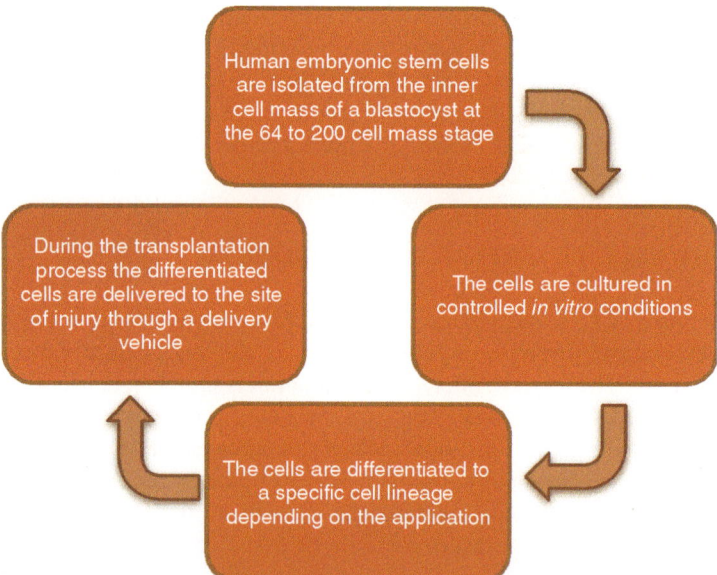

Figure 2.4 The Cell Transplantation Process—Human embryonic stem cells are differentiated to form different cell types that are delivered to the site of injury to support functional recovery.

cultured under controlled *in vitro* conditions. These cells are then differentiated to the specific cell lineage, which varies depending on application; for example, hES cells will need to be differentiated to cardiac myocytes for the treatment of myocardial infarction. The next step in the stem cell transplantation process is delivery of differentiated cells to the site of injury. Differentiated cells are maintained in 2D monolayer culture and will first need to be detached from the culture surface and suspended in a suitable buffer or cell culture media. Once suspended, the cells are placed within the delivery vehicle, with catheter-based delivery systems being commonly used. The differentiated cells are then delivered to the site of injury using these catheter-based delivery systems.

The cell transplantation process may appear to be a simple and straightforward process; however, this is not the case, and cell transplantation therapies are extremely complicated and are faced with many critical scientific and technological challenges.

Critical Challenges in Cell Transplantation—The isolation and culture of hES cells or any other stem cell requires trained technicians with a high degree of expertise. The researcher needs to identify the right culture conditions that will support the differentiation of hES to the required cell lineage. This is often done using a cocktail of chemical compounds and growth factors; the exact composition and concentration requires rigorous optimization, testing, and validation. In addition to using chemical compounds to drive the differentiation of hES to a specific lineage, other strategies have been used. These strategies include genetic engineering to manipulate the gene profile of cells, co-culturing with specific cell types, and electromechanical stimulation.

Another critical issue is the ability to maximize differentiation efficiency, which refers to the percentage of cells that differentiate to the desired cell lineage. The differentiation efficiency will never be 100%, and only a certain percentage of cells will be differentiated to any given lineage. One of the challenges in the field is the rigorous optimization of protocols to maximize differentiation efficiency. Another important challenge is the development of strategies to purify and separate differentiated cells from undifferentiated cells. At the end of any differentiation strategy, there will be a mixed cell population that contains differentiated and undifferentiated cells. The differentiated cells need to be separated from the undifferentiated cells prior to transplantation. Again, experimental strategies need to be developed for the separation of differentiated and undifferentiated cells, and this is no trivial task.

There is one more critical challenge associated with the differentiation of hES cells to specific lineages. *After differentiating hES cell to any given cell lineage, how effective is the differentiation strategy in producing functional cells?* For example, if the objective is to obtain cardiac myocytes, then we will develop and optimize differentiation strategies that drive the differentiation of hES cells to a cardiac myocyte lineage. Any given differentiation strategy will have finite differentiation efficiency, and a certain percentage of hES cells will form cardiac myocytes. Then the critical question becomes—*how close are the differentiated cardiac myocytes in form and function to mammalian cardiac myocytes?* In the case of cardiac myocytes, this question can be answered by obtaining functional and biological data and comparing the results to mammalian heart muscle. Functional tests will be designed to measure the contractile and electrical properties and the changes in calcium transients. In order to assess the biological properties of cardiac myocytes, researchers can undertake studies to measure changes in gene and protein expression using rt-PCR and Western blotting, respectively. In addition, information about the localization of specific proteins can be obtained by immunohistochemistry strategies, and 3D volumetric rendering can be accomplished using confocal imaging. Collectively, this data will provide a comprehensive assessment of cardiac myocytes' function. Similar studies can be conducted for any cell type. This data can be compared to published and experimental values for mammalian tissue to judge the effectiveness of hES differentiation. While the results vary significantly based on the tissue system, differentiation strategy, and between research laboratories, the functional performance of differentiated cells will not approach that of mammalian cells, and, in most cases, the functional performance will be significantly lower. Therefore, before these cells can be considered a true therapeutic option for patients, strategies will need to be developed to bridge the gap between the functional performance of differentiated and mammalian cells.

One of the most significant limitations of the cell transplantation strategy is low cell retention upon transplantation. *What percentage of total transplanted cells is actually retained at the site of injury?* This may come as somewhat of a surprise, but cell retention is extremely low and has been reported in the range of 1–2%. This means that for every one hundred cells that are transplanted to the site of injury, only 1–2 cells actually are retained at the site of delivery. This is a major limitation of cell transplantation as a potential therapeutic strategy, as a small population of cells cannot support tissue-level function and support recovery of injured tissue.

2.9 CELLS FOR CELL TRANSPLANTATION

A large number of cells have been evaluated as therapeutic agents for cell transplantation. Human embryonic stem cells, induced pluripotent stem cells, and bone marrow-derived mesenchymal stem cells have been used extensively for cell transplantation. Hematopoietic stem cells have been used as a source of blood cells and can be used to treat patients with blood disorders. MSCs derived from adipose tissue are also an attractive option, as adipose tissue is widely available, can easily be obtained from a tissue biopsy, and does not require any invasive procedures. Umbilical cord tissue and blood have received a lot of attention recently, as specimens can be preserved at birth for future use; cord tissue is a source for MSCs, while cord blood is a source for hematopoietic stem cells. Resident stem cells refer to stem cells that are present in tissue in a dormant state under normal physiological conditions; however, in times of injury, become activated and participate in the tissue repair process. Most mammalian tissues have resident stem cells, and as one common example, satellite cells are known to be resident stem cells, which are present in skeletal muscle tissue.

2.10 MODE OF ACTION OF CELLS DURING CELL TRANSPLANTATION

The purpose of cell transplantation is to elicit functional recovery of host tissue. Let us recall the process of cell transplantation—isolated cells are delivered to the site of injury to support functional recovery of host tissue. *What are some possible mechanisms by which this can occur?* There have been several hypotheses postulated, but not yet conclusively proven. There are four primary mechanisms by which transplanted cells can support functional recovery: trans-differentiation to host cells, secretion of paracrine factors, recruitment of circulating stem cells, and increase in vascularization of host tissue (Figure 2.5).

Trans-differentiation of Transplanted Cells—This hypothesis states that transplanted cells differentiate to form functional cells that support functional recovery of damaged or diseased tissue (118–122). The transplanted cells functionally interact with cells of host tissue, and cell-cell interactions lead to differentiation of transplanted cells. Stated another way, the transplanted cells senses the local environment and identifies the need to differentiate to form functional host cells. Stem cells are plastic and have the potential to become many different cell types. By sensing the local environment and interfacing with cells of the host tissue, the transplanted cells differentiate to the lineage of the host tissue.

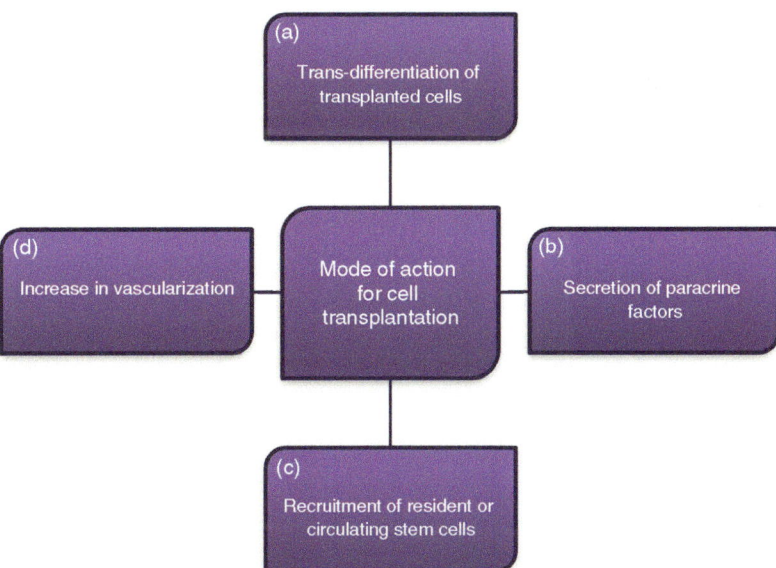

Figure 2.5 Mode of Action for Cell Transplantation—There have been four mechanisms postulated for the mode of action for cell transplantation. (a) Trans-differentiation of Transplantation Cells, (b) Secretion of Paracrine Factors, (c) Recruitment of Resident or Circulating Stem Cells, and (d) Increase in Vascularization.

Secretion of Paracrine Factors—The second potential mechanism is by secretion of paracrine factors—factors which are released into the local environment and act on neighboring cells (123–126). This hypothesis states that transplanted cells either secrete paracrine factors into the local environment or stimulate cells of the host tissue to secrete these factors. These paracrine factors then act by reducing cell apoptosis, increasing cell proliferation, and increasing angiogenesis and neovascularization of injured tissue. These actions lead to an increase in the functional performance of injured tissue.

Recruitment of Circulating Stem Cells—The third mechanism by which transplanted cells can improve functional performance of host tissue is by recruitment of resident or circulating stem cells (127–129). Resident stem cells are present within the tissue, while circulating stem cells refers to those present in the circulation, like hematopoietic stem cells. This hypothesis states that transplanted cells serve to recruit resident or circulating stem cells to the site of injury. Resident stem cells migrate from neighboring regions to the site of injury, while circulating stem cells can be recruited via capillaries. This process results in an increase in the number of stem cells present at the site of injury, which in turn leads to functional recovery of host tissue.

Increase in Vascularization of Host Tissue—The fourth mechanism by which transplanted stem cells can improve function is by promoting angiogenesis and neovascularization (130). Any increase in vascularization of host tissue will have a direct impact on tissue survival and will lead to an increase in cell viability, which in turn will lead to an increase in functional performance. Transplanted stem cells are postulated to support vascularization in one out of two possible ways; either by direct incorporation with host vasculature or by the release of angiogenic factors into the local environment. These two mechanisms will lead to a direct increase in host vascularization, which in turn will support functional recovery of the host tissue.

2.11 CELL TRANSPLANTATION AND TISSUE ENGINEERING

We will conclude our discussion of cell transplantation by presenting a comparison between cell transplantation and tissue engineering.

Which one is better, cell transplantation or tissue engineering? Let us begin by a brief discussion of the two fields. Cell transplantation is focused on delivery of isolated cells to the site of injury, while tissue engineering is focused on the fabrication of artificial tissue or entire organs that can be transplanted to the site of injury to support or replace lost tissue functionality. One of the most attractive features of cell transplantation is the simplicity of the approach, which utilizes isolated cells as a therapeutic option. From a methodological standpoint, isolated cells can easily be expanded and manipulated in culture and then delivered to the site of injury. There are, however, many limitations to the approach, the most significant of which is low cell retention, reported in the range of 1–2%.

Tissue engineering is focused on the design and fabrication of artificial tissue under controlled *in vitro* conditions, followed by transplantation of artificial tissue *in vivo* to support and/or replace lost tissue functionality. Tissue engineering addresses two significant drawbacks of cell transplantation. The first is the basic premise of the field, which is based on the hypothesis that lost tissue function can be recovered by transplantation of isolated cells. The field of tissue engineering takes this premise one step further and suggests that bioengineered tissue and/or organs would be better suited than isolated cells to repair, augment and/or replace lost tissue function. The second major drawback of cell transplantation is low cell retention; tissue engineering addresses the limitation of low cell retention by fabrication of artificial tissue/organs, which does not lead to significant loss of cells at the site of implantation. The reason for this is that the cells are tightly bound to other cells and to biomaterial and have come together to form complex 3D tissue under controlled *in vitro* conditions. The 3D architecture of bioengineered tissue and organs is maintained upon implantation *in vitro*.

The fields of cell transplantation and tissue engineering have made rapid progress during the last decade and both provide novel and innovative therapies to support lost tissue functionality. While the reader of this book is encouraged to formulate his/her opinion about the two fields, it is the view of the author that tissue engineering offers advantages over cell transplantation that make it a better therapeutic option to support lost tissue function.

2.12 THE CELL CULTURE PROCESS

Introduction—Isolated cells can be maintained in monolayer 2D cultures using a technique known as cell culture (131–143). The process of monolayer 2D cell culture has become very common and routine with very well-established protocols and standardized equipment. This can be appreciated by recognizing the abundance of cell culture laboratories in major research universities and medical centers, exceeding one hundred per institution. Cell culture is extremely critical to the development of cell transplantation and tissue engineering technologies. The application of cell culture during cell transplantation studies can easily be appreciated, as isolated cells need to be maintained, manipulated, and expanded in culture prior to implantation; at least fifty percent of cell transplantation studies are based on cell culture experiments. The same is true for tissue engineering studies; cells form the functional components of artificial tissue and, coupled with biomaterials and bioreactors, form one of the building blocks of 3D artificial tissue. The ability to isolate, maintain, manipulate, and expand cells is central to the tissue fabrication process. Development of tissue engineering technologies relies heavily on cell culture; at least fifty percent of tissue engineering research is based on cell culture experiments. Due to the essential role of cell culture techniques in tissue engineering, we will study some of the general principles of cell culture in this section.

The Cell Culture Process—The steps involved in the isolation, culture and expansion of cells are shown in Figure 2.6.

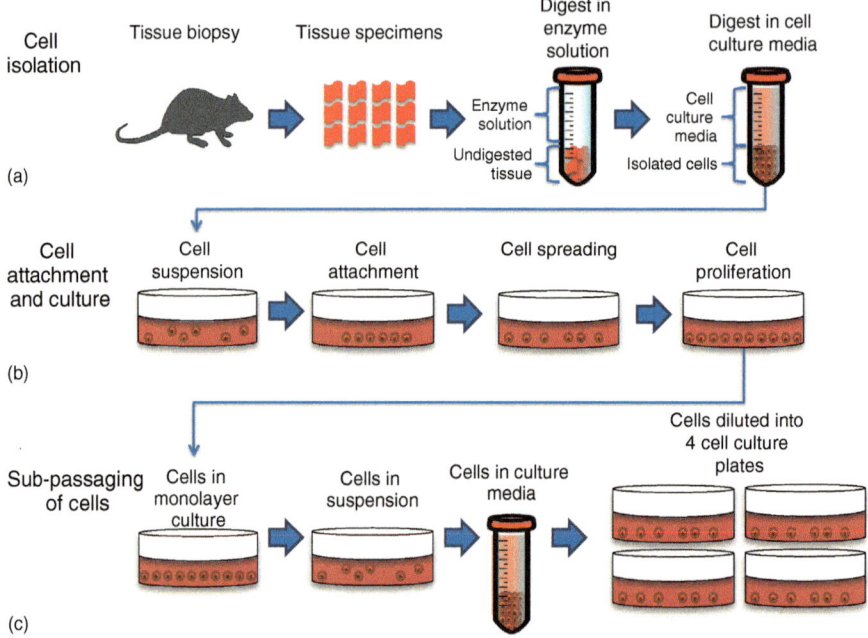

Figure 2.6 The Cell Culture Process—(a) Cell Isolation—A tissue specimen from a biopsy is cut into small pieces and subjected to an enzymatic digestion process to yield isolated cells. (b) Cell Attachment and Culture—The primary cells are cultured on the surface of tissue culture plates where they attach, spread, and proliferate. (c) Cell Expansion—Upon confluency, cells are subjected to treatment with trypsin and then replated to promote cell expansion.

Our discussion here is focused on primary cell culture, cells which are isolated from a tissue biopsy. There are several steps in the cell culture process: cell isolation, cell attachment, cell proliferation, and sub-passaging of cells. The first step in the process is the isolation of primary cells from a tissue biopsy. The tissue specimen is harvested either from animal or human tissue and is maintained in a buffered solution. The tissue specimen is cut into small pieces and suspended in digestion solution, which consists of digestive enzymes, with collagenase and trypsin being commonly used examples. The exact digestion time varies between tissue systems and is a variable that needs to be optimized and validated experimentally; digestion times of 30 to 60 minutes are common. The purpose of the digestive enzymes is to dissociate cells from the ECM, and over time, the cells are completely separated from the tissue. The isolated cells are transferred from the digestive solution to cell culture media. The tissue digestion process separates cells from the ECM, and the isolated cells are then maintained in culture and used for a variety of applications.

The next step in the cell culture process is cell attachment and culture on the surface of monolayer tissue culture plates. Most mammalian cells are anchorage-dependent, which means they need to be anchored to a substrate in order to support cell viability, growth and proliferation. Anchorage-dependent cells cannot maintain viability and function during suspension culture and will die over time. In order to support cell viability and growth, isolated cells are plated on the surface of tissue culture plates that have been specifically designed to support the attachment and growth of anchorage-dependent mammalian cells. The process of transferring isolated cells from suspension culture to the surface of a tissue culture plate is known as cell plating. The cells are plated at a specific density, which refers to the number of cells per milliliter of culture media or cells per unit area; in order to calculate the plating density, the number of cells will need to be determined using a hemocytometer or an automated cell counter.

Once the cells are plated, they attach to the culture surface. Often, the culture surface is coated with adhesion proteins like collagen, fibronectin, or laminin to promote specific integrin-mediated cell-matrix interactions. The choice of ECM protein and specific concentration depends on the cell type and cell plating density; the protein choice and its concentration needs to be experimentally tested and validated. Cells undergo morphological changes after attachment to the culture surface and start to spread out and occupy a large footprint on the culture surface. This process, referred to as cell spreading, is due to changes in cytoskeletal proteins. Cell spreading is common with mammalian cells and is indicative of a positive cell phenotype in response to cell attachment. If cell spreading occurs, it is considered to be a normal cellular response, while the lack of cell spreading indicates a problem with cell culture process.

So far, we have looked at cell attachment and cell spreading; the next step in the cell culture process is cell proliferation, which refers to the process by which cells divide and increase in number. Cells will only proliferate if they are cultured in the correct environment, including temperature, pH, nutrient composition, and adhesion matrix. Cell proliferation is an important part of the cell culture process and is essential in tissue engineering and cell transplantation studies, as a very large cell number is required for these studies. In the case of cell transplantation technologies, a large number of cells are required for *in vivo* implantation, whereas in the fabrication of 3D artificial tissue, a large number of cells are required to populate synthetic scaffolds.

The next step in the cell culture process is subpassaging. During the course of monolayer culture, attached cells continue to proliferate and increase in number. This process of cell proliferation continues as a function of time. However, there are certain limitations, and the cells cannot continue to proliferate indefinitely. The primary limitation is space—the culture surface has a finite area to support cell attachment; as the cells continue to proliferate, the culture surface gets covered with cells leaving no room for additional cells. The term confluency refers to the percentage of total culture surface that is covered by cells. An increase in cell confluency leads to contact inhibition, which refers to the process by which attached cells no longer proliferate due to constraints imposed by space limitations. As the cells no longer have space to grow and proliferate, they enter a decay or decline phase that leads to a reduction in cell number. This is not a desirable outcome, and prior to contact inhibition, the cells need to be subpassaged. Subpassaging refers to the process by which attached cells are detached from the culture surface, resuspended in cell culture media and re-plated at a lower cell density. Re-plating cells at a lower density reduces or eliminates confluency-induced contact inhibition and allows cells to proliferate and expand in number. Digestive enzymes like trypsin are used to detach cells from the culture surface, and once cells have detached, they are collected and diluted to the required cell concentration and then replated onto additional culture plates, commonly with a one to four dilution.

2.13 APPLICATIONS OF MONOLAYER 2D CELL CULTURE

Monolayer cell culture techniques have been used extensively in research for many different applications, some of which are summarized in Figure 2.7.

A significant amount of information regarding cell-cell interactions and cell-matrix interactions has been obtained from monolayer 2D cell cultures. Different cell types can be cultured under controlled *in vitro* conditions, and specific interactions between cells or with ECM components can be studied in isolation without the confounding effect of mammalian physiology. Therefore, specific cell-cell and cell-matrix interactions, along with subsequent signaling pathways, can be identified using isolated 2D monolayer cell culture systems. Intracellular signaling pathways,

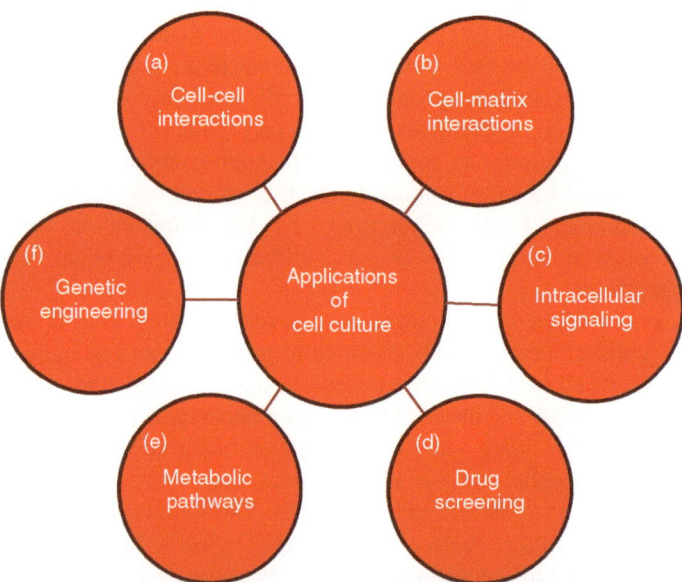

Figure 2.7 Applications of Cell Culture—Cell culture studies have been used for many different applications, some of which include: (a) Cell-Cell Interactions, (b) Cell-Matrix Interactions, (c) Intracellular Signaling, (d) Drug Screening, (e) Metabolic Pathways, and (f) Genetic Engineering.

either in response to cell-cell or cell-matrix interactions, or other modes of stimulation like growth factor conditioning, have been elucidated based on cell culture studies. Another area of research that has benefitted from cell culture studies is drug development. During the early stages of drug development, there are thousands of potential candidates that need to be narrowed down to a manageable number. Cell culture models provide a quick and cost-effective away of screening potential pharmacological compounds. Candidate compounds can be added to cell culture media and exposed to cells during monolayer culture; after a finite expose time, cellular response can be evaluated. Along the same line of thought, cell culture models have proven to be effective in toxicology studies and in the screening of environment toxins, agents that have a negative impact on cell and tissue function. Dose–response studies using cells in monolayer culture can be conducted to determine human exposure limits. Much of the information we know about metabolic pathways have been obtained from isolated cell culture studies. Advances in stem cell engineering, genetic engineering, and strategies to drive the differentiation fate of stem cells toward a cell specific lineage have been a result of cell culture studies.

2.14 CELL CULTURE VERSUS TISSUE ENGINEERING

A Typical Tissue Engineering Experiment—The first step of the process involves the isolation, culture, and expansion of primary cells. In most cases, a large number of cells are required to bioengineer artificial tissue, and several rounds of subpassaging are required. Once a sufficient number of cells have been obtained, cells are detached from the culture surface using digestive enzymes like trypsin, and the suspended cells are collected in a conical flask. These cells are then plated onto the surface of a biomaterial and maintained in culture. The cells attach to the biomaterial, proliferate to increase in cell number, and over time populate the entire scaffold. This process leads to the formation of bioengineered 3D artificial tissue. This example serves to illustrate the interplay between cell culture and tissue engineering and the relationship between the two fields.

Cell Culture versus Tissue Engineering—In the previous example, we looked at the relationship between cell culture and tissue engineering; techniques in cell culture are essential to obtain the large cell numbers required for tissue engineering and tissue fabrication. Cell culture, in a sense, feeds into tissue engineering. However, the two fields can be compared side by side as two separate areas of research. Cell culture techniques have been used extensively in basic research and have provided an understanding of cell behavior and function. In an earlier section, we identified a list of research areas in which cell culture has proven to be a valuable tool, like cell-cell interactions, cell-matrix interactions, intracellular signaling, and others. During cell culture, isolated cells are maintained in 2D culture and do not have the 3D geometry of mammalian tissue; therefore, information obtained from cell culture models is limited.

However, this information can be extended by using tissue engineering models, which provide complex 3D architecture, absent in 2D monolayer culture systems.

Researchers can extract considerably more information using tissue engineering models when studying cell-cell interactions and cell-matrix interactions, as the cells are now cultured in a complex 3D geometry.

Tissue engineering models have a tremendous role to play in basic research and in understanding cell and tissue organization, structure, and function.

2.15 INTRODUCTION TO STEM CELL ENGINEERING

Introduction—Stem cells have proven to be an attractive cell source for tissue engineering and tissue fabrication. Embryonic stem cells, induced pluripotent stem cells, and adult-derived mesenchymal stem cells have received significant attention in recent literature. Advances in stem cell engineering are based on manipulation of the genetic material of stem cells, and in order to understand this process, an understanding of the central dogma of molecular biology is important. We begin our discussion on stem cell engineering with a brief overview of the central dogma of molecular biology.

Central Dogma of Molecular Biology—The central dogma of molecular biology, states that DNA is transcribed to RNA, which is then translated to proteins (144–146). All cells of any given species have the same genetic material, which is stored in the nucleus of cells. The genetic material is stored in the form of DNA molecules, which have a double helical structure with repeating units of four nucleotides: adenine, thymine, guanine, and cytosine. Human DNA contains all the genetic information required to synthesize all proteins that are required for normal function. As such, DNA molecules do not leave the nucleus and are kept within the nucleus for safekeeping. The genetic information necessary for protein synthesis is transferred from the nucleus by messenger RNA (mRNA) molecules. The process by which information stored within segments of DNA molecules is converted to mRNA is known as transcription. Messenger RNA is single-stranded, compared to double-stranded DNA and is also composed of the same four nucleotides found in DNA. The process by which proteins are synthesized using the information within the mRNA molecules is known as translation. Ribosomes are the site of protein synthesis, and mRNA leaves the nucleus and binds to ribosomes, which leads to the formation of a specific peptide by the process known as translation. The central dogma states that DNA is converted to RNA and then to protein.

What exactly does the central dogma of molecular biology tell us regarding stem cell engineering? All mammalian cells have the necessary genetic information to produce all proteins required for normal function; the production of specific proteins distinguishes the function of one cell type from another cell type. For example, cardiac myocytes require the expression of contractile proteins in order to support function of heart muscle. Protein expression in turn is governed by the expression of mRNA, which is regulated by the process known as transcription. Therefore, transcriptional regulation is at the heart of cell specialization and is also a defining concept in the field of stem cell engineering.

Stem Cell Differentiation During Embryogenesis—During human development, fertilization of an ovulated oocyte by spermatozoa leads to the formation of a zygote. The zygote is a single cell, and early on during human development, it undergoes a series of rapid cell divisions to give rise to the 2-cell stage, 4-cell stage, 8-cell stage, and so forth (147–149). This process continues until the formation of a blastocyst, which contains an inner cell mass of embryonic stem cells. This blastocyst develops from an early stage of development known as an early blastocyst to a later stage of development known as a late blastocyst. The late blastocyst is implanted within the walls of the uterus and gives rise to an embryo. All cells within the embryo at this stage are the same and are known as human embryonic stem cells (hES). During the course of human development, hES cells give rise to all cells within the human body. This means that hES cells have the necessary genetic information to form all cells in the human body. *What causes a hES to become a cardiac myocyte versus a hepatocyte?* Different cell types have different functions that are performed by expression of cell specific proteins—cardiac myocytes exhibit contractile functions due to the presence of myosin heavy chain and other contractile proteins. The expression of different proteins is regulated by transcription of DNA to RNA; therefore, the regulation of gene expression is responsible for the expression of specific proteins (Figure 2.8). This means that hES can become any cell type in the human body by regulation of gene expression. This is the basic premise of stem cell engineering, which is also referred to as genetic engineering. The excitement in the field of stem cell engineering has been derived from the fact that a single cell type can, in theory, give rise to all cell types in the human body.

Human Embryonic Stem Cell and Tissue Engineering—In the previous section, we looked at the differentiation of human embryonic stem cells to any type in the

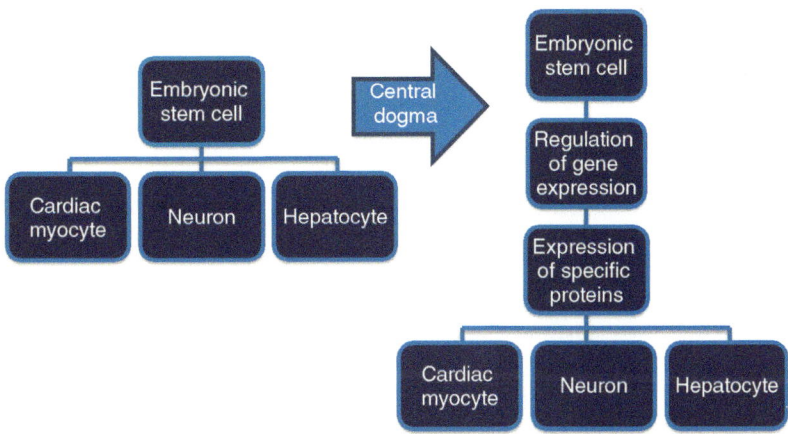

Figure 2.8 Differentiation of Human Embryonic Stem Cells—hES cells have the potential to become any cell type that is present in the human body. This is achieved by regulation of the genetic code, which in turn affects the expression of specific proteins, which in turn dictates cell function.

human body, at least in theory, by gene regulation. *What exactly does this mean for tissue engineering and why is this important?* In order to bioengineer 3D vascularized heart muscle, multiple cell types are required. Some of these required cell types which include cardiac myocytes for contractile function, cardiac fibroblasts for ECM production, and vascular cells, including endothelial cells and smooth muscle cells. Let us begin by using a scheme that relies on primary cells isolated from tissue biopsies. This scheme requires heart muscle specimens to isolate cardiac myocytes and cardiac fibroblasts. Similarly, vascular segments are required to isolate endothelial cells and smooth muscle cells. The isolated cells are then expanded in culture and combined with suitable scaffolds to fabricate 3D vascularized heart muscle.

Now let us look at the same scenario using hES cells. Once we have established isolation and culture conditions for hES, we can drive the differentiation of these cells toward all cell types required to fabricate vascularized heart muscle. This means that a single cell type can give rise to cardiac myocytes, cardiac fibroblasts, endothelial cells, and smooth muscle cells. These cells can then be combined with scaffolding material to fabricate 3D vascularized heart muscle. While controlling the differentiation fate of hES to multiple lineages is challenging and much more difficult than stated here, this example serves to illustrate the potential of hES cells during the tissue fabrication process.

Characteristics of Stem Cells—Stem cells have three important characteristics that distinguish them from other cell types: self-renewal, unspecialized function, and differentiation potential (150–153).

1. Self-renewal—This refers to the property of stem cells by which they are able to proliferate in order to maintain a constant supply of stem cells within any given tissue. We can illustrate this by looking at the case of satellite cells within skeletal muscle. Satellite cells are dormant under normal physiological conditions and become activated in response to muscle injury. The satellite cells proliferate and expand in number and then fuse to form myotubes. However, during this process, satellite cells are, in effect, used up. This means that the total number of satellite cells is reduced, limiting the ability to respond to the next muscle injury. Stem cells overcome this problem by the process of self-renewal. This means that every time a stem cell divides, one of the daughter cells is an identical copy of the parent and is maintained as a stem cell. The second copy may also be an identical copy, and if this occurs, the process is known as symmetrical division. Alternatively, the second daughter cell can differentiate to a specialized cell lineage, and in this case, the cell division process is known as asymmetrical division.
2. Unspecialized Function—Stem cells are unspecialized cells and do not perform any physiological function. Most cells in the human body are specialized and are programmed to undertake a specific function. For example, cardiac myocytes are specialized cells, and their primary function is to generate contractile force, which leads to the pumping action of the heart. In order to perform this specialized function, cardiac myocytes express specific contractile

proteins like actin, myosin heavy chain, tropomyosin, and troponin. However, stem cells do not perform any such specialized functions, and during growth and development, they are maintained in an unspecialized form. When the need arises, either due to disease or injury, stem cells have the potential to be differentiated to form specialized cells.

3. Differentiation Potential—Under controlled physiological conditions, stem cells have the potential to become specialized cells. This is the most talked-about property of stem cells and the most extensively studied as well. The ability of stem cells to become one or more specialized cells has broad-reaching applications in tissue engineering. Therefore, any cells that possess these three characteristics—self-renewal, unspecialized function, and differentiation potential—are referred to as stem cells.

Differentiation of Stem Cells—Early in the process of embryogenesis, embryonic stem cells are present and have the potential to give rise to any cell type. Embryonic stem cells give rise to cells of different lineages via a process known as cell differentiation, which refers to an increase in the degree of specialization of stem cells. As previously described, stem cells are unspecialized and do not perform any specialized function; however, stem cells gradually differentiate to form specialized cells. The process of cell differentiation transforms an unspecialized stem cell to a more specialized cell type. Cell differentiation is a result of regulation of gene expression, which in turn alters expression of specific proteins; these proteins undertake specialized functions in any given cell type. Therefore, the process of cell differentiation can be viewed as an increase in the degree of specialization of the cell brought about by expression of specific proteins that are specific to the cell and tissue.

Cell Potency—There are several terms that are used frequently in stem cell literature: cell potency, totipotent stem cells, multipotent stem cells, and unipotent stem cells (154–156) (Figure 2.9).

Cell potency refers to the differentiation potential of stem cells. Certain stem cells, like human embryonic stem cells, have the potential to differentiate to form all cell types, at least in theory, and are known as totipotent stem cells. In order to be truly considered totipotent stem cells, the differentiation potential of hES to all cell types will need to be experimentally validated. While this has not been done, hES are still considered to be totipotent, as it is known that these cells do give rise to all cell types in the human body during embryogenesis. Certain stem cells give rise to many different cell types, but not all, and are referred to as multipotent stem cells; hematopoietic stem cells are a classic example of multipotent stem cells, as they have the potential to become blood cells, but they cannot be differentiated into other cell types. Unipotent stem cells refer to stem cells that can only be differentiated into a single cell type. Most adult tissue has a population of resident stem cells, which are dormant under normal physiological conditions, but can be activated when the tissue is diseased, damaged, or injured. When resident stem cells are activated, they differentiate to form host tissue cells that aide the process of repair and recovery of lost tissue functionality. These resident stem cells are known as unipotent, as they

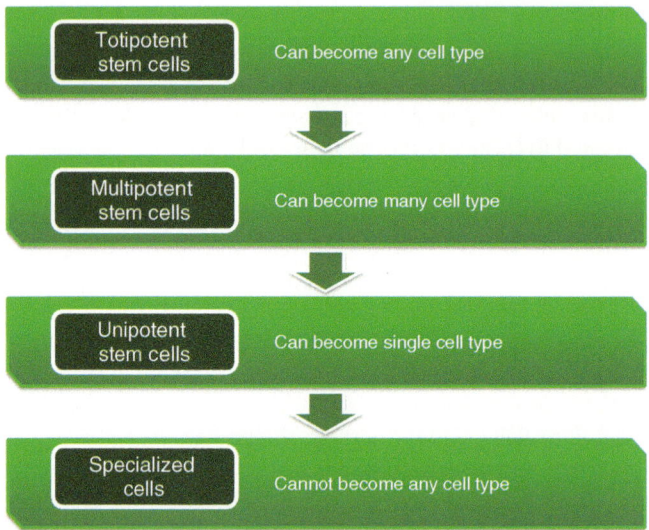

Figure 2.9 Differentiation Potential of Stem Cells—Totipotent stem cells have the potential to be differentiated to any cell type. Multipotent stem cells have the potential to be differentiated to form many cell types. Unipotent stem cells have the potential to be differentiated to form a single cell type. Specialized cells cannot be differentiated to form any cell types.

can only differentiate to a single cell type and cannot be differentiated to form any other cell type.

There have been three sources of stem cells that have been used extensively for tissue engineering studies: human embryonic stem cells (hES), which have been introduced before, induced pluripotent stem cells (iPS), and adult stem cells. We will discuss these three stem cell choices in the next three sections and also present a general scheme in Figure 2.10 outlining the potency of these cell types.

Human embryonic stem cells are totipotent and have the potential to become any cell type. Induced pluripotent stem cells are pluripotent, which means they have the capacity to become almost all types, but cannot be differentiated to form all cell types. Adult stem cells are either multipotent or unipotent, as they have the potential to differentiate only to a few cell types and often can only differentiate to form cells of the tissue in which they reside.

2.16 HUMAN EMBRYONIC STEM CELLS

Human embryonic stem cells hES are found very early during development and give rise to all cells in the body. Human embryonic stem cells are isolated from the inner mass of the human blastocyst and are cultured on tissue culture plates that have been coated with a feeder layer of mouse embryonic fibroblast cells (112,113).

INDUCED PLURIPOTENT STEM CELLS

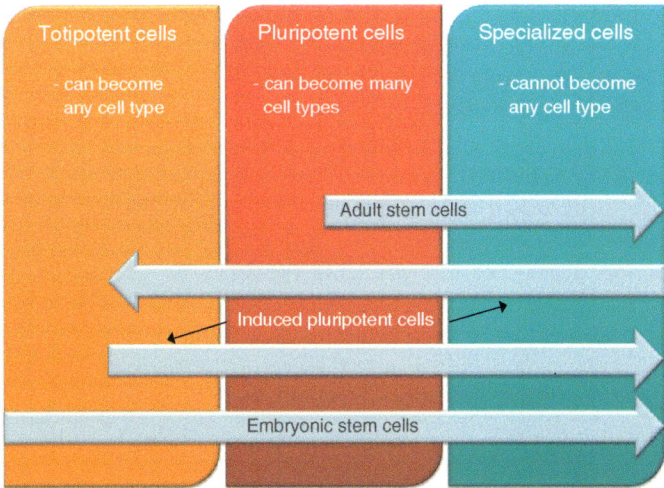

Figure 2.10 Stem Cell Sources for Tissue Engineering—There are three sources of stem cells that have been used extensively for tissue engineering studies: human embryonic stem cells, induced pluripotent stem cells, and adult stem cells.

After several days in culture, the inner cell mass divides and forms clusters of cells. Human embryonic stem cells are separated from the periphery of these clusters and mechanically or chemically dissociated, and then replated onto additional culture plates coated with feeder layers of fibroblast cells. As the cells continue to proliferate in culture, colonies of homogenous cells form and are mechanically/chemically selected and replated on a culture surface coated with feeder fibroblast cells. As this process continues, it leads to the formation of a continuous cell line. Human embryonic stem cells are desirable due to the ability of the cells to differentiate into all cell types, which can be used to support tissue fabrication technologies. However, the use of hES for tissue engineering or any other application has been surrounded by controversy, as the isolation of the cells requires destruction of human embryos.

2.17 INDUCED PLURIPOTENT STEM CELLS

Induced pluripotent stem cells have received a lot of attention in the recent literature and these cells have been extensively used for tissue engineering studies. The most attractive feature of iPS cells is their ability to differentiate into many different cell types without the need to destroy human embryos. Induced pluripotent stem cells are generated from differentiated adult cells, with skin fibroblasts being one common example. The specialized cells are reprogrammed to an embryonic state, and the reprogrammed cells, known as iPS cells, can be differentiated to form may cell types (157–160). The formation of induced pluripotent stem cells was first described in 2006, when it was demonstrated that retroviral overexpression of four transcription factors (octamer 3/4(Oct 4), SRY box-containing gene 2 (Sox2), Kruppel-like factor 4 (Klf4) and c-Myc) was sufficient to transform murine fibroblasts to an embryonic state (161). The main advantage of iPS cells is that terminally differentiated cells can be transformed to an embryonic state that can then be differentiated to multiple cell types. The use of iPS cells provides the same or comparable advantages as hES in terms of differentiation potential while eliminating the controversial issues associated with hES stem cells. In addition, iPS cells can be generated from terminally differentiated cells like fibroblasts, which can be isolated from a patient-derived biopsy, therefore providing a pathway for the development of autologous treatment strategies. The primary limitation of iPS cells has been the use of retroviral transduction of terminally differentiated cells, which can lead to random insertions and mutations in host genome. However, technology has been developed to address this limitation, and several methods have been developed to reprogram differentiated cells to form iPS cells using fewer transcription factors, viral integration followed by transgene removal, and chemical compounds.

2.18 ADULT STEM CELLS

The concept of adult stem cells or resident stem cells has been introduced earlier in this chapter; these cells refer to stem cells that are present within most tissues in a dormant state and become activated upon disease or injury. Adult stem cells are multipotent, as they can only be differentiated to a few cell types, and at times, they are unipotent and can only be differentiated to a single cell type (162). Adult stem cells isolated from the bone marrow, known as bone marrow-derived mesenchymal stem cells, have been studied extensively for tissue engineering applications (163,164). Bone marrow contains mesenchymal stem cells (MSCs) and hematopoietic stem cells (HSCs). Bone marrow-derived mesenchymal stem cells give rise to a variety of cell types, some of which include bone cells (osteocytes), cartilage cells (chondrocytes), fat cells (adipocytes), and other kinds of connective tissue cells, such as those in tendons. Hematopoietic stem cells give rise to all the types of blood cells: red blood cells, B lymphocytes, T lymphocytes, natural killer cells, neutrophils, basophils, eosinophils, monocytes, macrophages, and platelets.

SUMMARY

Current State of the Art—Cells are one of the building blocks of tissue engineering. In this chapter, we have provided an overview of cell biology and looked at the function of many cell organelles. We also looked at cell culture and examined how cell culture techniques are used to maintain and expand isolated cells. We also provided a comparison between monolayer cell culture and tissue engineering, and we looked at the relative advantages and disadvantages of each technique. In this chapter, we also studied cell transplantation and the use of isolated cells, including stem cells, to provide functional support for the repair of damaged and diseased tissue. In the concluding part of this chapter, we studied stem cell engineering and looked at several sources for stem cells to support the tissue fabrication process.

Thoughts for the Future—Significant research has been dedicated toward understanding the drivers of stem cells toward specific lineages. Many different strategies are currently under investigation to drive the differentiation fate of hES cells, iPS cells, and adult stem cells toward cell specific lineages. However, differentiation efficiency is generally low and differentiated cells do not fully replicate all functions of the cells that are necessary for therapeutic purposes. The ability to regulate the differentiation of stem cells toward a specific cell lineage, with high differentiation efficiencies to form highly functional cells, is a high priority research area.

PRACTICE QUESTIONS

1. Why are cells important for tissue engineering? How are cells used during the tissue fabrication process?
2. There are four aspects of cells that are important in tissue engineering: cell biology, cell culture, cell transplantation, and stem cell engineering. Explain how each of these four aspects relates to the fabrication of 3D artificial tissue.
3. Describe the process of nuclear import and nuclear export.
4. What is the role of the mitochondria in cell function?
5. Explain the process of phagocytosis and autophagy.
6. During the course of this chapter, we provided a description of several organelles and their role in maintaining the function of eukaryotic cells. How does this relate to tissue engineering, and why is it important during the tissue fabrication process?
7. Explain the following terms as they relate to cellular signaling: endocrine signaling, paracrine signaling, autocrine signaling, and direct cell-cell signaling.
8. Why is cellular signaling important for the fabrication of 3D artificial tissue?
9. Explain the following terms as they relate to cellular junctions: tight junction, adherens junction, desmosome junction, gap junction, and hemidesmosome junction.
10. What is the difference between structural and functional cellular junctions? Give examples of each.
11. Why are cellular junctions important during the tissue fabrication process?
12. Describe the steps involved in cell transplantation.
13. What does stem cell transplantation refer to?
14. What are some of the critical challenges in the field of cell transplantation?

15. Describe three mechanisms by which transplanted cells can improve the functional performance of damaged and/or diseased tissue.
16. Explain how cell therapy can be used as a treatment modality for acute myocardial infarction.
17. Compare stem cell therapy and tissue engineering as therapeutic modalities. Discuss the relative advantages and disadvantages of stem cell therapy and tissue engineering. Which one is better and why?
18. Describe the process of maintaining and expanding cells during monolayer 2D culture on the surface of a tissue culture plate. Explain the following steps in the process: cell isolation, cell attachment, cell proliferation, and sub-passaging of cells.
19. What does cell confluency refer to? How is it relevant to the cell culture process?
20. Cell culture and tissue engineering are related. Discuss the relationship between cell culture and tissue engineering. What are the relative advantages and disadvantages of each of these two processes?
21. What do the terms cell-cell and cell-matrix interactions mean? Why are cell-cell and cell-matrix interactions important for tissue engineering?
22. We discussed the following cell sources for tissue engineering: cell lines, animal derived cells, human embryonic stem cells, and adult stem cells. Pick any tissue fabrication application of your interest—which cell source will you use and why?
23. What are the relative advantages and disadvantages of human embryonic stem cells and induced pluripotent stem cells to support tissue fabrication?
24. We discussed cell culture, stem cell therapy, and tissue engineering. In order to support the fabrication of artificial heart muscle, discuss how each of these three strategies comes into play.
25. In your opinion, what is the most significant scientific challenge associated with cells as they relate to the tissue fabrication process? What can be done to overcome this scientific challenge?

REFERENCES

1. Yoshida M, Oh H. Stem cell engineering for cardiac tissue regeneration. Cardiology 2010;115(3):191–3.
2. Ashton RS, Keung AJ, Peltier J, Schaffer DV. Progress and prospects for stem cell engineering. Annu. Rev. Chem. Biomol. Eng. 2011;2:479–502.
3. Lim JM, Lee M, Lee EJ, Gong SP, Lee ST. Stem cell engineering: limitation, alternatives, and insight. Ann. N.Y. Acad. Sci. 2011 Jul;1229:89–98.

4. Ogle BM, Palecek SP. Editorial: stem cell engineering—discovery, diagnostics and therapies. Biotechnol. J. 2013 Apr;8(4):390–1.
5. Zhang Y, Rockett PI. A Generic multi-dimensional feature extraction method using multiobjective genetic programming. Evol. Comput. 2009;17(1):89–115.
6. Yang ZR, Lertmemongkolchai G, Tan G, Felgner PL, Titball R. A genetic programming approach for Burkholderia pseudomallei diagnostic pattern discovery. Bioinformatics. 2009 Sep 1;25(17):2256–62. PMCID:PMC2734322.
7. Mattick JS. Deconstructing the dogma: a new view of the evolution and genetic programming of complex organisms. Ann. N.Y. Acad. Sci. 2009 Oct;1178:29–46.
8. DePalma A. Enhancement of Cell Culture Techniques. Genetic Engineering & Biotechnology News 2009 Oct 15;29(18):46–8.
9. Helmrich A, Barnes D. Animal cell culture equipment and techniques. Methods in Cell Biology, Vol 57 1998;57:3–17.
10. Mozdziak PE, Petitte JN, Carson SD. An introductory undergraduate course covering animal cell culture techniques. Biochemistry and Molecular Biology Education 2004 Sep;32(5):319–22.
11. Murakami H. Basal Techniques for Serum-Free Animal-Cell Culture. Seikagaku 1988 Jun;60(6):450–3.
12. Dalle JH. Hematopoietic stem cell transplantation in SCD. C.R. Biol. 2013 Mar;336(3):148–51.
13. Haddad IY. Stem cell transplantation and lung dysfunction. Curr. Opin. Pediatr. 2013 Jun;25(3):350–6.
14. Lai PF, Panama BK, Masse S, Li G, Zhang Y, Kusha M, Farid TA, Asta J, Backx PH, Yau TM, et al. Mesenchymal Stem Cell Transplantation Mitigates Electrophysiological Remodeling in a Rat Model of Myocardial Infarction. J. Cardiovasc. Electrophysiol. 2013 Apr 8.
15. Maroof AM, Keros S, Tyson JA, Ying SW, Ganat YM, Merkle FT, Liu B, Goulburn A, Stanley EG, Elefanty AG, et al. Directed differentiation and functional maturation of cortical interneurons from human embryonic stem cells. Cell Stem Cell 2013 May 2;12(5):559–72.
16. Xie W, Schultz MD, Lister R, Hou Z, Rajagopal N, Ray P, Whitaker JW, Tian S, Hawkins RD, Leung D, et al. Epigenomic analysis of multilineage differentiation of human embryonic stem cells. Cell 2013 May 23;153(5):1134–48.
17. Tachibana M, Amato P, Sparman M, Gutierrez NM, Tippner-Hedges R, Ma H, Kang E, Fulati A, Lee HS, Sritanaudomchai H, et al. Human Embryonic Stem Cells Derived by Somatic Cell Nuclear Transfer. Cell 2013 May 15.
18. Bilokapic S, Schwartz TU. 3D ultrastructure of the nuclear pore complex. Curr. Opin. Cell Biol. 2012 Feb;24(1):86–91. PMCID:PMC3398480.
19. Grossman E, Medalia O, Zwerger M. Functional architecture of the nuclear pore complex. Annu. Rev. Biophys. 2012;41:557–84.
20. Maimon T, Elad N, Dahan I, Medalia O. The human nuclear pore complex as revealed by cryo-electron tomography. Structure. 2012 Jun 6;20(6):998–1006.
21. Raices M, D'Angelo MA. Nuclear pore complex composition: a new regulator of tissue-specific and developmental functions. Nat. Rev. Mol. Cell Biol. 2012 Nov;13(11):687–99.

22. Clever M, Mimura Y, Funakoshi T, Imamoto N. Regulation and coordination of nuclear envelope and nuclear pore complex assembly. Nucleus. 2013 Mar;4(2):105–14. PMCID:PMC3621742.
23. Sampathkumar P, Kim SJ, Upla P, Rice WJ, Phillips J, Timney BL, Pieper U, Bonanno JB, Fernandez-Martinez J, Hakhverdyan Z, et al. Structure, dynamics, evolution, and function of a major scaffold component in the nuclear pore complex. Structure 2013 Apr 2;21(4):560–71.
24. Imamoto N. Cargo recognition explains nuclear transport regulation induced by nuclear pore complex reorganization. J. Mol. Biol. 2013 Jun 12;425(11):1849–51.
25. Rothballer A, Kutay U. Poring over pores: nuclear pore complex insertion into the nuclear envelope. Trends Biochem. Sci. 2013 Jun;38(6):292–301.
26. Yang W. Distinct, but not completely separate spatial transport routes in the nuclear pore complex. Nucleus. 2013 May 1;4(3).
27. Strasser C, Grote P, Schauble K, Ganz M, Ferrando-May E. Regulation of nuclear envelope permeability in cell death and survival. Nucleus. 2012 Nov;3(6):540–51. PMCID:PMC3515537.
28. Rothballer A, Kutay U. SnapShot: the nuclear envelope II. Cell 2012 Aug 31;150(5):1084.
29. Korfali N, Wilkie GS, Swanson SK, Srsen V, de Las HJ, Batrakou DG, Malik P, Zuleger N, Kerr AR, Florens L, et al. The nuclear envelope proteome differs notably between tissues. Nucleus. 2012 Nov;3(6):552–64. PMCID:PMC3515538.
30. Barcena C, Osorio FG, Freije JM. Detection of nuclear envelope alterations in senescence. Methods Mol. Biol. 2013;965:243–51.
31. Cai S, Zhai Z. Relation between nuclear envelope and nuclear lamina in nuclear assembly in vitro. Sci. China C Life Sci. 1997 Dec;40(6):576–82.
32. Goldberg MW, Fiserova J, Huttenlauch I, Stick R. A new model for nuclear lamina organization. Biochem. Soc. Trans. 2008 Dec;36(Pt 6):1339–43.
33. Righolt CH, Raz V, Vermolen BJ, Dirks RW, Tanke HJ, Young IT. Molecular image analysis: quantitative description and classification of the nuclear lamina in human mesenchymal stem cells. Int. J. Mol. Imaging 2011;2011:723283. PMCID:PMC3065845.
34. Shevelyov YY, Nurminsky DI. The nuclear lamina as a gene-silencing hub. Curr. Issues Mol. Biol. 2012;14(1):27–38.
35. Bank EM, Gruenbaum Y. The nuclear lamina and heterochromatin: a complex relationship. Biochem. Soc. Trans. 2011 Dec;39(6):1705–9.
36. Barbi M, Mozziconacci J, Wong H, Victor JM. DNA topology in chromosomes: a quantitative survey and its physiological implications. J. Math. Biol. 2012 Nov 20.
37. Baumann K. Chromosomes: getting the architecture right. Nat. Rev. Mol. Cell Biol. 2013 Jan;14(1):2–3.
38. Bickmore WA, van SB. Genome architecture: domain organization of interphase chromosomes. Cell 2013 Mar 14;152(6):1270–84.
39. Donev R. Preface: organization of chromosomes. Adv. Protein Chem. Struct. Biol. 2013;90:vii–viii.
40. Cmarko D, Smigova J, Minichova L, Popov A. Nucleolus: the ribosome factory. Histol. Histopathol. 2008 Oct;23(10):1291–8.

41. Boulon S, Westman BJ, Hutten S, Boisvert FM, Lamond AI. The nucleolus under stress. Mol. Cell 2010 Oct 22;40(2):216–27. PMCID:PMC2987465.
42. Pederson T. The nucleolus. Cold Spring Harb. Perspect. Biol. 2011 Mar;3(3). PMCID:PMC3039934.
43. Hernandez-Verdun D. Assembly and disassembly of the nucleolus during the cell cycle. Nucleus. 2011 May;2(3):189–94. PMCID:PMC3149879.
44. Hernandez-Verdun D, Roussel P, Thiry M, Sirri V, Lafontaine DL. The nucleolus: structure/function relationship in RNA metabolism. Wiley. Interdiscip. Rev. RNA. 2010 Nov;1(3):415–31.
45. Ruggero D. Revisiting the nucleolus: from marker to dynamic integrator of cancer signaling. Sci. Signal. 2012;5(241):e38.
46. Panse VG, Johnson AW. Maturation of eukaryotic ribosomes: acquisition of functionality. Trends Biochem. Sci. 2010 May;35(5):260–6. PMCID:PMC2866757.
47. Gilbert WV. Functional specialization of ribosomes? Trends Biochem. Sci. 2011 Mar;36(3):127–32. PMCID:PMC3056915.
48. Pfeffer S, Brandt F, Hrabe T, Lang S, Eibauer M, Zimmermann R, Forster F. Structure and 3D arrangement of endoplasmic reticulum membrane-associated ribosomes. Structure. 2012 Sep 5;20(9):1508–18.
49. Schutz S, Panse VG. Getting ready to commit: ribosomes rehearse translation. Nat. Struct. Mol. Biol. 2012 Sep;19(9):861–2.
50. Nawy T. Ribosomes, start your engines. Nat. Methods 2012 Aug;9(8):780.
51. Liljas A, Moore P. Ribosomes—structure and function. Curr. Opin. Struct. Biol. 2012 Dec;22(6):730–2.
52. Barna M. Ribosomes take control. Proc. Natl. Acad. Sci. U.S.A 2013 Jan 2; 110(1):9–10. PMCID:PMC3538271.
53. Garcia ML, Fernandez A, Solas MT. Mitochondria, motor neurons and aging. J. Neurol. Sci. 2013 Apr 26.
54. Vannuvel K, Renard P, Raes M, Arnould T. Functional and morphological impact of ER stress on mitochondria. J. Cell Physiol 2013 Sep;228(9):1802–18.
55. Hwang AB, Jeong DE, Lee SJ. Mitochondria and organismal longevity. Curr. Genomics 2012 Nov;13(7):519–32. PMCID:PMC3468885.
56. Nezich CL, Youle RJ. Make or break for mitochondria. Elife. 2013;2:e00804. PMCID:PMC3654434.
57. Desler C, Rasmussen LJ. Mitochondria In Biology and Medicine—2012. Mitochondrion. 2013 May 29.
58. Jerome WG. Lysosomes, cholesterol and atherosclerosis. Clin. Lipidol. 2010 Dec 1; 5(6):853–65. PMCID:PMC3105626.
59. Kurz T, Eaton JW, Brunk UT. The role of lysosomes in iron metabolism and recycling. Int. J. Biochem. Cell Biol. 2011 Dec;43(12):1686–97.
60. Pryor PR. Analyzing lysosomes in live cells. Methods Enzymol. 2012;505:145–57.
61. Hamer I, Van BG, Arnould T, Jadot M. Lipids and lysosomes. Curr. Drug Metab 2012 Dec;13(10):1371–87.
62. Oczypok EA, Oury TD, Chu CT. It's a cell-eat-cell world: autophagy and phagocytosis. Am. J. Pathol. 2013 Mar;182(3):612–22. PMCID:PMC3589073.

63. Garg M, Chandawarkar RY. Phagocytosis: history's lessons. Conn. Med. 2013 Jan;77(1):23–6.
64. Braakman I, Hebert DN. Protein folding in the endoplasmic reticulum. Cold Spring Harb. Perspect. Biol. 2013;5(5).
65. Spang A. Retrograde traffic from the Golgi to the endoplasmic reticulum. Cold Spring Harb. Perspect. Biol. 2013;5(6).
66. Moseley JB. An expanded view of the eukaryotic cytoskeleton. Mol. Biol. Cell 2013 Jun;24(11):1615–8. PMCID:PMC3667716.
67. Brini M, Cali T, Ottolini D, Carafoli E. The plasma membrane calcium pump in health and disease. FEBS J. 2013 Feb 18.
68. Ueda Y, Makino A, Murase-Tamada K, Sakai S, Inaba T, Hullin-Matsuda F, Kobayashi T. Sphingomyelin regulates the transbilayer movement of diacylglycerol in the plasma membrane of Madin-Darby canine kidney cells. FASEB J. 2013 May 16.
69. Babcock JJ, Li M. Inside job: ligand-receptor pharmacology beneath the plasma membrane. Sin: Acta Pharmacol; 2013 May 20.
70. Rilla K, Pasonen-Seppanen S, Deen AJ, Koistinen VV, Wojciechowski S, Oikari S, Karna R, Bart G, Torronen K, Tammi RH, et al. Hyaluronan production enhances shedding of plasma membrane-derived microvesicles. Exp. Cell Res. 2013 May 31.
71. Volpato FZ, Fuhrmann T, Migliaresi C, Hutmacher DW, Dalton PD. Using extracellular matrix for regenerative medicine in the spinal cord. Biomaterials 2013 Jul;34(21):4945–55.
72. Trappmann B, Chen CS. How cells sense extracellular matrix stiffness: a material's perspective. Curr. Opin. Biotechnol. 2013 Apr 20.
73. Heinegard D. Extracellular matrix: pathobiology and signaling. Biol. Chem. 2013 Jun 1; 394(6):805–6.
74. Clause KC, Barker TH. Extracellular matrix signaling in morphogenesis and repair. Curr. Opin. Biotechnol. 2013 May 28.
75. Bennett JS. Integrin structure and function in hemostasis and thrombosis. Ann. N.Y. Acad. Sci. 1991;614:214–28.
76. Humphries MJ. Integrin structure. Biochem. Soc. Trans. 2000;28(4):311–39.
77. Shimaoka M, Takagi J, Springer TA. Conformational regulation of integrin structure and function. Annu. Rev. Biophys. Biomol. Struct. 2002;31:485–516.
78. Arnaout MA, Mahalingam B, Xiong JP. Integrin structure, allostery, and bidirectional signaling. Annu. Rev. Cell Dev. Biol 2005;21:381–410.
79. Arcangeli A, Becchetti A. Integrin structure and functional relation with ion channels. Adv. Exp. Med. Biol 2010;674:1–7.
80. Campbell ID, Humphries MJ. Integrin structure, activation, and interactions. Cold Spring Harb. Perspect. Biol. 2011 Mar;3(3). PMCID:PMC3039929.
81. Ruoslahti E, Hayman EG, Engvall E, Cothran WC, Butler WT. Alignment of biologically active domains in the fibronectin molecule. J. Biol. Chem. 1981 Jul 25;256(14):7277–81.
82. Venyaminov SY, Metsis ML, Chernousov MA, Koteliansky VE. Distribution of secondary structure along the fibronectin molecule. Eur. J. Biochem. 1983 Oct 3;135(3):485–9.
83. Hunyady L. Cellular signaling in health and disease. Mol. Cell Endocrinol. 2012 Apr 28;353(1–2):1–2.

84. Walker C, Ahmed SA, Brown T, Ho SM, Hodges L, Lucier G, Russo J, Weigel N, Weise T, Vandenbergh J. Species, interindividual, and tissue specificity in endocrine signaling. Environ. Health Perspect. 1999 Aug;107 Suppl 4:619–24. PMCID:PMC1567505.

85. Gough NR. Focus issue: endocrine signaling from clinic to cell. Sci. Signal. 2010;3(143):eg9.

86. Dietze GJ, Taegtmeyer H. Introduction: autocrine and paracrine signaling between contracting myocardium and coronary endothelium during myocardial ischemia: effects of insulin resistance. Am. J. Cardiol. 1997 Aug 4;80(3A):1A-2A.

87. Wang N, De BM, Decrock E, Bol M, Gadicherla A, Vinken M, Rogiers V, Bukauskas FF, Bultynck G, Leybaert L. Paracrine signaling through plasma membrane hemichannels. Biochim. Biophys. Acta 2013 Jan;1828(1):35–50. PMCID:PMC3666170.

88. Mann DA, Oakley F. Serotonin paracrine signaling in tissue fibrosis. Biochim. Biophys. Acta 2013 Jul;1832(7):905–10.

89. Mureli S, Gans CP, Bare DJ, Geenen DL, Kumar NM, Banach K. Mesenchymal stem cells improve cardiac conduction by upregulation of connexin 43 through paracrine signaling. Am. J. Physiol Heart Circ. Physiol. 2013 Feb 15;304(4):H600-H609. PMCID:PMC3566487.

90. Wilson KJ, Mill C, Lambert S, Buchman J, Wilson TR, Hernandez-Gordillo V, Gallo RM, Ades LM, Settleman J, Riese DJ. EGFR ligands exhibit functional differences in models of paracrine and autocrine signaling. Growth Factors 2012 Apr;30(2):107–16.

91. Yi EH, Lee CS, Lee JK, Lee YJ, Shin MK, Cho CH, Kang KW, Lee JW, Han W, Noh DY, et al. STAT3-RANTES autocrine signaling is essential for tamoxifen resistance in human breast cancer cells. Mol. Cancer Res. 2013 Jan;11(1):31–42.

92. Leibiger B, Moede T, Muhandiramlage TP, Kaiser D, Vaca SP, Leibiger IB, Berggren PO. Glucagon regulates its own synthesis by autocrine signaling. Proc. Natl. Acad. Sci. U.S.A 2012 Dec 18;109(51):20925–30. PMCID:PMC3529083.

93. Datta S, Sarvetnick N. Lymphocyte proliferation in immune-mediated diseases. Trends Immunol 2009 Sep;30(9):430–8.

94. Zagon IS, Donahue RN, Bonneau RH, McLaughlin PJ. B lymphocyte proliferation is suppressed by the opioid growth factor-opioid growth factor receptor axis: Implication for the treatment of autoimmune diseases. Immunobiology 2011 Jan;216(1–2):173–83.

95. Luckerath K, Kirkin V, Melzer IM, Thalheimer FB, Siele D, Milani W, Adler T, Aguilar-Pimentel A, Horsch M, Michel G, et al. Immune modulation by Fas ligand reverse signaling: lymphocyte proliferation is attenuated by the intracellular Fas ligand domain. Blood 2011 Jan 13;117(2):519–29.

96. Abraham R, Verfaillie CM. Neural differentiation and support of neuroregeneration of non-neural adult stem cells. Prog. Brain Res. 2012;201:17–34.

97. Vilchez D, Boyer L, Lutz M, Merkwirth C, Morantte I, Tse C, Spencer B, Page L, Masliah E, Travis BW, et al. FOXO4 is necessary for neural differentiation of human embryonic stem cells. Aging Cell 2013 Mar 7;

98. Shen L. Tight junctions on the move: molecular mechanisms for epithelial barrier regulation. Ann. N.Y. Acad. Sci. 2012 Jul;1258:9–18.

99. Cording J, Berg J, Kading N, Bellmann C, Tscheik C, Westphal JK, Milatz S, Gunzel D, Wolburg H, Piontek J, et al. In tight junctions, claudins regulate the interactions between occludin, tricellulin and marvelD3, which, inversely, modulate claudin oligomerization. J. Cell Sci. 2013 Jan 15;126(Pt 2):554–64.
100. Sapra B, Jindal M, Tiwary AK. Tight junctions in skin: new perspectives. Ther. Deliv. 2012 Nov;3(11):1297–327.
101. Campbell M, Humphries P. The blood-retina barrier: tight junctions and barrier modulation. Adv. Exp. Med. Biol. 2012;763:70–84.
102. Franke WW. Discovering the molecular components of intercellular junctions–a historical view. Cold Spring Harb. Perspect. Biol. 2009 Sep;1(3):a003061. PMCID:PMC2773636.
103. Penn EJ, Hobson C, Rees DA, Magee AI. Structure and assembly of desmosome junctions: biosynthesis, processing, and transport of the major protein and glycoprotein components in cultured epithelial cells. J. Cell Biol. 1987 Jul;105(1):57–68. PMCID:PMC2114930.
104. Wheeler GN, Parker AE, Thomas CL, Ataliotis P, Poynter D, Arnemann J, Rutman AJ, Pidsley SC, Watt FM, Rees DA, et al. Desmosomal glycoprotein DGI, a component of intercellular desmosome junctions, is related to the cadherin family of cell adhesion molecules. Proc. Natl. Acad. Sci. U.S.A 1991 Jun 1;88(11):4796–800. PMCID:PMC51753.
105. Kleber AG. Gap junctions and conduction of cardiac excitation. Heart Rhythm. 2011 Dec;8(12):1981–4.
106. Saffitz JE, Kleber AG. Gap junctions, slow conduction, and ventricular tachycardia after myocardial infarction. J. Am. Coll. Cardiol. 2012 Sep 18; 60(12):1111–3.
107. Li MW, Mruk DD, Cheng CY. Gap junctions and blood-tissue barriers. Adv. Exp. Med. Biol. 2012;763:260–80.
108. Nielsen MS, Nygaard AL, Sorgen PL, Verma V, Delmar M, Holstein-Rathlou NH. Gap junctions. Compr. Physiol 2012 Jul 1;2(3):1981–2035.
109. Manz BN, Groves JT. Spatial organization and signal transduction at intercellular junctions. Nat. Rev. Mol. Cell Biol. 2010 May;11(5):342–52.
110. Niethammer D, Bader P, Handgretinger R, Klingebiel T. Stem Cell Transplantation. Klin. Padiatr. 2013 May;225(S 01):94–6.
111. Maqsood MI, Matin MM, Bahrami AR, Ghasroldasht MM. Immortality of cell lines: Challenges and advantages of establishment. Cell Biol.Int. 2013 May 30.
112. Amit M. Sources and derivation of human embryonic stem cells. Methods Mol. Biol. 2013;997:3–11.
113. Kaur J, Tilkins ML. Methods for culturing human embryonic stem cells on feeders. Methods Mol. Biol. 2013;997:93–113.
114. Cao Q, Whittemore SR. Cell transplantation: stem cells and precursor cells. Handb. Clin. Neurol. 2012;109:551–61.
115. Segers VF, Lee RT. Stem-cell therapy for cardiac disease. Nature 2008 Feb 21;451(7181):937–42.
116. Teng M, Geng Z, Huang L, Zhao X. Stem cell transplantation in cardiovascular disease: an update. J. Int. Med. Res. 2012;40(3):833–8.

117. Leor J, Gerecht S, Cohen S, Miller L, Holbova R, Ziskind A, Shachar M, Feinberg MS, Guetta E, Itskovitz-Eldor J. Human embryonic stem cell transplantation to repair the infarcted myocardium. Heart 2007 Oct;93(10):1278–84. PMCID:PMC2000918.
118. Terai S, Sakaida I, Yamamoto N, Omori K, Watanabe T, Ohata S, Katada T, Miyamoto K, Shinoda K, Nishina H, et al. An in vivo model for monitoring trans-differentiation of bone marrow cells into functional hepatocytes. J. Biochem. 2003 Oct;134(4):551–8.
119. Choi KS, Shin JS, Lee JJ, Kim YS, Kim SB, Kim CW. In vitro trans-differentiation of rat mesenchymal cells into insulin-producing cells by rat pancreatic extract. Biochem. Biophys. Res. Commun. 2005 May 20;330(4):1299–305.
120. Hombach-Klonisch S, Panigrahi S, Rashedi I, Seifert A, Alberti E, Pocar P, Kurpisz M, Schulze-Osthoff K, Mackiewicz A, Los M. Adult stem cells and their trans-differentiation potential—perspectives and therapeutic applications. J. Mol. Med. (Berl) 2008 Dec;86(12):1301–14. PMCID:PMC2954191.
121. Laco F, Grant MH, Flint DJ, Black RA. Cellular trans-differentiation and morphogenesis toward the lymphatic lineage in regenerative medicine. Stem Cells Dev. 2011 Feb;20(2):181–95.
122. Joo KM, Jin J, Kang BG, Lee SJ, Kim KH, Yang H, Lee YA, Cho YJ, Im YS, Lee DS, et al. Trans-differentiation of neural stem cells: a therapeutic mechanism against the radiation induced brain damage. PLoS One 2012;7(2):e25936. PMCID:PMC3277599.
123. Li H, Zuo S, He Z, Yang Y, Pasha Z, Wang Y, Xu M. Paracrine factors released by GATA-4 overexpressed mesenchymal stem cells increase angiogenesis and cell survival. Am. J. Physiol Heart Circ. Physiol 2010 Dec;299(6):H1772-H1781. PMCID:PMC3006287.
124. Burdon TJ, Paul A, Noiseux N, Prakash S, Shum-Tim D. Bone marrow stem cell derived paracrine factors for regenerative medicine: current perspectives and therapeutic potential. Bone Marrow Res. 2011;2011:207326. PMCID:PMC3195349.
125. Xu S, Zhu J, Yu L, Fu G. Endothelial progenitor cells: current development of their paracrine factors in cardiovascular therapy. J. Cardiovasc. Pharmacol. 2012 Apr;59(4):387–96.
126. Bell GI, Meschino MT, Hughes-Large JM, Broughton HC, Xenocostas A, Hess DA. Combinatorial human progenitor cell transplantation optimizes islet regeneration through secretion of paracrine factors. Stem Cells Dev. 2012 Jul 20;21(11):1863–76.
127. de FP, Gonzalez M, Meloni G, De Propris MS, Bellucci R, Cordone I, Gozzer M, Leone G, Mandelli F. Monitoring of CD34+ cells during leukapheresis allows a single, successful collection of hemopoietic progenitors in patients with low numbers of circulating stem cells. Bone Marrow Transplant. 1999 Jun;23(12):1229–36.
128. Hillebrands JL, Klatter FA, Rozing J. Origin of vascular smooth muscle cells and the role of circulating stem cells in transplant arteriosclerosis. Arterioscler. Thromb. Vasc. Biol. 2003 Mar 1;23(3):380–7.
129. Hennessy B, Korbling M, Estrov Z. Circulating stem cells and tissue repair. Panminerva Med. 2004 Mar;46(1):1–11.
130. Seebach C, Henrich D, Wilhelm K, Barker JH, Marzi I. Endothelial progenitor cells improve directly and indirectly early vascularization of mesenchymal stem cell-driven bone regeneration in a critical bone defect in rats. Cell Transplant. 2012;21(8):1667–77.
131. Walker BA, Brown BB, Krohmer JS, Bonte FJ. Adaptation of disposable plastics to quantitative mammalian cell culture. Tex. Rep. Biol. Med 1962;20:686–92.

132. Perlman D. Value of mammalian cell culture as a biochemical tool. Science 1968 Apr 5; 160(3823):42–6.
133. Gospodarowicz D, Moran JS. Growth factors in mammalian cell culture. Annu. Rev. Biochem. 1976;45:531–58.
134. Oxender DL, Lee M, Cecchini G. Regulation of transport in mammalian cell culture. Prog. Clin. Biol. Res. 1976;9:41–7.
135. Tolbert WR, Schoenfeld RA, Lewis C, Feder J. Large-scale mammalian cell culture: Design and use of an economical batch suspension system. Biotechnol. Bioeng. 1982 Jul;24(7):1671–9.
136. Harakas NK, Lewis C, Bartram RD, Wildi BS, Feder J. Mammalian cell culture: technology and physiology. Adv. Exp. Med. Biol 1984;172:119–38.
137. Linhardt RJ. Mammalian cell culture. Patents and literature. Appl. Biochem. Biotechnol. 1986 Oct;13(2):167–74.
138. Nilsson K. Mammalian cell culture. Methods Enzymol. 1987;135:387–93.
139. Finter NB, Garland AJ, Telling RC. Large-scale mammalian cell culture: a perspective. Bioprocess. Technol. 1990;10:1–14.
140. McKeehan WL, Barnes D, Reid L, Stanbridge E, Murakami H, Sato GH. Frontiers in mammalian cell culture. In Vitro Cell Dev. Biol. 1990 Jan;26(1):9–23.
141. Hu WS, Piret JM. Mammalian cell culture processes. Curr. Opin. Biotechnol. 1992 Apr;3(2):110–4.
142. Reiter M, Bluml G. Large-scale mammalian cell culture. Curr. Opin. Biotechnol. 1994 Feb;5(2):175–9.
143. Hu WS, Aunins JG. Large-scale mammalian cell culture. Curr. Opin. Biotechnol. 1997 Apr;8(2):148–53.
144. Crick F. Central dogma of molecular biology. Nature 1970 Aug 8;227(5258):561–3.
145. Cooper S. The central dogma of cell biology. Cell Biol. Int. Rep. 1981 Jun;5(6):539–49.
146. Thieffry D, Sarkar S. Forty years under the central dogma. Trends Biochem. Sci. 1998 Aug;23(8):312–6.
147. Dvash T, Ben-Yosef D, Eiges R. Human embryonic stem cells as a powerful tool for studying human embryogenesis. Pediatr. Res. 2006 Aug;60(2):111–7.
148. Vaillancourt C, Lafond J. Human embryogenesis: overview. Methods Mol. Biol. 2009;550:3–7.
149. Yi H, Xue L, Guo MX, Ma J, Zeng Y, Wang W, Cai JY, Hu HM, Shu HB, Shi YB, et al. Gene expression atlas for human embryogenesis. FASEB J. 2010 Sep;24(9):3341–50. PMCID:PMC2923361.
150. Kim ND, Oberley TD, Yasukawa-Barnes J, Clifton KH. Stem cell characteristics of transplanted rat mammary clonogens. Exp. Cell Res. 2000 Oct 10;260(1):146–59.
151. Sottile V, Halleux C, Bassilana F, Keller H, Seuwen K. Stem cell characteristics of human trabecular bone-derived cells. Bone 2002 May;30(5):699–704.
152. Miki T, Lehmann T, Cai H, Stolz DB, Strom SC. Stem cell characteristics of amniotic epithelial cells. Stem Cells 2005 Nov;23(10):1549–59.
153. Pfeiffer MJ, Schalken JA. Stem cell characteristics in prostate cancer cell lines. Eur. Urol. 2010 Feb;57(2):246–54.

REFERENCES

154. Jiao J, Milwid JM, Yarmush ML, Parekkadan B. A mesenchymal stem cell potency assay. Methods Mol. Biol. 2011;677:221–31.
155. Polejaeva I, Mitalipov S. Stem cell potency and the ability to contribute to chimeric organisms. Reproduction 2013 Mar;145(3):R81–R88.
156. Romli F, Alitheen NB, Hamid M, Ismail R, Abd Rahman NM. Current techniques in reprogramming cell potency. J. Cell Biochem. 2013 Jun;114(6):1230–7.
157. Ferreira LM, Mostajo-Radji MA. How induced pluripotent stem cells are redefining personalized medicine. Gene. 2013 May 10;520(1):1–6.
158. Scott CW, Peters MF, Dragan YP. Human induced pluripotent stem cells and their use in drug discovery for toxicity testing. Toxicol. Lett. 2013 May 10;219(1):49–58.
159. Garate Z, Davis BR, Quintana-Bustamante O, Segovia JC. New frontier in regenerative medicine: site-specific gene correction in patient-specific induced pluripotent stem cells. Hum. Gene Ther. 2013 May 15.
160. Takahashi K, Yamanaka S. Induced pluripotent stem cells in medicine and biology. Development 2013 Jun;140(12):2457–61.
161. Takahashi K, Yamanaka S. Induction of pluripotent stem cells from mouse embryonic and adult fibroblast cultures by defined factors. Cell 2006 Aug 25;126(4):663–76.
162. Gonzalez MA, Bernad A. Characteristics of adult stem cells. Adv. Exp. Med. Biol. 2012;741:103–20.
163. Kang Y, Kim S, Khademhosseini A, Yang Y. Creation of bony microenvironment with CaP and cell-derived ECM to enhance human bone-marrow MSC behavior and delivery of BMP-2. Biomaterials 2011 Sep;32(26):6119–30. PMCID:PMC3130069.
164. Pelekanos RA, Li J, Gongora M, Chandrakanthan V, Scown J, Suhaimi N, Brooke G, Christensen ME, Doan T, Rice AM, et al. Comprehensive transcriptome and immunophenotype analysis of renal and cardiac MSC-like populations supports strong congruence with bone marrow MSC despite maintenance of distinct identities. Stem Cell Res. 2012 Jan;8(1):58–73.

3

BIOMATERIALS FOR TISSUE ENGINEERING

Learning Objectives

After completing this chapter, students should be able to:

1. Provide a definition for biomaterials.
2. Describe a process flow sheet for biomaterial development and describe the following terms as they relate to biomaterial development: biocompatibility, mechanical properties, biomimetic properties, and material degradation.
3. Discuss the historical development of biomaterials, including examples of dental implants and prosthetic implants.
4. Describe the tensile properties of a material, including sample preparation, components of a mechanical testing system, and components of the stress–strain curve.
5. Describe methods that can be used to improve the tensile properties of biomaterials.
6. Discuss the role of biodegradation in tissue engineering.
7. Describe the role of biocompatibility in tissue engineering.
8. Define biomimetic activity as it relates to biomaterials and tissue fabrication.

Introduction to Tissue Engineering: Applications and Challenges, First Edition. Ravi Birla.
© 2014 The Institute of Electrical and Electronics Engineers, Inc. Published 2014 by John Wiley & Sons, Inc.

DEFINITION OF BIOMATERIALS   85

9. Describe the differences between natural and synthetic materials, degradable and nondegradable materials, and metals, ceramics, and polymers.
10. Discuss the following biomaterial platforms: hydrogels, acellular scaffolds, polymeric scaffolds, and self-organization strategies.
11. Explain the concept of smart materials and give examples of smart materials in tissue engineering.
12. Describe the composition and function of the mammalian extracellular matrix and how it relates to the tissue fabrication process.
13. Explain the concept of an idealized biomaterial.

CHAPTER OVERVIEW

In this chapter, we will study the role of biomaterials in tissue engineering. We begin this chapter by providing a definition of biomaterials and then provide a general scheme for the development of biomaterials to support the tissue fabrication process. We then provide a brief description of the historical relevance of biomaterials and the way in which biomaterials have been used over the centuries. After these introductory sections, we provide a discussion of biomaterial properties that are important for tissue engineering: tensile properties, degradation kinetics, biocompatibility, and biomimetic properties. We then provide a classification scheme for biomaterials using three categories: natural vs. synthetic, degradable vs. nondegradable, and metals, ceramics, or polymers. The next section is devoted to a discussion of biomaterial platforms, which include polymeric scaffolds, biodegradable hydrogels, acellular matrices, and scaffold-free platforms. We next focus our discussion on the evolution of smart materials and the development of these materials for tissue engineering. We end this chapter with a discussion of the mammalian ECM and provide insights for the development of an idealized biomaterial.

3.1 DEFINITION OF BIOMATERIALS

In this section, we will look at several widely used definitions of biomaterials (1–6). We will then use this information to formulate a definition of biomaterials that is tailored to tissue engineering and tissue fabrication processes.

According to the National Institutes of Health (NIH), a biomaterial is defined as (7) *"any substance (other than a drug) or combination of substances synthetic or natural in origin, which can be used for any period of time, as a whole or part of a system which treats, augments, or replaces tissue, organ, or function of the body."* There are two important components of this definition that require further explanation. First, a biomaterial may be *"synthetic or natural in origin"*, thereby providing a broad classification of the sources of biomaterials. This provides a simple classification scheme that is discussed later in this chapter. However, at this stage, it can be appreciated that biomaterials can either be synthetic, which means that the biomaterial is synthesized in the laboratory using controlled conditions; or the biomaterial can be natural, which means that it is extracted from tissue specimens like collagen, which is extracted from the tail of rats (referred to as rat tail collagen). The second component of the definition—*"treats, augments, or replaces tissue, organ, or function of the body"*—provides a direct statement of the potential application of the biomaterial. This statement illustrates the potential application of the biomaterial in the medical field as a therapeutic option. The definition provided by NIH suggests that the biomaterial can be used as a whole or as a part of a system for therapeutic purposes. The utilization of a biomaterial as a whole for medical applications means that the biomaterial itself is the therapeutic agent; an example of this application is when biomaterials are sutured onto left ventricular tissue after a myocardial infarction to limit cardiac hypertrophy and support functional remodeling. The second application of a biomaterial is as part of a system; this application refers to biomaterials that are used the fabrication of devices like stents or pacemakers. The use of biomaterials as part of a system applies to tissue engineering, where biomaterials are used for fabrication of 3D artificial tissue.

Additional definitions for biomaterials have been proposed. Clemson University has played a significant role in development of the field of biomaterials. During one of the annual biomaterials symposia at Clemson University, the Sixth Annual International Biomaterials Symposium in April 20–24th 1974, Clemson's Advisory Board for Biomaterials provided the following definition (8): *"a biomaterial is a systematically, pharmacological inert substance designed for implantation within or incorporation with a living system."* Similar to the NIH definition, this one provided by Clemson's Advisory Board can be broken down into two components. The first part of the definition—*"systematically, pharmacological inert substance"*—refers to a specific property of biomaterials, inertness of the material, which is important for any given biological application. The second part of the definition—*"designed for implantation within or incorporation with a living system"*—refers to the application of the material for *in vivo* applications as a direct therapeutic modality. This definition was conceived in 1974, and at the time, the field of tissue engineering was not very well-developed, and therefore, the definition does not discuss the incorporation of materials for artificial tissue fabrication.

We have looked at two definitions of biomaterials: one provided by NIH and one provided by Clemson's Advisory Board for Biomaterials. If we compare the two definitions, there is a common theme—the definition always comprises two parts, the first part focused on material classification or property, and the second part focused on application of the biomaterial.

A third definition is provided by J. Black in a 1982 publication (9): *"a biomaterial is any pharmacological inert material, viable or non-viable, natural product or man-made, that is a part of or is capable of interacting in a beneficial way within a living organism."* The definition provided by J. Black has also been broken into two components, just as the case with the definition provided by NIH and by Clemson's Advisory Board on Biomaterials. The first part of the definition—*"is any pharmacological inert material, viable or non-viable, natural product or man-made"*—talks about biomaterial properties and classification schemes. The second part of the definition describes potential applications of the biomaterial by stating that it is *"capable of interacting in a beneficial way within a living organism."*

Continuing with this discussion o the definition of biomaterials, let us look at one final definition provided by CP Sharma (8): *"biomaterials are materials designed for interfacing and/or interacting with a living system, inducing no adverse reaction at the site of implantation in vivo or ex vivo and systematically."* As we have seen with the three definitions of biomaterials that have been presented before, the definition by CP Sharma is also composed of two parts, though in this case the application is stated first and is followed by the material properties. In his definition, CP Sharma states that the applications of biomaterials are *"materials designed for interfacing and/or interacting with a living system."* This part of the definition alludes to the use of biomaterials as a therapeutic modality. In the second part of the definition, CP Sharma states an important property of biomaterials: *"inducing no adverse reaction."* This phrase alludes to the biocompatibility of the material and is indeed an important property for any biomaterial.

We have looked at four definitions of biomaterials and seen a general theme. All four definitions can be viewed as two-part definitions, the first of which is focused on a specific material property or classification scheme and the second part focused on *in vivo* application or interfacing with living systems. However, none of these definitions adequately represent the use of biomaterials for tissue engineering and for the tissue fabrication process. In the field of tissue engineering, biomaterials are used specifically to support fabrication, culture, and maturation of 3D artificial tissue by providing functional integration at the cell-material interface. Keeping with the theme of the biomaterial definitions provided by experts in the field, we present our definition of a biomaterial as it applies to tissue engineering and the tissue fabrication process (Figure 3.1): *a biomaterial is any substance that simulates the extracellular matrix by functionally interacting with isolated cells to support fabrication and maturation of 3D artificial tissue.*

Let us discuss this definition. The first point to note is the elimination of any classification schemes and any reference to material properties. We do believe that classification schemes and materials properties are critical in the development of biomaterials for any application; however, we do not believe this information has to be incorporated within the definition. Rather, we will discuss biomaterial classification schemes and biomaterial properties in a later section. Our definition is focused on the application of biomaterials to interface with isolated cells in the context of artificial tissue fabrication: *"a biomaterial is any substance which simulates the extracellular matrix."* In this sense, the biomaterial simulates the roles of the mammalian extracellular matrix. This is an important criterion for any biomaterial used for fabrication of 3D artificial tissue; the role of biomaterials is indeed to simulate properties of the extracellular matrix.

We stated that a biomaterial is any substance that stimulates the extracellular matrix. Implied in this definition is that biomaterials will functionally interact with isolated cells based on specific cell-matrix interactions mediated between cell surface integrins and specific binding sites on the surface of biomaterials. Cell-matrix interactions are known to initiate a complex set of intracellular signaling pathways,

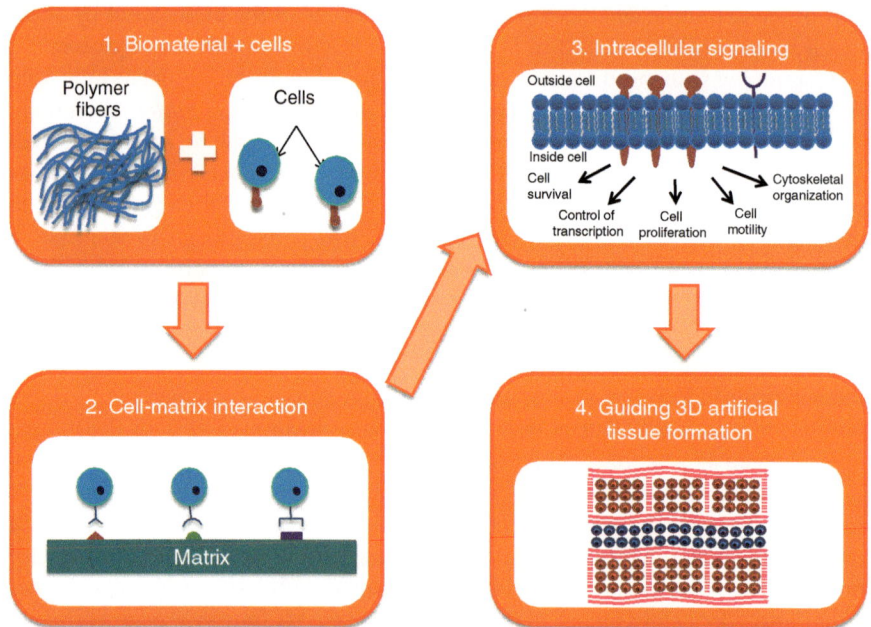

Figure 3.1 Definition of Biomaterials—The definition of biomaterials is presented in four parts. The first part shows a biomaterial that has specific binding sites for integrins, along with cell surface integrins. The second part shows functional coupling between biomaterials and cells, known as cell-matrix interaction. The specific cell-matrix interaction leads to a sequence of intracellular signaling events that support growth, remodeling, and health of cells and the tissue, as shown in the third part of the figure. Finally, fabrication of functional 3D artificial tissue is a result of intracellular signaling events, as can be seen in the fourth and final part of the definition.

which modulate cellular and molecular behavior and phenotype. This is embedded in the second part of our definition of a biomaterial: *"functionally interacting with isolated cells."*

The third and final part of our definition of a biomaterial is: *"support the fabrication and maturation artificial 3-dimensional tissue."* The objective of tissue engineering is to fabricate 3D artificial tissue. Specific functional interactions at the cell-matrix interface will support the fabrication of 3-dimensional tissue, and this interrelationship between biomaterials and isolated cells will continue as artificial tissue matures during controlled *in vitro* culture.

3.2 SCHEME FOR BIOMATERIAL DEVELOPMENT

Figure 3.2 shows a generic scheme for biomaterial development for applications in tissue engineering.

SCHEME FOR BIOMATERIAL DEVELOPMENT

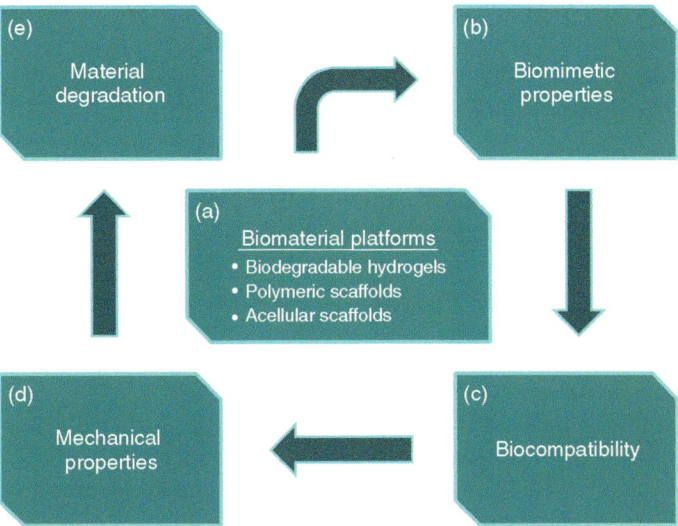

Figure 3.2 Biomaterial Development for Tissue Engineering—(a) Biomaterial Platforms—The first step in biomaterial development involves selection of a suitable platform. Options for biomaterial platforms include biodegradable hydrogels, polymeric scaffolds, and acellular scaffolds. (b) Biomimetic Properties—The second step in the process refers to biomimetic properties, cell-matrix interactions between specific binding sites on the surface of biomaterials and cell surface integrins. (c) Biocompatibility—Biomaterials elicit a foreign body response upon implantation, and biocompatibility refers to the host response to the biomaterial. (d) Mechanical Properties—Depending on application, biomaterials will be exposed to different mechanical conditions. The mechanical properties of biomaterials need to be optimized to match the requirements for any given application. (e) Degradation Kinetics—The rate of material degradation is an important variable that needs to be controlled and optimized when working with degradable biomaterials.

Researchers are often confronted with the difficult task of selecting a biomaterial platform for their specific application, and for sake of simplicity, we can say they have three platforms to choose from: polymeric scaffolds, biodegradable hydrogels, and acellular scaffolds (Figure 3.2a). Once a biomaterial platform has been selected, there are several processing conditions that need to be optimized, and these conditions depend on the platform selected. Fabrication of each of the three platforms requires optimization of processing conditions that vary based on the platform selected.

The second stage of biomaterial development is the incorporation of biomimetic functionality to support cell-matrix interactions (Figure 3.2b). Cells have specific cell surface receptors known as integrins, which bind to specific binding sites on the extracellular matrix, like the RGD site on fibronectin; this process is known as cell-matrix interaction. By definition, in order for a material to be considered a biomaterial, it must have biomimetic activity; however, the material properties may need to be modified to improve the number of binding sites, distribution of these sites, and binding strength at the cell-material interface, along with a host of other variables.

The third stage of biomaterial development is biocompatibility (Figure 3.2c), which refers to the ability of the biomaterial and/or artificial tissue fabricated using the biomaterial to maintain functionality upon implantation. When a biomaterial or 3D artificial tissue is implanted *in vivo*, the host's immune system undergoes a foreign body reaction to minimize detrimental effects of this foreign body to the host. This process involves infiltration of neutrophils and macrophages to the site of implantation, followed by formation of a fibrotic capsule to seal the foreign body from host cells. Many of the biomaterials utilized for tissue engineering are based on synthetic polymers and, upon implantation, elicit a foreign body response. In order to prevent this, properties of the biomaterial need to be modified to minimize host rejection upon implantation.

The fourth stage in biomaterial development is assessment and optimization of mechanical properties (Figure 3.2d). Depending on the application, artificial tissue will be subjected to various biomechanical forces, and biomaterials need to withstand these forces. The mechanical properties of a biomaterial are commonly characterized based on tensile properties, which are measured by stretching the material to the point of failure. The tensile properties of a material are used to plot a stress–strain graph, which is then utilized to obtain several variables that are used to assess tensile properties of the material.

The fifth and final stage of the biomaterial development process is degradation kinetics (Figure 3.2e). There are many applications in which the biomaterial acts as a temporary scaffold to support the initial homing of cells, which then synthesizes the extracellular matrix. As cells produce ECM, the biomaterial is degraded, and the rate of material degradation is controlled to match the rate of new tissue formation. As such, the degradation properties of the biomaterial need to be characterized and optimized. Often, the rate of degradation of a material can be controlled by changing the material synthesis process; for example, the addition of a cross-linking agent during polymer synthesis can stabilize the scaffold and delay the rate of material degradation.

In this section we presented a generic scheme for biomaterial development as it applies to tissue engineering for the fabrication of 3D artificial tissue. We introduced the role of the biomaterial platform, biomimetic properties, biocompatibility, mechanical properties, and material degradation for biomaterial development. This section serves to familiarize the reader with these concepts, and in subsequent sections in the chapter, we will provide additional details on each of these topics.

3.3 HISTORICAL PERSPECTIVE ON BIOMATERIALS

Biomaterial science has advanced to a well-defined scientific discipline with broad applications for surgical reconstruction. However, the use of materials to restore lost functionality is not a new concept, and specific cases have been cited hundreds and even thousands of years ago (10–13). Some relevant examples can be seen in the history of dental implants, which dates back thousands of years, and in the history of prosthetics, which dates back hundreds of years. In this section, we provide a brief historical overview of the materials used in dental implants, prosthetics, and other significant developments in the field.

One of the earliest documented cases of a dental implant took place about 3000 years ago (1000 BC), where a copper stud was implanted into a patient in Egypt, using nails to secure the implant (14–16). Another early example of a dental implant dates back about 1400 years ago (600 AD), where pieces of shell were implanted into sockets of three incisor teeth of a young woman. This incident happened in the Mayan civilization, in what is today known as Honduras. In addition to copper and shells, ivory was often used as a dental implant in both Egyptian and South American cultures (14–16). Gold and platinum were also used as dental implants, with early use dating back to 1809 and 1887, respectively. In 1937, vitallium, an alloy consisting of cobalt, chromium, and molybdenum, was used in patients at Harvard University (17,18). More recently, in 1965, titanium was first used as a dental implant in a patient in Sweden (19). Copper, shells, ivory, gold, platinum, and titanium are all examples of early biomaterials that were used for reconstructive applications in humans. These materials were not developed for dental applications, but rather used due to accessibility. The success rate was limited, with immune tolerance being low and likelihood of rejection being very high. Nonetheless, these early examples serve to illustrate the utility of biomaterials to restore lost functionality in humans.

Artificial limbs have a long history filled with several innovations spanning several centuries (20–24). Although limb prosthetics are commonly used in cases of birth defects, accidents, and in cases of amputation required due to cancer or infection the long-standing history of prosthetic limbs has been brought about by the loss of limbs during warfare. The earliest documentation of the use of a prosthetic limb was about 5500 years ago (3500 BC) in India, where Queen Vishpla used an iron limb in battle after losing her own leg. Another early documentation of the use of a prosthetic limb was about 2500 years ago (484 BC), when a Persian solder lost his leg while escaping from enemy captivity and later used a wooden support as a prosthetic. The earliest prosthetic limb that has been discovered was from about 2300 years ago (300 BC) in Capau Italy and was fabricated from copper and wood, while more recent prosthetic limbs in the 15th and 16th centuries were fabricated from iron. Notably, in the 19th century, prosthetic limbs were primarily fabricated from wood, likely due to their light weight compared with the heavier metallic compounds.

The advent of modern biomaterial science can be traced back to early work conducted by Sir Nicholas Harold Lloyd Ridley in the development of intraocular lenses to restore vision in patients affected by cataracts (25–30). Under normal conditions, the eye has a specialized structure known as the crystalline lens, which together with the cornea, refracts light to the retina leading to the formation of an image and, therefore, vision. A cataract is a medical condition that results in clouding of the crystalline lens and alteration of the ability to refract light on the retina and leads to total loss of vision. Intraocular lenses were developed to replace crystalline lenses that were affected by cataracts. This work was pioneered by Sir Nicholas Harold Lloyd Ridley and has been instrumental in defining the field of biomaterial science. Sir Nicholas Harold Lloyd Ridley was an English ophthalmologist who was working with the Royal Air Force treating patients during World War II. He found splinters of acrylic plastic from aircraft cockpit canopies in the eyes of wounded pilots and made the observation that the material did not trigger any host immune response in the eyes of these pilots. This led him to conclude that the material that was used to fabricate the canopies, polymethyl methacrylate (PMMA), would serve as a good biomaterial for fabrication of intraocular lenses for treatment of cataracts, as it is not rejected by the host. He went on to develop intraocular lenses, and the first artificial lens was implanted in a patient on November 29th, 1949, at St. Thomas Hospital in London. This technology proved to be very successful, and intraocular lens are now implanted in over 10 million patients per year worldwide as a treatment modality for cataracts. The work conducted by Sir Nicholas Harold Lloyd Ridley pioneered artificial intraocular lens implantation as a corrective surgery for patients with cataracts This laid the foundation for biomaterial science as we know it today.

The history of Biomaterials as a scientific discipline can be traced back to 1969 when Clemson University hosted the first symposium of Biomaterials, which was later known as the Annual International Biomaterials Symposium (7). At this inaugural meeting, 17 research papers were presented and there were approximately 100 participants. There were two hallmarks of this meeting that provided the foundation for the evolution of Biomaterials as a scientific discipline. First, the participants of this meeting represented an interdisciplinary group of scientists and physicians, thereby actively engaging a broad spectrum of thoughts and opinions from basic material design, fabrication, and properties all the way to potential applications in very specific clinical situations. Second, discussions at this inaugural meeting provided the impetus for the formation of the Society of Biomaterials, the leading authority in the field, which was founded in 1975.

The field of biomaterials has been a part of mankind for thousands of years and has slowly evolved into a knowledge-based scientific discipline. Although the field had modest beginnings, biomaterial science has grown at an exponential rate during the last few decades, and there are entire academic departments in major research universities dedicated to the advancement of this field. Our understanding of biomaterials, including design considerations, functional assessment, and host response upon implantation has been greatly improved over the last couple of years. Some of this information will be presented in subsequent sections of this chapter.

3.4 TENSILE PROPERTIES

The mechanical properties of a material are important in determining function in any given application and are commonly assessed based on tensile properties. The tensile properties of a material are used very frequently in engineering design as an important criterion for material selection. The tensile properties of a material provide information about the strength of the material, its ability to withstand a particular load, and information about elastic properties (31–33). All of these properties are extremely important for material selection during tissue fabrication, and depending on the application, certain properties will be desirable: material strength is an important design criterion for load-bearing applications like bone tissue engineering, while elasticity becomes important in valve tissue engineering.

We begin with a description of tensile testing of biomaterials and follow this with a discussion of strategies to change the tensile properties of biomaterials. For the first section—tensile properties of biomaterials—we will discuss sample preparation, mechanics of tensile testing, and data acquisition and interpretation.

Introduction to Tensile Testing—The deformation of a material in response to a load provides information about tensile properties (31–33). The tensile properties of a material are extrapolated from a stress–strain plot, which is obtained after a load-deflection test (also known as a stress–strain test or tensile test). The test material is clamped in a mechanical testing system with one end being held stationary while the other end is subjected to a load, which stretches the material to the point of failure. The force applied to the material, along with material deformation, is recorded and used to plot a stress (force per unit area) versus strain (deformation) curve; the stress–strain plot is used to obtain specific variables that provide information about the mechanical properties.

Sample Preparation for Tensile Testing—Sample preparation is very specific for tensile testing; the material is prepared in a dog bone shape, as shown in Figure 3.3a. The ends of the materials are referred to as the shoulder and are thicker compared to the rest of the material to allow gripping during tensile testing. The gage length is the center region of the material; it is thinner than the shoulder and is part of the material for which tensile properties are determined.

Testing Apparatus—Tensile testing is conducted in a specialized apparatus known as a mechanical testing system (MTS) (34,35), which consists of the following major components: a fixed member, a movable member, a set of grips, a drive mechanism, a load indicator, a crosshead extension indicator, and an extensometer. The fixed and movable members position the grips in place, and the specimen is secured to these grips by attachment at the shoulder region of the sample. The drive mechanism applies a uniform, controlled velocity to the movable member resulting in a tensile load on the sample. The load indicator shows the total tensile load carried by the test specimen, while the extensometer is used to determine the distance between two designated points on the test specimen. A crosshead extension indicator shows change in the separation of the grips. ASTM International, which develops international standards for materials, products, systems, and services used in construction, manufacturing, and transportation, has defined the specific requirements for the testing apparatus.

The Tensile Test—The specimen is secured between the two grips, and the movable grip applies a tensile load that causes deformation of the material. Tensile load is applied until the material fails (Figure 3.3b), and material failure is accompanied by a change in length and diameter (Figure 3.3c).

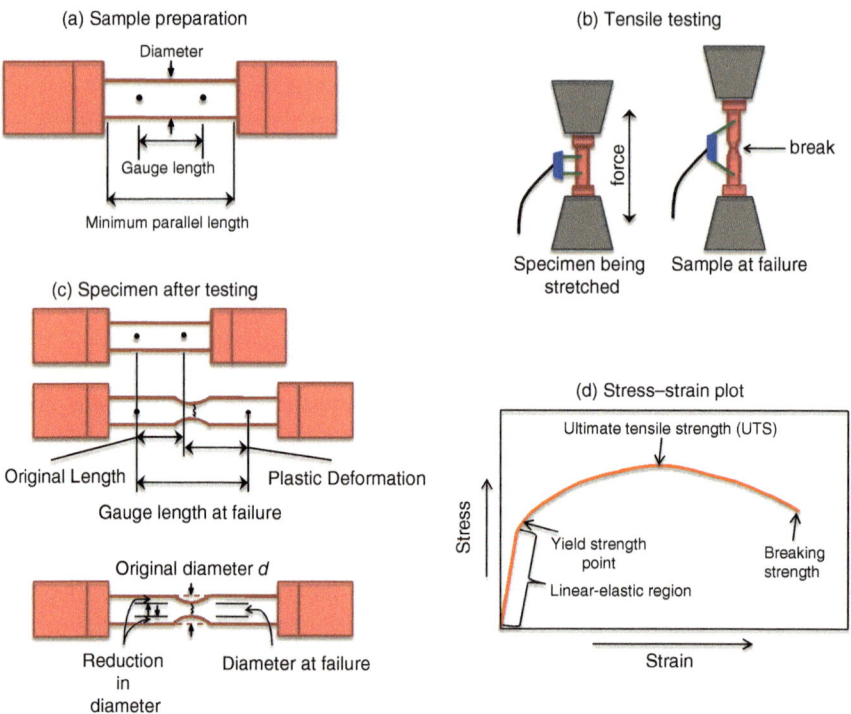

Figure 3.3 Tensile Properties of Materials—(a) Sample Preparation—Dog bone-shaped materials are used for tensile testing. The edges of the material are known as the shoulder and are designed for ease of gripping in a mechanical testing system. The gauge length is the region in the center of the material. **(b) Tensile Testing**—During tensile testing, the specimen is secured between two grips, and a tensile load is applied to the specimen until the material reaches a point of failure. **(c) Specimen after Testing**—As a result of tensile testing, the material undergoes a change in geometry, becomes elongated, and is accompanied by a change in diameter. **(d) Stress–strain Plot**—The slope of the linear region of the stress–strain curve is known as the elastic modulus. The yield strength is the point at which the material transitions from the linear region to plastic deformation. The tensile strength of a material is the maximum force that the material can withstand without failure.

Stress–strain Curve—The stress versus strain curve for most materials is divided into two components; a linear region and a region in which the material exhibits plastic deformation (Figure 3.3d). The linear region is characterized by a linear relationship between strain and stress, and for every unit increase in stress, there is one unit increase in strain or deformation of the material. This behavior is characterized by Hooke's law, and the slope of the linear region, obtained by dividing tensile stress by tensile strain, is known as Young's modulus, also known as elastic modulus or modulus of elasticity (36–39). Many materials, known as brittle, fracture upon application of any additional stress beyond this linear region.

However, ductile materials do not fracture and exhibit plastic deformation, which means that application of additional stress results in greater material deformation. The material deformation is not fully reversible, as the material will not return to its original state upon removing external stresses. The point at which a ductile material transitions from linear stress–strain behavior to plastic deformation is known as yield strength. The toughness (of either brittle or ductile materials) is obtained by the area under the stress–strain curve, and is a measure of the work required to deform a material until it reaches a point of failure. Finally, tensile strength, also known as ultimate tensile strength, is an important material property and measures the maximum force the material can withstand without failure. The tensile strength can be obtained from the stress versus strain curve by determining the maximum value of stress on this curve.

Tensile Properties and Tissue Engineering—We have seen that the tensile property of a material is an important criterion used to aide material selection in any engineering design problem; this argument also applies to the tissue fabrication process (40–44). During tissue fabrication, the function of a biomaterial is to support formation of 3D artificial tissue and guide tissue development and maturation. In order to satisfy this requirement, the biomaterial needs to satisfy a set of design constraints, which include constraints on tensile properties. For any given tissue engineering application, mechanical properties of the artificial tissue need to match mechanical properties of mammalian tissue. The tensile properties of most mammalian tissue have been published, and values can be obtained from the literature. During the tissue fabrication process, the objective is to fabricate 3D artificial tissue that has the same or similar tensile properties to mammalian tissue. While this may sound like a trivial task, this is not one that has been accomplished with a high degree of success in the recent tissue engineering studies. This is due to the inherent lack of mechanical compatibility between engineered naturals and those found in nature. In order to address this limitation, numerous strategies have been implemented to improve tensile properties of engineered materials, some of which are discussed in the following section.

3.5 MODULATION OF TENSILE PROPERTIES

In this section we examine strategies to modulate tensile properties of biomaterials. As we saw in the concluding paragraph of the previous section, for any tissue engineering application, it is important for the tensile properties of biomaterials to match those of mammalian tissue. However, this similarity is not always the case, and various strategies have been developed to change and improve the tensile properties of biomaterials to bridge the gap with mammalian tissue. Let us look at some of these strategies in this section.

Polylactic acid (PLA) has been used extensively in tissue engineering for many different applications, including the development of bone and heart muscle. For bone tissue engineering applications, the tensile properties of PLA do not match that of normal bone tissue. This incongruence can be seen by looking at one particular metric, the Young's modulus, also known as the modulus of elasticity or the elastic modulus. As we have seen in the previous section, the elastic modulus is a measure of the deformation of the material in response to stress and is measured by the slope of the stress–strain curve in the linear region. For bone tissue engineering, the Young's modulus should be in the range of 3–30 GPa, while for PLA, the Young's modulus has been reported to be in the range 2–7 GPa. Therefore, the tensile properties of PLA are not suitable for bone tissue engineering, and strategies need to be developed to bridge this gap. *How exactly can this be achieved?*

In one study, hydroxyapatite (HA) was used to increase the Young's modulus of PLA fibers (45). HA is a naturally occurring mineral that is present in human bone and teeth and has calcium and phosphate as components. The hypothesis is that addition of HA to PLA during the fabrication process will enhance the tensile properties of the scaffold, thereby making it suitable for bone tissue engineering. In this study, PLA fibers were fabricated with the addition of varying amounts of HA, from 0 to 70 wt. % (45). There was a linear relationship between HA percentage and Young's modulus, which increases to approximately 12 GPa with incorporation of 70% HA (45).

This study showed a clear relationship between the composition of HA and tensile properties of PLA fibers. Utilization of an additive is commonly used to modulate the tensile properties of biomaterials, and has been used extensively for tissue engineering applications. The choice of additive and composition at which it is used, are important design variables and must becarefully chosen.

In the previous example, we looked at the use of additives to modulate tensile properties of PLA fibers. Another commonly used strategy to improve tensile properties of biomaterials is the use of cross-linking agents to stabilize fibers. Let us look at this from a conceptual standpoint first; then we will provide a specific example. Cross-linking is the process by which polymer chains are linked together by ionic or covalent bonds, or by the use of specific probes which form a bridge between polymer chains. Irrespective of the method used for cross-linking, the objective is always the same—to stabilize the material. Intuitively, this would translate to an increase in tensile properties of the material, as stabilization of polymer chains would result in an increase in mechanical properties. Glutaraldehyde is a commonly used cross-linking agent and has been used extensively in biological sciences for cross-linking of many proteins, including collagen.

Cross-linking of protein molecules has been used as a strategy to improve tensile properties of biomaterials. In one study, glutaraldehyde was used as cross-linking agent to stabilize gelatin molecules in an attempt to improve tensile properties of the scaffold (46). Gelatin films were fabricated in a petri dish by solvent evaporation followed by air-drying. Different concentrations of glutaraldehyde solution were used, in the range 0.125–2.5 (w/w), to cross-link gelatin films, and tensile properties were evaluated (46). Glutaraldehyde cross-linking proved to be an effective strategy to increase the Young's modulus of gelatin films (46).

Cross-linking of polymer fibers has been used extensively to improve the tensile properties of biomaterials and proves to be an effective method. During the design process, the choice of cross-linking agent, the concentration of cross-linking agent, and processing conditions necessary for efficient cross-linking of polymer fibers need to be addressed and optimized.

In the two previous examples, tensile properties were modulated by use of an additive or cross-linking agent. In both cases, properties of an existing material were modified. However, a completely different and novel approach is to engineer custom materials by polymerization of predefined monomer units under optimized processing condition. This strategy is very different from the first two, as this strategy involves custom fabrication of the biomaterial, which allows greater flexibility for tuning tensile and other properties.

In one study, researchers set out to fabricate titin-mimicking artificial elastomeric proteins or, in other words, artificial proteins that mimic the properties of the naturally occurring protein titin (47). Titin is one of the largest proteins known; it can be found in muscle tissue of humans, and it provides passive elasticity and acts as a molecular spring and scaffolding protein. In order to develop artificial equivalents of titin, protein domains GB1 and resilin were used to mimic titin immunoglobulin domains, and the resulting polymer was fabricated into a ring configuration (47). Based on the proportion of the two components used for biomaterial synthesis, Young's modulus could be changed. Utilization of urea as a denaturing agent also had a significant impact on tensile properties, as can be seen by changes in Young's modulus (47).

The third strategy for the synthesis and modulation of tensile properties is very complex; only highly specialized research laboratories have the necessary technological capabilities to undertake such an endeavor. However, the strategy of building tailor-made biomaterials is novel and provides great promise for the future of biomaterials for tissue engineering.

3.6 MATERIAL DEGRADATION

Introduction—Material degradation refers to the loss of integrity and molecular organization, which in turn affects function of the material for any given application (48–52). Most, if not all, materials are subjected to degradation, although in some cases the rate of degradation may be too slow to be observed or measured in any meaningful way. We can think about material degradation from an engineering standpoint, where degradation of the material can have a catastrophic effect and lead to structural failure. Loss of functionality of materials like stainless steel can lead to instability in major structures like bridges. In engineering design, material degradation is a negative result; it is something that needs to be reduced, eliminated, or managed. Another important application of material degradation can be found in human physiology. Turnaround of proteins in the human body is a normal part of homeostasis, and based on the physiological state of the person, proteins are either degraded or synthesized; this is important for normal human function. Material degradation during the tissue fabrication process is an important property of the biomaterial. The purpose of the biomaterial is to support culture and remodeling of isolated cells to form functional 3D artificial tissue. Degradation of the biomaterial during early stages of tissue fabrication, or any stage, as a matter of fact, will have a significant impact on functional performance of artificial tissue.

Definition of Material Degradation—The definition of material degradation varies depending on the application and context in which it is used (53–57). The word "biodegradation" has often been used to refer to degradation that occurs in a biological environment and has been defined as "*gradual breakdown of a material mediated by a specific biological activity.*" Protein degradation in the human body will fall within this category and can be referred to as biodegradation. Material degradation from an engineering standpoint has a completely different meaning and has been defined as "*a simple definition of materials degradation is that it is the consequence of a wide range of physical processes; it is almost universal in occurrence and is a major engineering problem.*" When we compare the two definitions, we can easily see that from a biological perspective, degradation is considered to be a natural process and the objective is not to control or regulate the process, but to understand it. From an engineering standpoint, material degradation is an undesirable outcome, and the objective is to develop countermeasures to limit any accompanying adverse effects. Applied to tissue engineering, neither of the two definitions is adequate; the definition of biodegradation only provides a platform to start with. Biodegradation has been defined as the "*gradual breakdown of a material mediated by a specific biological activity.*" This process restricts material breakdown to a specific biological activity, which is the case during human physiology but not always the case in tissue engineering. For tissue engineering and the tissue fabrication process, degradation can be defined as "*gradual breakdown of a biomaterial mediated in a controlled manner to support the fabrication of 3-dimensional artificial tissue.*" This definition removes any restriction of material degradation by biological activity and expands the scope to include 3D tissue fabrication.

Biomaterial Degradation and Tissue Engineering—Many materials used in tissue engineering, particularly for soft tissue applications (for example, in the cardiovascularsystem), are biodegradable. Biodegradable materials are those materials that undergo a significant change in chemical structure under specific conditions, and these changes result in a loss of physical and mechanical properties (58–62). The rationale for using a biodegradable material for tissue engineering is the ability to provide a temporary scaffold to support tissue fabrication and remodeling. During the early stages of 3D tissue formation, the biodegradable scaffold acts as the extracellular matrix, providing structural and functional support during tissue formation (Figure 3.4).

As the 3D tissue develops and matures, extracellular matrix components are generated by cells, and the temporary scaffold is no longer required. At this stage, it is desirable for the biomaterial to be completely degraded (Figure 3.4). In order for a biodegradable material to be suitable for tissue engineering, two requirements need to be met. First, the degradation kinetics needs to be tunable, so that the rate of biomaterial degradation is balanced by the rate of extracellular matrix formation by cells. Second, the degradation products need to be nontoxic to the cells, and if

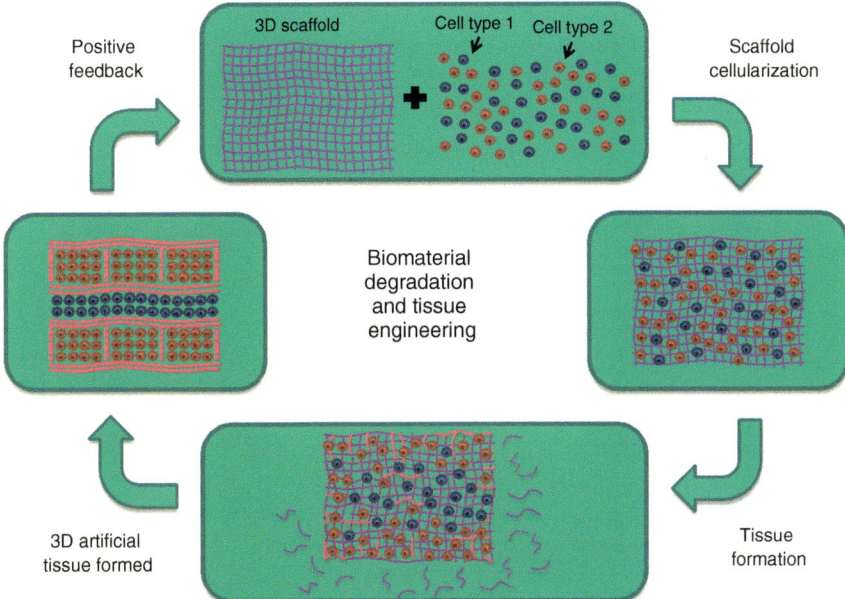

Figure 3.4 Biomaterial Biodegradation—During initial stages of tissue fabrication, the 3D scaffold provides structural support. Gradually, cells produce their own extracellular matrix, and as this process continues, the cells are less dependent on the scaffold for structural support; scaffold degradation can be initiated. The process of scaffold degradation and extracellular matrix production continues until the entire biomaterial has been degraded and replaced by extracellular matrix produced by the cells. This results in the formation of 3D artificial tissue.

the material is used *in vivo*, the degradation products need to be safely eliminated from the body.

Mode of Action for Biomaterial Degradation—Material degradation can occur as a result of physical processes, chemical reactions, or biological activity. Physical degradation refers to material degradation, which occurs in response to physical processes like heat and does not require any chemical reaction or biological intervention (63). Chemical degradation occurs in response to a chemical reaction like oxidation and results in an organized change in the 3D architecture of the biomaterial (64–66). Chemical processes that lead to material degradation include hydrolysis (mediated by water) or enzymatic activity. In addition to the mode of action, the chemical degradation of materials varies based on the mechanism of degradation, cleave of crosslinks between polymer chains, transformation/cleavage of side chains, and/or cleave of backbone linkages between polymer repeating units. Degradation due to biological activity is caused by enzymatic reactions involving specific protein interactions and highly orchestrated sequences of events (67–70). Examples of material degradation by physical processes, chemical reactions, and biological processes are as follows:

- *Physical*—heat, wear, fracture and fatigue, impact fracture, creep, radioactivity, sorption, swelling, softening, dissolution, mineralization, extraction, crystallization, decrystallization, stress cracking
- *Chemical*—aqueous corrosion, solvation by liquid metals, reaction with organic solvents, thermolysis, oxidation, solvolysis, photolysis, radiolysis, fracture-induced radical reactions
- *Biological*—enzymatic

Selection Criteria for Degradation Strategy—The choice of degradation strategy must be carefully selected to support fabrication of artificial tissue and must satisfy the following design requirements: 1) the rate of degradation should be tunable and user-defined on requirements of the specific application. The rate of degradation can vary from days to months, depending on application. 2) The mode of action of degradation should not be harmful to cells. While there are many options for material degradation, some of these will not be suitable for tissue engineering, as they can induce damage to the cells. Degradation strategies that involve fracture, fatigue and corrosion may not be suitable for tissue engineering studies. 3) The degradation products should not be harmful to cells. As polymers degrade, they are broken down into simpler monomer units; these monomer units should not damage the cells in any way. The degradation products will be affected by the composition of the polymer itself along with the degradation strategy employed.

3.7 BIOCOMPATIBILITY

Introduction—Under normal physiological conditions, the human body has a host of defense mechanisms that work in tandem to protect against a variety of treats from the environment. The foreign body response, complement activation, and thrombosis are examples of host defense systems that work to protect against foreign bodies in the environment. These systems are very sophisticated and remarkable in their ability to provide a host defense mechanism. These systems also come into play when a foreign body is implanted for therapeutic purposes, which can be in a medical device, a biomaterial, isolated cells during cell transplantation, or 3D artificial tissue (71). The human body considers these therapeutic agents to be foreign bodies and unleashes its defenses to limit the effect of these agents on human physiology. This reaction to therapeutic agents, in turn, limits the therapeutic benefit from the implanted agent and negates any intended beneficial effects. In order for any implanted biomaterial or artificial tissue to have a beneficial effect, it has to interface with the host immune system in a way that allows it to be accepted by the host. Functional integration needs to take place at the host-implant interface. In general terms, the property of an implanted agent to be accepted by the host immune system is referred to as biocompatibility.

Definition of Biocompatibility—As we have seen throughout this book, a specific definition that relates biocompatibility and tissue engineering does not exist.

While many definitions have been presented in the literature, a commonly utilized one is *"biocompatibility refers to the ability of a material to perform with an appropriate host response in a specific situation"* (72). Unfortunately, this definition is very general and does little to relate or provide a specific interpretation of biocompatibility to tissue engineering. As we have done in several instances throughout this book, not by choice but rather by need, we will develop a working definition of biocompatibility as it relates to tissue engineering. Prior to providing a definition of biocompatibility, let us take a step back and revisit the tissue fabrication process. In tissue engineering, our objective is to fabricate 3D artificial tissue, which, upon implantation, serves to augment, repair, and/or restore lost tissue function. In order for artificial tissue to function in the host environment, it has to be accepted and tolerated by the host immune system, which includes a large number of defense mechanisms, including the foreign body reaction, complement system, and thrombosis. The argument that we have just presented will form the basis for biocompatibility as it applies to tissue engineering—*"The ability of 3-dimensional artificial tissue to be accepted by host defense mechanisms upon implantation, while maintaining functional capacity, is known as biocompatibility."* This definition focuses on tissue engineering and refers to the host defense mechanisms in their entirety, rather than specifying one particular mechanism. The definition also refers to the functional capacity of artificial tissue, which is necessary for the implanted tissue to serve as a therapeutic agent to recover and/or restore lost functionality of host tissue.

Biocompatibility and Tissue Engineering—From a tissue engineering or tissue fabrication standpoint, biocompatibility refers to the ability of the implanted tissue to be accepted by host defense mechanisms, which include the foreign body reaction, complement activation, and the coagulation pathway (67–70,72,73) (Figure 3.5).

In subsequent sections, we provide a brief description of these pathways and show how they function to protect against foreign pathogens. A similar response takes place when artificial tissue is implanted, and biocompatibility refers to the ability to design artificial tissue that minimizes these reactions.

Foreign Body Reaction—When a foreign body is implanted, it elicits the foreign body response which consists of the following steps (71,74–77): 1) injury, 2) acute inflammation, 3) chronic inflammation, 4) granulation tissue, 5) foreign body reaction, and 6) fibrosis.

Injury—In order to implant any biomaterial at a functional site, invasive procedures are often required, which result in perturbation of the host tissue, thereby leading to some degree of injury (71). Implantation of a biomaterial requires severing skin tissue, disturbing existing musculature, and excising vasculature, all of which lead to some form of tissue injury. In addition, the physical positioning of 3D artificial tissue at the site of injury requires physical contact with host tissue, particularly to secure the implanted tissue using surgical sutures. All steps in the implantation cascade can lead to tissue injury. In response to injury, a cascade of molecular events is triggered; this cascade begins with acute inflammation and can

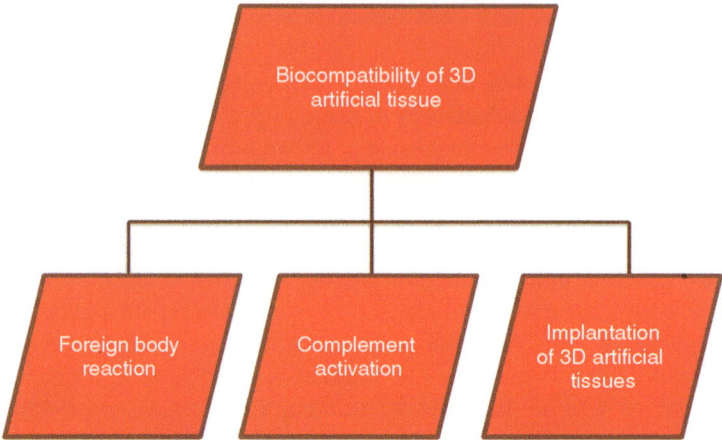

Figure 3.5 Biomaterial Biocompatibility—Biocompatibility refers to the ability of 3D artificial tissue to functionally interface with host defense mechanisms, which include foreign body reactions, complement systems, and coagulation pathways.

lead to fibrotic encapsulation of the implanted graft. While under physiological conditions, the tissue response to injury is designed to contain the injury, limit infection, and reduce the adverse effects on tissue function This response can reduce functionality of artificial tissue. Therefore, strategies need to be designed to minimize these effects. The host response to injury is dependent upon the extent of injury and the properties of the implanted biomaterial or artificial tissue.

Acute Inflammation—Inflammation is defined as the reaction of vascularized living tissue to local injury and is designed to contain, neutralize or dilute the injurious agent or process (71,78–80). Acute inflammation occurs within the first few hours or the first few days of tissue injury and is characterized by leukocyte accumulation at the site of injury. Neutrophils are generally the first cells to be found at the site of injury and are then replaced by monocytes, which later differentiate to form macrophages. The primary function of the neutrophils is to phagocytize microorganisms and foreign materials.

Chronic Inflammation—Chronic inflammation occurs over time, with a time horizon ranging from weeks to years. Although multiple cell types mediate the chronic inflammatory response, macrophages are central due to the large number of compounds that can be secreted by these cells, including neutral proteases, chemotactic factors, arachidonic acid metabolites, reactive oxygen metabolites, complement components, coagulation factors, growth-promoting factors, and cytokines (71,81–83). Some of the tissue responses associated with chronic inflammation include the proliferation of fibroblasts, vascularization of the injured tissue, and regeneration of epithelial cells.

Granulation Tissue—The time frame for the onset of granulation tissue depends on the extent of injury, and upon biomaterial implantation, it can be seen within 3–5 days. Granulation tissue consists of connective tissue and cells like fibroblasts that are known to produce extracellular matrix components (71,84). Granulation tissue is also vascularized and contains vascular cells like endothelial cells. Due to the presence of vasculature, granulation tissue has a pinkish color. Granulation tissue serves as a protective barrier, preventing the infiltration of pathogens at the site of injury.

Foreign Body Reaction—Foreign body giant cells are the products of macrophage fusion and are a hallmark of the foreign body reaction. When macrophages encounter a foreign object too large to be phagocytized, such as an implant, it is thought that the macrophages experience "frustrated phagocytosis." They fuse to form larger foreign body giant cells composed of up to a few dozen individual macrophages (71,85,86). Giant cells secrete degradative agents such as superoxides and free radicals, causing localized damage to implants and other foreign bodies. Currently, little is known of the role of foreign body giant cells and it is hard to say whether they are "more or less inflammatory" than a collection of macrophages. Macrophages and foreign body giant cells tend to remain at the surface of an implant for the duration of its residence.

Fibrosis—Fibrosis is the final stage of the host response to a foreign material, like an implanted biomaterial graft or artificial tissue. Fibrosis involves encapsulating the biomaterial with fibrous tissue, which is about 50–200 μm thick and consists of an abundance of collagen (71,87). The primary rationale for the fibrotic response is to separate the foreign body from the host and minimize any adverse effects to the host.

Complement Activation—The complement system is a part of our immune system and provides a defense mechanism against a host of pathogens that we are constantly exposed to, including bacteria, viruses, and fungi (88–90). While the complement system can act alone and serve to recognize and destroy pathogens that would otherwise have adverse effects on health, it can also serve as an intermediary to tag pathogens for phagocytosis; the latter process is known as opsonization. There is more than one molecular pathway by which our complement system works, one of which is the classical pathway. There is a cascade of events that takes place in the classical pathway. The trigger for initiation of the classical pathway is a foreign antigen on a microbe, followed by binding of host antibody to the foreign antigen (88–90). Formation of this antigen-antibody complex is followed by activation of the protein C1, which then leads to cleavage of the protein C4 to form C4a and C4b, with C4b binding to the surface of the pathogen. This is followed by cleavage of C2 to form C2a and C2b, with C2a binding to C4b on the pathogen surface and C2b acting on C3 to form C3b and C3a. Formation of C3b can have one of two possible outcomes. First, C3b can bind to cell surface receptors on macrophages, promoting opsonization. Second, C3b can bind to the C4b and C2a complex, leading to formation of a C5 convertase; this in turn can lead to formation of C5b, which results in the formation of a membrane attack complex (MAC) (91,92). The MAC is formed in the cell membrane of pathogens, acting as a transmembrane channel leading to disruption of cell activity and function and eventually leading to cell death. Therefore, the classical pathway of the complement system can lead to direct destruction of pathogens by formation of the MAC or can act as an intermediary by formation of C3b to promote recognition and binding by macrophages followed by phagocytosis.

Platelet Activation and Blood Coagulation—Thrombosis is the process by which platelets, in combination with fibrin, form a blood clot designed to plug injured blood vessels, thereby limiting the loss of blood and containing the site of injury (93–95). This is a very important hemostasis mechanism during normal function and serves to regulate blood loss after injury. When a blood vessel is damaged or injured, collagen in the endothelium layer is exposed and serves as a trigger for platelet adhesion and activation. Activation of platelets leads to a cascade of signaling events culminating in the conversion of prothombin to thrombin, which then acts to convert fibrinogen to fibrin and promotes formation of a blood clot that plugs the injured vessel.

3.8 BIOMIMETIC BIOMATERIAL

Introduction—Naturally occurring biomaterials like collagen have functional sites that support cellular interactions, leading to enhanced cell-material interactions. On the other hand, synthetic polymers like PLA and PGA do not possess functional interaction sites for cells; these materials can be modified by introducing functional sites within the polymeric structure. Proteins and peptides are commonly used as linking agents to modify the biomaterial properties, and enzymes, antibodies, antibiotics, and cell adhesion molecules are commonly employed. While covalent bonding has been extensively used to link functional molecules to polymer chains, physical methods like adsorption or electrostatic attractions have also been employed. Biomimetic activity refers to the "cell-friendliness" of the biomaterial, and in order for a biomaterial to be cell-friendly, it must possess functional binding sites for cell surface integrins (96–107).

Definition of Biomimetic Biomaterial—A two-part definition of biomimetic biomaterials has been provided in a recent article (108):

1. *The development of biomaterials for tissue engineering applications has recently focused on the design of biomimetic materials that are able to interact with surrounding tissues by biomolecular recognition.*
2. *The design of biomimetic materials is an attempt to make the materials such that they are capable of eliciting specific cellular responses and directing new tissue formation mediated by specific interactions, which can be manipulated by altering design parameters; instead of by non-specifically adsorbed ECM proteins.*

In the article, the two-part definition was used as a discussion point to describe the nature of biomimetic biomaterials, rather than using the statements as a definition. However, the explanation provided in the article fits into the role of defining biomimetic biomaterials and has therefore been used in this manner and will be retained in this book. Simply stated, a biomimetic biomaterial is one that elicits specific cell-matrix interactions that guide intracellular signaling pathways thereby regulating cellular and molecular responses; this in turn dictates 3D tissue formation and function. Another way to state this is: biomimetic biomaterials are designed to resemble mammalian extracellular matrix, which itself functionally interacts with cells to support 3D remodeling and dynamic tissue formation.

Fabrication of Biomimetic Biomaterials—There is an abundance of examples of biomimetic biomaterials in nature and almost all, if not all, components of the extracellular matrix are considered to be biomimetic biomaterials. The objective of tissue engineering is to replicate biomimetic properties of naturally occurring biomaterials to support fabrication of functional 3D artificial tissue. *How can this be achieved?* Let us take a closer look at naturally occurring biomaterials using a commonly utilized one in tissue engineering, fibronectin (109–113). Fibronectin is a large extracellular glycoprotein with a very complex 3D architecture; however, a sequence of three amino acids, Arg-Gly-Asp (RGD) is known to be critical for cell adhesion (Figure 3.6). Cells interface with the RGD sequence on fibronectin through specific cell surface integrins, $\alpha_5\beta_1$. The $\alpha_5\beta_1$-RGD interaction is an example of a specific cell-matrix interaction and is one of the factors responsible

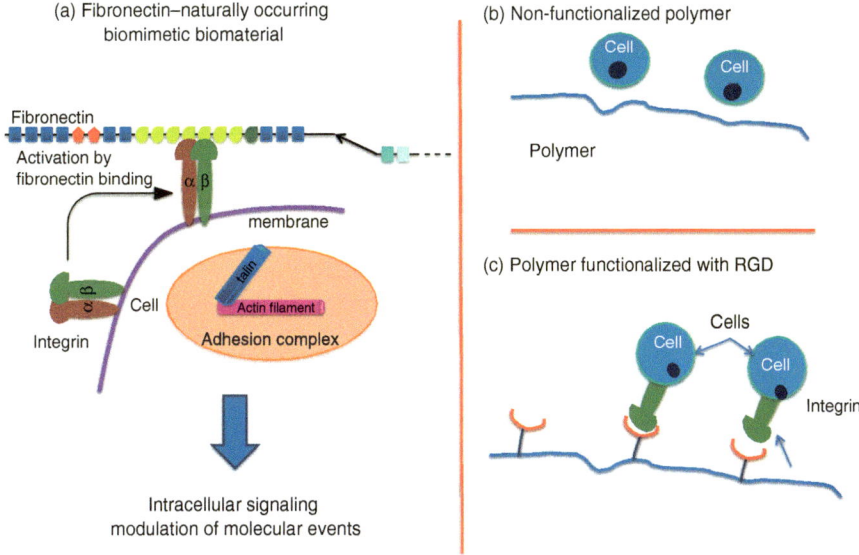

Figure 3.6 Biomimetic Biomaterials—(a) Fibronectin (FN)—Naturally Occurring biomimetic biomaterial—FN has a specific sequence of three amino acids that bind to specific alpha-5 beta-1 integrins on the surface of cells. Binding of alpha-5 beta-1 to the RGD site on FN leads to a cascade of intracellular signaling events that result in cellular and molecular changes. (b) Non-Functionalized Polymer—Cells passively interact with a polymer that does not possess any functional sites to support cell-matrix interaction. (c) Polymer Functionalized with RGD—A RGD sequence can be linked to a polymer that can then support functional interaction with alpha5beta1 integrins on the surface of cells.

for the biomimetic activity of fibronectin; this in turn leads to specific intracellular signaling events that regulate molecular and cellular behavior.

How do we use this information in the design of synthetic biomimetic biomaterials? Biomaterials have been synthesized by functionalizing the RGD sequence, chemically linking this sequence to the polymer backbone, thereby providing a binding site for cells that express the $\alpha_5\beta_1$ integrin. Linking the RGD sequence to polymer backbones has shown to significantly increase cell adhesion and functionality of 3D artificial tissue.

3.9 CLASSIFICATION OF BIOMATERIALS

Several schemes have been used to classify biomaterials, and the most common ones are discussed here. Biomaterials are frequently classified based on source (natural and synthetic), based on degradation (biodegradable and non-biodegradable), and based on interatomic bonding forces (metals, polymers, and ceramics) (114–116), as seen in Figure 3.7. These terms appear frequently throughout this chapter and the remainder of this book; therefore, we take a moment to discuss their meaning and relevance to tissue engineering.

Natural versus Synthetic Materials—This classification scheme is based on the source of the material. It is very simple and self-explanatory, and its relevance to tissue engineering is paramount; therefore, frequent discussion of this topic can be found in many scientific forums. Simply stated, naturally derived materials are obtained from natural sources, while synthetic materials are synthesized in the laboratory. One common example of a naturally occurring material is collagen, which is frequently extracted from rat tails and used extensively for tissue engineering and other medical applications. As example of synthetic material used for tissue engineering and other medical applications is the aliphatic polymer poly(glycolic acid), PGA. PGA has found extensive application in resorbable sutures and as scaffolding material to support the fabrication of 3-dimensional tissue constructs.

There are clear advantages of natural or synthetic materials in any given tissue engineering application. Natural materials have anatomically matched 3-dimensional architecture and are biologically active, thereby supporting functional interaction with isolated cells. However, the main disadvantage of natural materials is batch variability inherently due to differences in isolation efficiency. Synthetic materials have the advantage of reproducibility, as synthesis is tightly regulated, and tenability, as materials of different properties and functionality can be fabricated. However, the main disadvantage of synthetic materials is the lack of biological functionality, as many synthetic materials do not have functional binding sites for isolated cells.

Degradable versus Nondegradable—The degradation kinetics of a material define the rate at which the material disintegrates or loses structural stability as a function of time. On the surface, material degradation may appear to be a nondesirable material property, as loss of structural integrity can lead to catastrophic effects. This is indeed the case for numerous medical applications, as

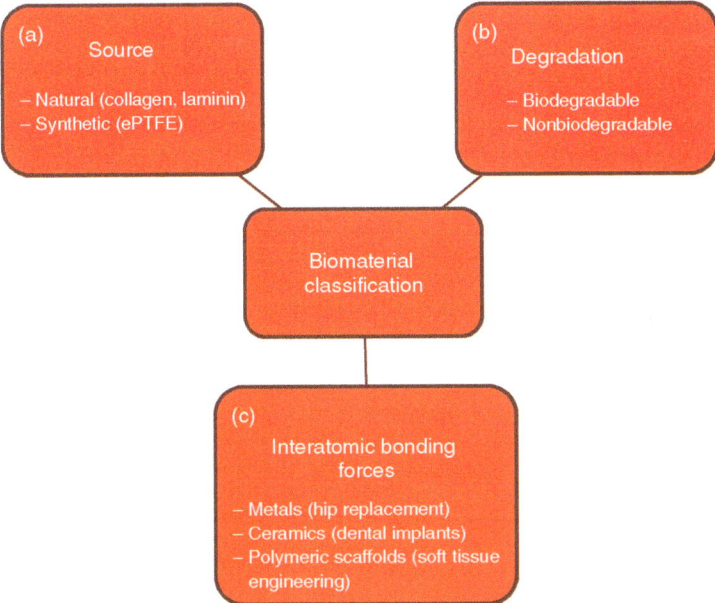

Figure 3.7 Classification of Biomaterials—(a) Natural versus Synthetic—Natural biomaterials are found in nature and are extracted from mammalian tissue to support fabrication of 3D artificial tissue. Examples of naturally occurring biomaterials include collagen, fibronectin, and laminin. Synthetic biomaterials are synthesized in the laboratory under controlled reaction conditions. (b) Degradable versus Nondegradable—Degradable materials have a measurable change in weight over a specific time frame while nondegradable materials maintain a constant weight over any given time frame. (c) Metals, Ceramics, or Polymers—Biomaterials are classified as metals, ceramics, or polymers, each of which has very different applications in tissue engineering.

in the case of knee and hip replacements, which are often fabricated using metallic components like stainless steel, and in which longevity of the implant is a critical functional determinant. Metallic materials fall into the category of nondegradable materials where long-term structural stability is essential for function. In the case of tissue engineering, a biodegradable material is molded into a scaffold and then populated with isolated cells. The cells use the scaffold as a temporary support matrix, and during the culture period, extracellular matrix components are fabricated by cells. During the tissue fabrication process, the material degrades and is replaced by extracellular matrix fabricated by the cells. In this case, material degradation is an important property required to support 3D tissue formation. The degradation kinetics of scaffolds is an important material property and is the focus of many tissue engineering studies.

Metals versus Ceramics versus Polymers—Metals, ceramics, and polymers represent a large group of materials that are used frequently for medical and tissue engineering applications. Some examples of metals used as implants include titanium and its alloys, and stainless steel. These materials have been used in hip and knee replacement implants and in bone applications, including bone plates, screws, pins, and rods. Titanium screws or posts are also used in dental implants to anchor prosthetic teeth. The titanium implant is placed in the bone socket of the missing tooth, and within a couple of weeks, the jawbone and implant form a functional bond. An abutment is then attached to the titanium implant as an intermediary between the implant and prosthetic tooth. Artificial teeth, also known as dentures, are fabricated from acrylic resins like PMMA—polymethyl methacrylate. In addition to PMMA, ceramics like porcelain are also used in dental implants and have also found applications in hip and joint replacement implants. Metals and ceramics have traditionally been used for hard tissue applications like orthopedic and dental, and most materials in these categories are known to be nondegradable. Metals are naturally occurring, while ceramics are synthetic. Polymers have been used extensively for soft tissue engineering and can be derived from natural sources or can be synthesized in the laboratory; degradable and nondegradable polymers are both used in tissue engineering, although the former are more common.

3.10 BIOMATERIAL PLATFORMS

One of the early decisions that need to be made when considering the use of biomaterials for any tissue engineering application is the biomaterial platform to be implemented. There are four platforms that have been widely used for tissue engineering applications: polymeric scaffolds, biodegradable hydrogels, acellular matrices, and self-organization strategies (Figure 3.8). In this section, we provide an overview of these platforms, describe their properties, identify advantages/disadvantages of each, and provide examples of tissue engineering applications where each platform has been successfully utilized.

Decellularized Matrices—This strategy is based on the utilization of naturally occurring extracellular matrix as the scaffolding material for 3D tissue formation (117–122). Tissue specimens are obtained from cadaveric or xenogeneic sources, and cells are completely removed using one of several potential strategies. Removal of cellular components from tissue specimens is known as decellularization, and the material that is obtained after removal of the cells is known as an acellular scaffold. Removal of cells does not distort the extracellular matrix components in any significant manner. The composition and 3-dimensional organization of individual components of the extracellular matrix remains intact. As a result, the acellular scaffold retains the physiological architecture of the extracellular matrix and therefore can serve as a scaffold for tissue engineering. In addition, removal of cells reduces the immunogenicity of the scaffold, as the immunogenic response is primarily a cellular response. Some examples of tissue engineering applications where decellularized matrices have been used include heart valves, blood vessels, skeletal muscle, skin, nerves, tendons, and ligaments.

Several approaches have been implemented to decellularize tissue specimens. These approaches have been categorized as physical, chemical, and enzymatic, with

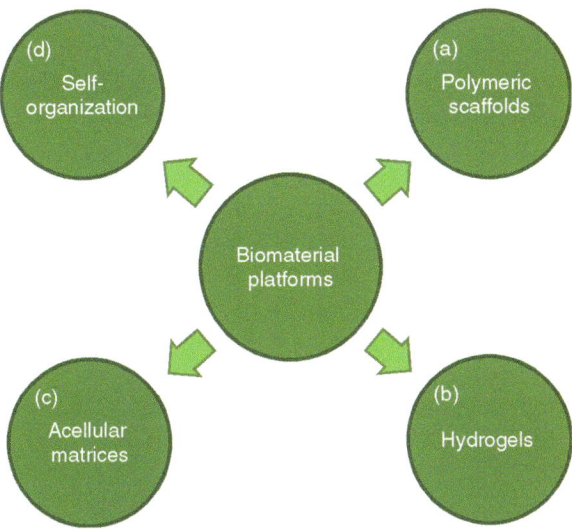

Figure 3.8 Biomaterial Platforms—(a) Polymeric Scaffolds—Polymerization of individual monomer units leads to formation of long chain structures known as polymers. Polymer chains can be fabricated into 3D scaffolds that can then be used to support tissue fabrication. (b) Hydrogels—Materials that have a high water content are known as hydrogels and are used extensively in tissue engineering. Many materials can be formed into 3D hydrogels, including collagen, chitosan, and fibrinogen. (c) Acellular Matrices—Mammalian tissue is subjected to a decellularization protocol that completely removes all cellular components leaving an intact ECM that can be used to support 3D tissue fabrication. (d) Self-Organization—Refers to scaffold-free technology to fabricate 3D artificial tissue; extracellular matrix is produced by the cells and used to support the fabrication of 3D artificial tissue.

some protocols based on a combination of these three broad classifications. Decellularization methods that rely on physical techniques for cell removal include the use of repeated freeze thaw cycles or the use of pressure, sonication, and/or mechanical agitation, all of which function to disrupt the cell membrane, with subsequent washing required to remove cellular components. The primary advantage of physical treatment methods is the ease of implementation. However, physical treatment methods are not always sufficient for complete removal of cellular components and can lead to damage of the extracellular matrix.

A wide variety of chemical compounds have been used to decellularize tissue specimens. Alkalinic and acidic solutions, detergents like Triton X-100 and sodium dodecyl sulfate (SDS), hypertonic/hypertonic solutions, and chelating agents like ethylene glycol tetraacetic acid (EGTA) or ethylenediaminetetraacetic acid (EDTA) have all been used. The mechanism by which these compounds act is very different. Alkalinic/acidic solutions are known to solubilize cytoplasmic components of the cells, detergents can disrupt the cell membrane and denature proteins, and hypertonic/hypertonic solutions result in changes in osmotic gradients causing cells to swell and eventually burst. In addition, chelating agents like EGTA/EDTA disrupt cellular interactions with other cells and the extracellular matrix by binding divalent ions like calcium. Chemical methods have been very efficient in the removal of cells with limited disruption of the extracellular matrix.

Enzymatic methods have commonly employed the use of trypsin, a protein known to cleave peptide chains at the carboxyl side of lysine and arginine by a process known as trypsin proteolysis. In addition, endonucleases and exonucleases, which degrade DNA and RNA, have been used in decellularization protocols. While the efficiency of cell removal is high with enzymatic methods, degradation of extracellular matrix can occur.

Hydrogels—The term hydrogel is composed of "hydro" (water) and "gel," and refers to aqueous (water-containing) gels; to be more precise, it refers to polymer networks that are insoluble in water; they swell to an equilibrium volume but retain their shapes. Specifically, the water content of hydrogels is greater than 30% on a weight basis. Polymers that form hydrogels are known to have specific chemical residues within their 3-dimensional lattice structure, some of which include hydroxylic (−OH), carboxylic (−COOH), amidic (−CONH−), primary amidic (−CONH$_2$), and sulfonic (−SO$_3$H) groups.

Hydrogels are classified using several different schemes, although two are commonly seen in tissue engineering: natural or synthetic and biodegradable or nonbiodegradable. Natural occurring hydrogels are based on polymers that are found in nature. These polymers include agarose, alginate, chitosan, collagen, fibrin, gelatin, and hyaluronic acid (123–128). Collagen in particular has been used extensively in tissue engineering due to its prominent role in modulating mammalian physiology. Synthetic hydrogels are formed using polymers that are synthesized in the laboratory and include compounds like poly(ethyleneoxide) (PEO), poly(vinyl alcohol) (PVA), poly(acrylicacid) (PAA), poly(propylene furmarate-co-ethylene glycol) (P(PF-co-EG)). Some hydrogels retain structural stability over time and are known as nondegradable, while other hydrogels degrade over time, result in loss of structurally stability, and are known as degradable.

The interaction of water with the polymer chains plays an integral role in hydrogel formation and subsequent functionality. The water content of hydrogels is divided into two components: the total bound water and free or bulk water. The total bound water is the amount of water that is functionally interacting with hydrophilic and hydrophobic groups in the polymer lattice. The free or bulk water is any additional amount of water present within the hydrogel lattice occupying spaces between polymer chains.

Hydrogels have been used extensively for tissue engineering, for example in the fabrication of blood vessels, heart muscle, tri-leaflet heart valves, and skeletal muscle (123–128). The advantages of hydrogels include the aqueous environment that spatially separates polymer fibers, thereby increasing the material porosity and supporting nutrient delivery throughout the 3D structure. The main disadvantage is the lack of mechanical strength due to the high porosity and water content, making hydrogels difficult to use for load bearing applications.

Polymeric Scaffolds—Polymers have been used extensively for tissue engineering applications. Before proceeding to discuss the role of polymers in tissue engineering, we introduce the definition of a polymer in terms of its constitutional units and macromolecules, followed by the definition of polymerization in terms of converting monomer molecules (129).

- A constitutional unit is an atom or group of atoms (with pendant atoms or groups, if any) comprising a part of the essential structure of a macromolecule.
- A macromolecule is a molecule of high relative molecular mass, the structure of which essentially comprises the multiple repetitions of units derived, actually or conceptually, from molecules of low relative molecular mass.
- A polymer is a substance composed of macromolecules.
- Polymerization is the process of converting a monomer or a mixture of monomers into a polymer.
- A monomer molecule is a molecule that can undergo polymerization, thereby contributing constitutional units to the essential structure of a macromolecule.
- A monomer is a substance composed of monomer molecules.

While it is important to use the right terminology to define polymers, the definition presented can be conceptually simplified, and polymers can be viewed as molecules of a high molecular weight that are composed of repeating monomer units.

Some of the earliest work in the field of tissue engineering was based on polymeric scaffolds. As discussed in Chapter 1, 3D liver constructs were fabricated by culturing primary hepatocytes on three different polymeric scaffolds. Since this initial application, utilization of polymeric scaffolds has increased considerably to support different tissue engineering applications, including applications in the cardiovascular space. Although there are several reasons for this exponential growth, the ability to synthesize polymers with different properties makes it possible to customize materials to suit different tissue engineering applications. The bioactivity, degradation kinetics, mechanical stretch, immunogenicity, and surface properties can be controlled by varying the polymer composition and processing conditions. This provides a very high degree of freedom and has been one of the reasons for such a high degree of interest in polymers (130–136).

Polymers used for tissue engineering applications are conveniently classified based on their source as natural or synthetic polymers. Natural polymers are derived from nature, and some common examples include the polysaccharides alginic acid, hyaluronic acid, chitin, chitosan, and collagen. The primary advantage of using naturally occurring polymers is the physiological role these polymers play in mammalian function. For example, collagen is the most abundant protein in humans and provides structural support for tissue formation and maturation; therefore, the rationale for using collagen in tissue engineering is to mimic its anatomical role. However, the main disadvantage of using natural polymers is the high degree of variability between independent batches, thereby making process control and scale-up challenging. Synthetic polymers are synthesized in the laboratory by polymerization of monomer units, some examples of which include poly(methyl methacrylate), poly(ethylene terephthalate), poly(dimethylsiloxane), poly(tetrafluoroethylene), polyethylene, and polyurethane. The primary advantage of using synthetic polymers is the reproducibility by which the polymers can be synthesized in the laboratory and the ability to tune material properties, allowing functional modifications to match the requirements for any given tissue engineering application. However, synthetic polymers do have some disadvantages, as they are not anatomically matched and may not exhibit biomimetic activity and may be rejected by the host upon implantation.

Self-Organization Strategies—Self-organization is prevalent in biological systems; it involves the physical interaction of molecules in a steady-state structure (137–142). In a broad sense, self-organization can be viewed as a process that occurs in the absence of any constraining conditions, thereby providing a greater degree of freedom and flexibility. Prior to evaluating the role of self-organization strategies in the development of 3D tissue constructs, we provide a working definition, along with specific requirements that need to be satisfied in order for a system to be considered self-organized. Self-organization in the context of cell biology can be defined as the capacity of a macromolecular complex or organelle to determine its own structure based on functional interactions of its components. In a self-organizing system, the interactions of its molecular parts determine its architectural and functional features. The processes that occur within a self-organized structure are not underpinned by a rigid architectural framework; rather, the components of the structure define its organization. For self-organization to act on macroscopic cellular structures, three requirements must be fulfilled: the cellular structure must be dynamic, material must be continuously exchanged, and an overall stable configuration must be generated from dynamic components.

Self-organization strategies are focused on the assembly of structures based on internal dynamics without significant external control. It is important to explore this phenomenon from a biomaterials and tissue engineering standpoint. Thus far, we have evaluated the role of acellular matrices, hydrogels, and polymeric scaffolds in the fabrication of 3D tissue constructs. The concept of self-organization is somewhat tangential and one may question its relevance to tissue engineering. Tissue engineering strategies are focused on the fabrication of 3D tissue, often by culturing cells within a support matrix; this process was introduced in Chapter 1 and is common in tissue engineering studies. However, there is some interest in exploring the ability of isolated cells to self-organize into functional tissue constructs without the need of external scaffolding. This viewpoint is based on the assumption that cells have all the information required for tissue formation; isolated cells interact with other biological components, like other cells or the extracellular matrix, in order to support tissue formation under normal mammalian function.

Based on the requirements put forth for self-organization to occur, cells must be dynamic, material must be exchanged, and an overall stable configuration must be the end result of the process. These requirements are satisfied by isolated cells, and therefore, these cells can be said to have the properties required to participate in a self-organization process. Using self-organization strategies, isolated cells participate in 3D tissue formation by interacting with other cells and extracellular matrices that are generated by the cells; the process of tissue formation is independent of any external scaffolding material. The ability to form 3D tissue without the need for any external scaffolding is central to the self-organization process.

The main advantage of self-organization strategies is the ability to minimize external constraints and therefore allowing the cells to govern the process of 3D tissue formation. With less constraints and a high degree of freedom, the probability of fabricating functionally and anatomically matched constructs increases. Culturing isolated cells on a prefabricated scaffold defines boundaries for tissue formation and growth, while self-organization processes remove or reduce these boundaries and allow cells to determine the best course of action. While self-organization strategies have advantages, they are also faced with challenges. Self-organization strategies can require long culture times and can be difficult processes to control.

3.11 SMART MATERIALS

Tissue engineering has traditionally focused on the use of scaffolds to support 3D tissue formation. The role of the scaffold has been to provide temporary structural support for cells. During tissue formation, extracellular matrix components are generated by cells, and as this happens, the temporary scaffold is degraded and replaced by newly formed extracellular matrix. The result is a 3D tissue construct that only consists of biological components. The first generation of scaffolds used for tissue fabrication was designed to provide passive support during tissue fabrication. The next generation of biomaterials was designed to be "cell-friendly" or biomimetic; these materials were fabricated by incorporation of biological activity within the scaffold, which provided functional attachment sites for cells (143–150). Covalent linking of the amino acid sequence RGD is one example used extensively to support fabrication of 3D tissue constructs. The most recent generation of biomaterials has been designed to respond to changes in the cellular environment; these materials, known as smart materials, are receptive to changes in the physiological environment and are adaptive to changes in the degree of tissue maturation (143–150).

We can illustrate the concept of smart biomaterials with one specific hypothetical example. In this case, the biomaterial is fabricated to deliver a specific target to mammalian tissue; the smart biomaterial is fabricated with specifically targeted cell attachment sites and targets for cleavage by changes in physiological environment, all linked to a polymer backbone containing internalized growth factors. The objective is to utilize the cell attachment site to deliver biomaterials to specific targets, while changes in the physiological state of cells promote the release of growth factors in the local environment. Although this concept is at an early stage of development, there have been several examples of the development of smart materials using several different stimuli like compression, oxidation state, pH, and MMP cleavage activity.

In one example, VEGF was embedded within a 3D alginate gel, and the rate of release was regulated by application of compressive loads (using a mechanical testing system), both during *in vitro* culture and upon subcutaneous implantation into the dorsal region of severe combined immunodeficient (SCID) mice (151). Application of compressive load resulted in an increase in the rate of release of VEGF (determined by a radioactive tracer) during *in vitro* culture; during *in vivo* culture, this led to an increase in neovascularization, as determined by vessel count. Compressive loads are commonly observed during normal mammalian bone function, and using this stimulus to regulate the functionality of biomaterials provides an excellent platform for novel tissue engineering therapies for development of bone grafts.

In another example, changes in oxidation state were used to modify material properties, leading to release of embedded biomolecules. A tri-block copolymer (ABA) was fabricated with the A block consisting of hydrophilic poly(ethylene glycol) and B block consisting of hydrophobic poly(propylene sulfide) (PPS) (152). Upon exposure of the ABA copolymer to an oxidative environment, the sulfide moieties of the PPS were oxidized to sulfoxides and then to sulfones, which changed the properties of the PPS from hydrophobic to hydrophilic. This resulted in destabilization of the copolymer material, and as a result, embedded biomolecules were released within the culture environment (153,154).

A similar concept has been developed by utilization of changes in pH upon cellular endocytosis of biomaterials (155). A polymer was designed by conjugating a biomolecule to a PEG copolymer using disulphide bonds and then forming a complex between the PEG copolymer (conjugated with a biomolecule) to a backbone or carrier polymer using pH-sensitive acetyl linkers. When the biomaterial is implanted, it is internalized within the cell endosome via endocytosis. The pH within the endosome is acidic; this acidity provides a mechanism to cleave pH-sensitive acetyl linkers, thereby separating the PEG copolymer from the carrier polymer (155). The biomolecule that is bound to the PEG copolymer is separated from the PEG copolymer and released from the endosome to the cytoplasm. In the cytoplasm, the biomolecule can perform its biological function.

The ability of matrix metalloproteinases (MMPs) to recognize and cleave specific amino acid sequences makes them a suitable mechanism for functionality of smart biomaterials. In one example, a polymer was synthesized by cross-linking PEG copolymers with the peptide gly-pro-gln-gly-lle-trp-gly-gln, a substrate which contains a cleavage site for MMP-2 (156). The polymer was embedded with VEGF and used to support 2D culture of endothelial cells. MMP-2- mediated release of VEGF during 2D culture was identified using a fluorescent tag; the release resulted in an increase in the rate of endothelial cell proliferation.

3.12 THE DYNAMIC EXTRACELLULAR MATRIX

Introduction—Mammalian tissue consists of two components—cells and the extracellular matrix. Cells provide functionality, while the extracellular matrix provides structural support. In the case of heart muscle tissue, cardiac myocytes are the functional cells, and cytoskeletal proteins within the myocytes generate contractile force in response to elevated intracellular intracellular calcium; this in turn results in pumping capacity of the heart. Heart muscle tissue also contains the extracellular matrix, which anchors cardiac myocytes and provides mechanical support during continuous contractions of the heart. The extracellular matrix has numerous functions and plays a critical role in tissue formation and function. In terms of tissue engineering, tissue fabrication technology consists of coupling functional cells with biomaterials, and the purpose of the biomaterial is to mimic functionality of the extracellular matrix. It is important to gain an appreciation of the composition, organization, and function of the extracellular matrix and the way in which extracellular components interact with cells to support tissue formation and function. In this section, we will look at the role of the ECM in the formation and function of mammalian tissue.

Components of the ECM—The extracellular matrix consists of proteins, glycoproteins, glycosaminoglycans (GAGs), and proteoglycans (157,158). Some examples of proteins in the extracellular matrix are collagen, laminin, fibronectin, and elastin, each of which has very distinct functions within the tissue, including mechanical and tensile properties and binding specificity for cells. GAGs are long chain polysaccharides, some examples of which are chondroitin sulfates, dermatan sulfates, heparan sulfates, heparin, keratan sulfates, and hyaluronic acid. Glycoproteins are proteins that are covalently linked to a carbohydrate; they serve many functions, including stabilizing protein molecules. Proteoglycans are glycosylated proteins, which mean that proteins are conjugated to specific types of GAGs; examples of proteoglycans based on the GAG chondroitin sulfate are decorin and versican. In mammalian tissue, every component of the extracellular matrix has a specific role in providing structural support or functional interaction with cells. The composition of ECM components varies from one tissue to another and changes during development and in response to change in the physiological or pathological state of the tissue. Hence, the ECM is considered to be dynamic in nature as it constantly changes in composition in response to its environment.

Functions of the ECM—The ECM has many diverse functions, ranging from mechanical support for tissue formation and function to regulation of cell behavior by specific cell matrix interaction. Functions of the ECM include (157,158): 1) *Structural integrity for 3D tissue*—The ECM provides scaffolding to confer strength to the tissue, and mechanical properties of tissue are a direct result of properties of ECM components. 2) *Attachment sites for cells*—Extracellular matrix proteins like fibronectin, laminin, and collagen have specific binding sites for cell surface integrins. The binding of cell surface integrins with ECM proteins is known as cell-matrix interaction and is responsible for initiating a cascade of intracellular signaling events that support cell proliferation, viability, and functionality. 3) *Binding site for growth factor*—The ECM binds to growth factors like BMPs and FGFs and acts as a reservoir of these factors. 4) *Serve as mechanosensitive receptors*—The ECM responds to changes in the biomechanical environment, including changes in the stretch profile, via mechanosensitive receptors. These signals are then transmitted to the intracellular environment, initiating a cascade of signaling events that allow cells to modify specific molecular events to accommodate and adjust to the changes in external biomechanical environment.

ECM and Tissue Engineering—The ECM is a complex structure consisting of proteins and many other molecules. It supports 3D tissue fabrication, organization, and function. During the tissue fabrication process, the properties of the extracellular matrix are replicated by biomaterials. The composition of the ECM depends on the tissue system under consideration, and biomaterials are designed to match these tissue-specific properties. Once fabricated, the properties of the biomaterials are compared with those of the mammalian extracellular matrix by assessing the tensile properties, biocompatibility, and biomimetic function. The objective in tissue engineering is to fabricate biomaterials that closely replicate the properties of the mammalian extracellular matrix.

3.13 IDEALIZED BIOMATERIAL

An idealized biomaterial needs to replicate properties of the mammalian extracellular matrix both in terms of form and function. Many of the biomaterials currently under development are uniform in composition, which means the entire scaffold has the same properties. On the other hand, mammalian ECM consists of a diverse array of proteins and related compounds that support many functions of the ECM. Therefore, the first requirement of an idealized biomaterial is that it must have fiber composition that mimics the composition of mammalian ECM. We looked at several classification schemes earlier in this chapter and studied natural versus synthetic biomaterials. An idealized biomaterial will need to resolve problems associated with both naturally occurring (batch variability) and synthetic biomaterials (lack of biological activity) which necessities synthetic fabrication strategies leading to biologically active biomaterials. This means that an idealized biomaterial will need to replicate many, if not all, properties of mammalian ECM, and the process for material fabrication would need to be carefully optimized and regulated. This is depicted in Figure 3.9a.

The second component of our idealized biomaterial, shown in Figure 3.9b, is the ability to selectively bind specific cell types using specific binding sites or receptors for selectivity. As we have seen before, cell-matrix interactions are critical for tissue formation and function, and in the case of mammalian tissue, cell surface integrins recognize specific amino acid sequences on ECM proteins (like RGD for fibronectin). Cell surface integrins bind amino acid sequences, and this leads to a series of intracellular signaling events that regulate cellular and molecular events and tissue function. In our idealized biomaterial, we have specific binding sequences for selectively binding specific cell types. In Figure 3.9b, four binding sites are shown with specificity for functional cells, cells that provide structural support, cells for vasculature formation, and stem cells. If we relate this general scheme to heart muscle, the four cell types would be cardiac myocytes, fibroblasts, endothelial cells, and cardiac stem cells.

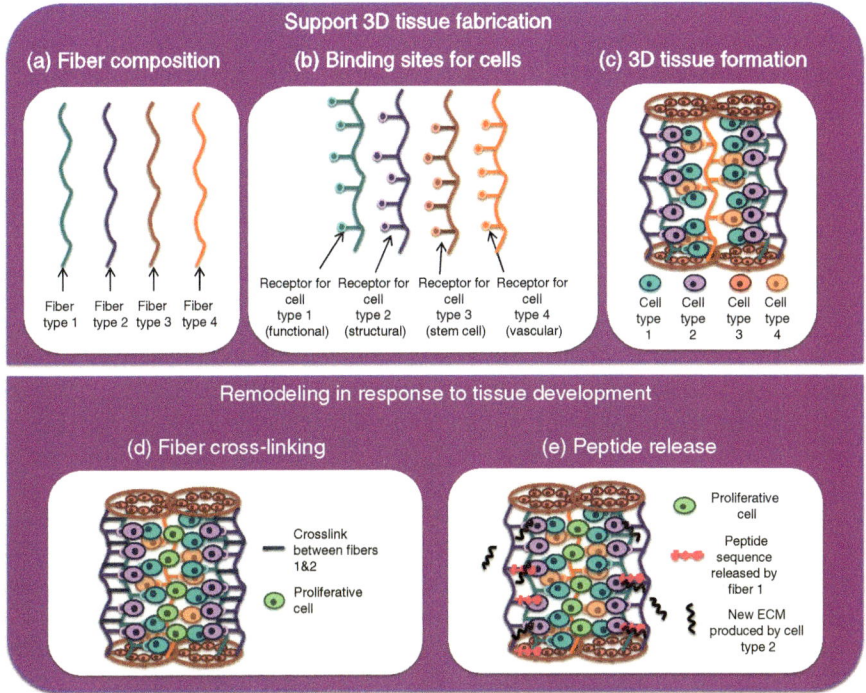

Figure 3.9 Idealized Biomaterial—(a) Fiber Composition—An idealized biomaterial will be composed of multiple fiber types, each with distinct properties. In the example presented here, the idealized biomaterial is shown to consist of 4 different fiber types. **(b) Binding Sites for Cells**—Individual fibers within the idealized biomaterial will have specific binding sites for cell surface integrins. **(c) 3D Tissue Formation**—Cells attach to specific binding sites on fibers of the idealized biomaterial to support 3D formation. **(d) Fiber Cross-Linking**—In response to cell proliferation, an idealized biomaterial will promote cross-linking between individual fibers to increase stability of the scaffold. **(e) Peptide Release**—In response to an increase in cell proliferation, peptide sequences can be cleaved from fibers within an idealized biomaterials. The peptide sequences act on structural cells, increasing the rate of ECM productions by these cells. This in turn has the effect of stabilizing the scaffold to support new tissue formation.

The third phase during the development of an idealized biomaterial is scaffold fabrication, as illustrated in Figure 3.9c. Important parameters in scaffold design and fabrication include fiber orientation and alignment, selective attachment of specific cell types to promote 3D tissue formation, and the ability to support functional organization of vasculature. As shown in Figure 3.9c, in order to support 3D tissue formation, the functional and structural cells are organized and aligned in a specific manner that allows cell-cell interaction. In the case of 3D heart muscle, functional coupling of cardiac myocytes is a prerequisite for tissue formation and can be achieved by aligning the fibers longitudinally; for another tissue system, a different organization of fibers may be optimal. Similarly, the organization of structural cells like fibroblasts is an important determinant of tissue function, and in our idealized biomaterials, the structural cells have been organized in parallel with functional cells; again, the specific organization will vary from application to application. However, the idealized biomaterial will have the capacity to selectively position structural cells in any orientation relative to functional cells. Finally, specific fibers within the idealized biomaterial have been fabricated in a circular pattern to support adhesion of vascular cells (endothelial cells) to promote capillary formation.

Collectively, we have discussed the ability of our idealized biomaterial to support 3D tissue formation by replicating many of the properties of mammalian ECM. These properties include composition and alignment of proteins and other related molecules, presence of cell specific binding sites, and the ability to orient fibers of the idealized biomaterials in customized patterns to support localization of specific cell types. The second function of the idealized biomaterial is to support maturation and development of 3D artificial tissue and modulate material properties in response to the phenotypic state of the tissue.

During the development and maturation of artificial tissue, traditional materials play a passive role; they do not respond to changes in cell behavior and/or phenotype. Cell proliferation is an important prerequisite for development of 3D tissue, and during normal mammalian tissue growth, this is accompanied by changes in the ECM to support increase in cell number. Many of the materials used in tissue engineering do not respond to these changes in cell number. However, in an idealized case, biomaterials will have the capacity to respond to changes in tissue growth, particularly an increase in cell number, as shown in Figure 3.9d–e. In the first case, an increase in cell proliferation leads to conformational changes in ECM proteins, which in turn results in cross-bridge formation between neighboring protein molecules. This in turn leads to stabilization of the biomaterial, providing additional structural stability. In the second case, an increase in cell proliferation leads to cleavage of a specific amino acid sequence from the biomaterial fibers. This amino acid sequence acts on structural cells, stimulating production of new extracellular matrix components. The newly formed ECM integrates with existing fibers to strengthen, stabilize, and expand the scaffold; this integration in turn serves to support the increase in cell number and tissue development and maturation.

In the two examples, the biomaterial responds to changes in the physiological state of cells. During early stages of tissue fabrication, the idealized biomaterial serves to support attachment of cells. This attachment leads to the formation of artificial 3D tissue. During later stages of tissue development, the idealized biomaterial responds to changes in the cellular environment to accommodate tissue maturation.

SUMMARY

Current State of the Art—Biomaterials are integral for the fabrication of artificial tissue and serve several functions during the tissue fabrication process. The field of biomaterial science is very well developed and extremely mature. It has provided the impetus for the development of biomaterials that can be used to support 3D artificial tissue. There are three biomaterial classification schemes: natural versus synthetic, degradable versus nondegradable, and metals, ceramics, and polymers. There are four biomaterial platforms: polymeric scaffolds, biodegradable hydrogels, acellular matrices, and scaffold-free models. Several properties of biomaterials are important during tissue fabrication: mechanical properties, degradation kinetics, biocompatibility, and biomimetic properties.

Thoughts for Future Research—One area of research that needs to be developed further is the use of scaffold-free technologies to fabricate 3D artificial tissue. Scaffold-free technologies are based on the premise that cells can generate their own ECM and that this ECM will be closer in form and function to mammalian ECM than synthetic biomaterials. In addition, it is hypothesized that ECM generated by cells will be superior to synthetic biomaterials. In order to achieve this objective, strategies need to be optimized to guide ECM production by cells by regulating the microenvironment and delivering controlled physiological stimuli. The development of scaffold-free technologies to support tissue fabrication is challenging; there are several hurdles that need to be overcome. The two most significant technological hurdles are: 1) guidance to regulate ECM production by cells, and 2) guidance to regulate tissue fabrication using newly synthesized ECM.

PRACTICE QUESTIONS

1. Based on your understanding of biomaterials, provide a general discussion of the role of biomaterials in tissue engineering. How are biomaterials used during the tissue fabrication process? What are the functions of the biomaterial during the fabrication of 3D artificial tissue?

2. Provide a definition for a biomaterial.

3. Provide a general scheme for biomaterial development. Explain the following terms: biomaterial platforms, biomimetic properties, biocompatibility, and mechanical properties.

4. During our discussion of biomaterial development, the following concepts were described: biomaterial platforms, biomimetic properties, biocompatibility, and mechanical properties. Pick any tissue fabrication application and explain how these concepts apply to the selected application.

5. Describe historical applications of biomaterials. Provide examples that are not discussed in the chapter.

6. Explain what the tensile properties of a material refer to. Draw the stress–strain curve and explain what different regions of the curve mean.

7. Why are the tensile properties of a biomaterial important for tissue engineering?

8. Describe one method that can be used to modify tensile properties of a biomaterial. Explain the impact of this modification on the biomaterial using the stress–strain curve.

9. Why are the degradation kinetics of biomaterials important for tissue engineering?

10. Pick any tissue engineering application. For your selected application, do you believe that a degradable or nondegradable biomaterial would be more suitable? Explain your selection.

11. Define biomaterial biocompatibility.

12. Explain why the biocompatibility of biomaterials is important for tissue engineering.

13. Describe the foreign body response, complement activation, and the coagulation pathway.

14. Define biomimetic biomaterials.

15. Explain why the biomimetic properties of a biomaterial are important for tissue engineering.

16. Provide one example of how you would synthesize a biomimetic biomaterial to fabricate 3D artificial heart muscle.

17. There are several classification schemes for biomaterials: natural versus synthetic, degradable versus nondegradable, and metals, ceramics and polymers. Explain these classification schemes and discuss the relative advantages and disadvantages of each.

18. As described in Question 17, there are several classification schemes for biomaterials. Pick any tissue engineering application and explain which group of biomaterials would be most suited for the selected application.

19. Scaffold-free methods rely upon cells to generate ECM to support tissue fabrication. Explain the concept of scaffold-free technology. Do you think that it is suitable for tissue engineering? Identify important design considerations for the development of scaffold-free technology.

20. Acellular matrices have received significant attention in recent literature. Explain the concept of acellular matrices. How is an acellular matrix fabricated? What are the relative advantages and disadvantages of acellular matrices? Do you believe that acellular matrices can be used clinically in patients?

21. We discussed four biomaterial platforms: polymeric scaffolds, biodegradable hydrogels, acellular tissue, and scaffold-free technology. Explain these classification schemes and discuss the relative advantages and disadvantages of each.

22. In Question 21, we discussed four biomaterial platforms. Which one of these four is best suited to bioengineer artificial heart muscle and why?

23. In Question 21, we discussed four biomaterial platforms. Pick any one tissue engineering application. Which one of these four is best suited for the selected application and why?

24. Describe the concept of smart materials—what exactly is a smart material? Pick any one tissue engineering application and design a smart material for the selected application.

25. What is the most significant challenge in the development of biomaterials to support the fabrication of artificial tissue? What steps will you take to overcome this challenge?

REFERENCES

1. Kosuge D, Khan WS, Haddad B, Marsh D. Biomaterials and scaffolds in bone and musculoskeletal engineering. Curr. Stem Cell Res. Ther. 2013 May;8(3): 185–91.
2. Zhang Z, Gupte MJ, Ma PX. Biomaterials and stem cells for tissue engineering. Expert. Opin. Biol. Ther. 2013 Apr;13(4):527–40. PMCID:PMC3596493.
3. D'Addessi A, Vittori M, Sacco E. An introduction to biomaterials in urology. Urologia, 2013 Apr 10;80(1):20–8.
4. Bhat S, Kumar A. Biomaterials and bioengineering tomorrow's healthcare. Biomatter. 2013 Apr 1; 3(2).
5. Williams D. Growth in the biomaterials market: the nature of growth factors. Med. Device Technol. 1998 Sep;9(7):6–11.
6. Allan B. Closer to nature: new biomaterials and tissue engineering in ophthalmology. Br. J. Ophthalmol. 1999 Nov;83(11):1235–40. PMCID:PMC1722846.
7. Williams DF. On the nature of biomaterials. Biomaterials. 2009 Oct;30(30): 5897–909.
8. Sharma CP. Biomaterials and artificial organs: few challenging areas. Trends in Biomaterials and Artificial Organs 2005;18:148.
9. Black J. The education of the biomaterialist: Report of a survey, 1980ΓÇô81. J. Biomed. Mater. Res. 1982 Mar 1;16(2):159–67.
10. Pariente JL, Conort P. [History of materials: from the stone age to the age of plastics]. Prog. Urol. 2005 Nov;15(5):863–4.
11. Ball P. Material witness: the materials of history. Nat. Mater. 2007 Nov;6(11):801.
12. Randall RC, Wilson NH. Clinical testing of restorative materials: some historical landmarks. J. Dent. 1999 Nov;27(8):543–50.
13. Belkin NL. A historical review of barrier materials. AORN J. 2002 Oct;76(4):648–53.
14. Katagiri M. [Dental implants. History and tissue reactions of implants]. Shigaku. 1989 Oct;77(SPEC):1152–61.
15. Ring ME. A thousand years of dental implants: a definitive history--part 1. Compend. Contin. Educ. Dent. 1995 Oct;16(10):1060, 1062, 1064.

16. Ring ME. A thousand years of dental implants: a definitive history--part 2. Compend. Contin. Educ. Dent. 1995 Nov;16(11):1132, 1134, 1136.
17. Lubit EC, Rappaport BA. Vitallium implantation. Oral Implantol. 1971;1(3):200–8.
18. Gore D, Frazer RQ, Kovarik RE, Yepes JE. Vitallium. J. Long Term Eff. Med. Implants 2005;15(6):673–86.
19. Nakajima H, Okabe T. Titanium in dentistry: development and research in the U.S.A. Dent. Mater. J. 1996 Dec;15(2):77–90.
20. Wilson AB, Jr. The modern history of amputation surgery and artificial limbs. Orthop. Clin. North Am. 1972 Jul;3(2):267–85.
21. Orr JF, James WV, Bahrani AS. The history and development of artificial limbs. Eng Med. 1982 Oct;11(4):155–61.
22. Mustapha NM. Artificial limbs, past, present and future. NATNEWS. 1985 Feb;22(2):suppl-20.
23. Marks LJ, Michael JW. Science, medicine, and the future: Artificial limbs. BMJ 2001 Sep 29;323(7315):732–5. PMCID:PMC1121287.
24. Putti V. Historic artificial limbs. 1933. Clin. Orthop. Relat Res. 2003 Jul;(412):4–7.
25. Apple DJ, Schmidbauer JM. [Sir Nicholas Harold Lloyd Ridley: pioneer of intraocular lens]. Klin. Monbl. Augenheilkd. 2001 Sep;218(9):583–5.
26. Escobar-Gomez M, Apple DJ, Vargas LG. [Tribute to Sir Nicholas Harold Ridley: inventor of intraocular lenses]. Arch. Soc. Esp. Oftalmol. 2001 Nov;76(11): 687–8.
27. Auffarth GU, Schmidbauer J, Apple DJ. [The life work of Sir Nicholas Harold Lloyd Ridley]. Ophthalmologe 2001 Nov;98(11):1012–6.
28. Apple DJ, Trivedi RH. Sir Nicholas Harold Ridley, Kt, MD, FRCS, FRS: contributions in addition to the intraocular lens. Arch. Ophthalmol. 2002 Sep;120(9):1198–202.
29. Trivedi RH, Apple DJ, Pandey SK, Werner L, Izak AM, Vasavada AR, Ram J. Sir Nicholas Harold Ridley. He changed the world, so that we might better see it. Indian J. Ophthalmol. 2003 Sep;51(3):211–6.
30. Apple DJ. Sir Nicholas Harold Lloyd Ridley: 10 July 1. Biogr. Mem. Fellows. R Soc. 2007;53:285–307.
31. Soden PD, Kershaw I. Tensile testing of connective tissues. Med. Biol. Eng 1974 Jul;12(4):510–8.
32. Wright TM, Hayes WC. Tensile testing of bone over a wide range of strain rates: effects of strain rate, microstructure and density. Med. Biol. Eng 1976 Nov;14(6): 671–80.
33. Jonas J, Burns J, Abel EW, Cresswell MJ, Strain JJ, Paterson CR. A technique for the tensile testing of demineralised bone. J. Biomech. 1993 Mar;26(3):271–6.
34. Perrott DH, Rahn B, Wahl D, Linke B, Thuruller P, Troulis M, Glowacki J, Kaban LB. Development of a mechanical testing system for a mandibular distraction wound. Int. J. Oral Maxillofac. Surg. 2003 Oct;32(5):523–7.
35. Nazarian A, Stauber M, Muller R. Design and implementation of a novel mechanical testing system for cellular solids. J. Biomed. Mater. Res. B Appl. Biomater. 2005 May;73(2):400–11.
36. Gilbert R, Eich RH, Auchincloss JH, Jr. Application of Hooke's law to the elastic properties of the lung. Am. Rev. Tuberc. 1958 May;77(5):863–6.

37. Zhang W, Wang C, Kassab GS. The mathematical formulation of a generalized Hooke's law for blood vessels. Biomaterials 2007 Aug;28(24):3569–78.
38. Wang C, Zhang W, Kassab GS. The validation of a generalized Hooke's law for coronary arteries. Am. J. Physiol Heart Circ. Physiol 2008 Jan;294(1):H66–H73.
39. Giuliodori MJ, Lujan HL, Briggs WS, Palani G, DiCarlo SE. Hooke's law: applications of a recurring principle. Adv. Physiol Educ. 2009 Dec;33(4):293–6.
40. Torzilli PA, Takebe K, Burstein AH, Zika JM, Heiple KG. The material properties of immature bone. J. Biomech. Eng 1982 Feb;104(1):12–20.
41. Jain MK, Chernomorsky A, Silver FH, Berg RA. Material properties of living soft tissue composites. J. Biomed. Mater. Res. 1988 Dec;22(3 Suppl):311–26.
42. Ferguson SJ, Bryant JT, Ito K. The material properties of the bovine acetabular labrum. J. Orthop. Res. 2001 Sep;19(5):887–96.
43. Ritchie J, Jimenez J, He Z, Sacks MS, Yoganathan AP. The material properties of the native porcine mitral valve chordae tendineae: an in vitro investigation. J. Biomech. 2006;39(6):1129–35.
44. Kemper AR, McNally C, Manoogian SJ, Duma SM. Tensile material properties of human tibia cortical bone effects of orientation and loading rate. Biomed. Sci. Instrum. 2008;44:419–27.
45. Kasuga T, Ota Y, Nogami M, Abe Y. Preparation and mechanical properties of polylactic acid composites containing hydroxyapatite fibers. Biomaterials 2001 Jan;22(1):19–23.
46. Bigi A, Cojazzi G, Panzavolta S, Rubini K, Roveri N. Mechanical and thermal properties of gelatin films at different degrees of glutaraldehyde crosslinking. Biomaterials 2001 Apr;22(8):763–8.
47. Lv S, Dudek DM, Cao Y, Balamurali MM, Gosline J, Li H. Designed biomaterials to mimic the mechanical properties of muscles. Nature 2010 May 6;465(7294):69–73.
48. Pidaparti RM, Merril BA, Downton NA. Fracture and material degradation properties of cortical bone under accelerated stress. J. Biomed. Mater. Res. 1997 Nov;37(2):161–5.
49. Chaturvedi TP, Upadhayay SN. An overview of orthodontic material degradation in oral cavity. Indian J. Dent. Res. 2010 Apr;21(2):275–84.
50. Hjalmarsson L, Smedberg JI, Wennerberg A. Material degradation in implant-retained cobalt-chrome and titanium frameworks. J. Oral Rehabil. 2011 Jan;38(1):61–71.
51. Bawolin NK, Li MG, Chen XB, Zhang WJ. Modeling material-degradation-induced elastic property of tissue engineering scaffolds. J. Biomech. Eng 2010 Nov;132(11):111001.
52. Ng AH, Ng NS, Zhu GH, Lim LH, Venkatraman SS. A fully degradable tracheal stent: in vitro and in vivo characterization of material degradation. J. Biomed. Mater. Res. B Appl. Biomater. 2012 Apr;100(3):693–9.
53. Ito T, Nakamura T, Takagi T, Toba T, Hagiwara A, Yamagishi H, Shimizu Y. Biodegradation of polyglycolic acid-collagen composite tubes for nerve guide in the peritoneal cavity. ASAIO J. 2003 Jul;49(4):417–21.
54. Zhijiang C. Biocompatibility and biodegradation of novel PHB porous substrates with controlled multi-pore size by emulsion templates method. J. Mater. Sci. Mater. Med. 2006 Dec;17(12):1297–303.

55. Cortizo MS, Molinuevo MS, Cortizo AM. Biocompatibility and biodegradation of polyester and polyfumarate based-scaffolds for bone tissue engineering. J. Tissue Eng Regen. Med. 2008 Jan;2(1):33–42.
56. Kean T, Thanou M. Biodegradation, biodistribution and toxicity of chitosan. Adv. Drug Deliv. Rev. 2010 Jan 31;62(1):3–11.
57. McBane JE, Sharifpoor S, Cai K, Labow RS, Santerre JP. Biodegradation and in vivo biocompatibility of a degradable, polar/hydrophobic/ionic polyurethane for tissue engineering applications. Biomaterials 2011 Sep;32(26):6034–44.
58. Wan Y, Yu A, Wu H, Wang Z, Wen D. Porous-conductive chitosan scaffolds for tissue engineering II. in vitro and in vivo degradation. J. Mater. Sci. Mater. Med. 2005 Nov;16(11):1017–28.
59. Liu H, Slamovich EB, Webster TJ. Less harmful acidic degradation of poly(lacticco-glycolic acid) bone tissue engineering scaffolds through titania nanoparticle addition. Int. J. Nanomedicine. 2006;1(4):541–5. PMCID:PMC2676635.
60. Xin AX, Gaydos C, Mao JJ. In vitro degradation behavior of photopolymerized PEG hydrogels as tissue engineering scaffold. Conf. Proc. IEEE Eng Med. Biol. Soc. 2006;1:2091–3.
61. Yixiang D, Yong T, Liao S, Chan CK, Ramakrishna S. Degradation of electrospun nanofiber scaffold by short wave length ultraviolet radiation treatment and its potential applications in tissue engineering. Tissue Eng Part A 2008 Aug;14(8):1321–9.
62. Habraken WJ, Wolke JG, Mikos AG, Jansen JA. PLGA microsphere/calcium phosphate cement composites for tissue engineering: in vitro release and degradation characteristics. J. Biomater. Sci. Polym. Ed 2008;19(9):1171–88.
63. Shen H, Zhu L, Castillon A, Majee M, Downie B, Huq E. Light-induced phosphorylation and degradation of the negative regulator PHYTOCHROME-INTERACTING FACTOR1 from Arabidopsis depend upon its direct physical interactions with photoactivated phytochromes. Plant Cell 2008 Jun;20(6):1586–602. PMCID:PMC2483374.
64. Lu M, Wu X, Wei X. Chemical degradation of polyacrylamide by advanced oxidation processes. Environ. Technol. 2012 Apr;33(7–9):1021–8.
65. Tanimoto S, Takahashi D, Toshima K. Chemical methods for degradation of target proteins using designed light-activatable organic molecules. Chem. Commun. (Camb.) 2012 Aug 11;48(62):7659–71.
66. Toshima K. Chemical biology based on target-selective degradation of proteins and carbohydrates using light-activatable organic molecules. Mol. Biosyst. 2013 May;9(5):834–54.
67. Liang SL, Yang XY, Fang XY, Cook WD, Thouas GA, Chen QZ. In vitro enzymatic degradation of poly (glycerol sebacate)-based materials. Biomaterials 2011 Nov;32(33):8486–96.
68. Habraken GJ, Peeters M, Thornton PD, Koning CE, Heise A. Selective enzymatic degradation of self-assembled particles from amphiphilic block copolymers obtained by the combination of N-carboxyanhydride and nitroxide-mediated polymerization. Biomacromolecules. 2011 Oct 10;12(10):3761–9.
69. Nilasaroya A, Martens PJ, Whitelock JM. Enzymatic degradation of heparin-modified hydrogels and its effect on bioactivity. Biomaterials 2012 Aug;33(22):5534–40.
70. Laffleur F, Hintzen F, Rahmat D, Shahnaz G, Millotti G, Bernkop-Schnurch A. Enzymatic degradation of thiolated chitosan. Drug Dev. Ind. Pharm. 2012 Oct 12.

71. Anderson JM, Rodriguez A, Chang DT. Foreign body reaction to biomaterials. Semin. Immunol. 2008 Apr;20(2):86–100. PMCID:PMC2327202.
72. Williams DF. On the mechanisms of biocompatibility. Biomaterials 2008 Jul;29(20):2941–53.
73. Sun J, Zheng Q, Wu Y, Liu Y, Guo X, Wu W. Biocompatibility of KLD-12 peptide hydrogel as a scaffold in tissue engineering of intervertebral discs in rabbits. J. Huazhong. Univ Sci. Technolog. Med. Sci. 2010 Apr;30(2):173–7.
74. van Luyn MJ, Plantinga JA, Brouwer LA, Khouw IM, de Leij LF, van Wachem PB. Repetitive subcutaneous implantation of different types of (biodegradable) biomaterials alters the foreign body reaction. Biomaterials 2001 Jun;22(11):1385–91.
75. Puolakkainen P, Bradshaw AD, Kyriakides TR, Reed M, Brekken R, Wight T, Bornstein P, Ratner B, Sage EH. Compromised production of extracellular matrix in mice lacking secreted protein, acidic and rich in cysteine (SPARC) leads to a reduced foreign body reaction to implanted biomaterials. Am. J. Pathol. 2003 Feb;162(2):627–35. PMCID:PMC1851143.
76. Jones JA, McNally AK, Chang DT, Qin LA, Meyerson H, Colton E, Kwon IL, Matsuda T, Anderson JM. Matrix metalloproteinases and their inhibitors in the foreign body reaction on biomaterials. J. Biomed. Mater. Res. A 2008 Jan;84(1):158–66.
77. Liu L, Chen G, Chao T, Ratner BD, Sage EH, Jiang S. Reduced foreign body reaction to implanted biomaterials by surface treatment with oriented osteopontin. J. Biomater. Sci. Polym. Ed 2008;19(6):821–35.
78. Liddiard K, Rosas M, Davies LC, Jones SA, Taylor PR. Macrophage heterogeneity and acute inflammation. Eur. J. Immunol. 2011 Sep;41(9):2503–8.
79. Calvo JA, Meira LB, Lee CY, Moroski-Erkul CA, Abolhassani N, Taghizadeh K, Eichinger LW, Muthupalani S, Nordstrand LM, Klungland A, et al. DNA repair is indispensable for survival after acute inflammation. J. Clin. Invest 2012 Jul 2;122(7):2680–9. PMCID:PMC3386829.
80. Sousa LP, Alessandri AL, Pinho V, Teixeira MM. Pharmacological strategies to resolve acute inflammation. Curr. Opin. Pharmacol. 2013 Apr 8.
81. Caielli S, Banchereau J, Pascual V. Neutrophils come of age in chronic inflammation. Curr. Opin. Immunol. 2012 Dec;24(6):671–7. PMCID:PMC3684162.
82. Murakami M, Hirano T. The molecular mechanisms of chronic inflammation development. Front Immunol. 2012;3:323. PMCID:PMC3498841.
83. Naylor AJ, Filer A, Buckley CD. The role of stromal cells in the persistence of chronic inflammation. Clin. Exp. Immunol. 2013 Jan;171(1):30–5. PMCID:PMC3530092.
84. Makela J, Yannopoulos F, Ylitalo K, Makikallio T, Lehtonen S, Lappi-Blanco E, Dahlbacka S, Rimpilainen E, Kaakinen H, Juvonen T, et al. Granulation tissue is altered after intramyocardial and intracoronary bone marrow-derived cell transfer for experimental acute myocardial infarction. Cardiovasc. Pathol. 2012 May;21(3):132–42.
85. Hu D, Cross JC. Development and function of trophoblast giant cells in the rodent placenta. Int. J. Dev. Biol. 2010;54(2–3):341–54.
86. Holt DJ, Grainger DW. Multinucleated giant cells from fibroblast cultures. Biomaterials 2011 Jun;32(16):3977–87. PMCID:PMC3071287.
87. Honda E, Park AM, Yoshida K, Tabuchi M, Munakata H. Myofibroblasts: biochemical and proteomic approaches to fibrosis. Tohoku J. Exp. Med. 2013;230(2):67–73.

88. Bosmann M, Haggadone MD, Hemmila MR, Zetoune FS, Sarma JV, Ward PA. Complement activation product C5a is a selective suppressor of TLR4-induced, but not TLR3-induced, production of IL-27(p28) from macrophages. J. Immunol. 2012 May 15;188(10):5086–93. PMCID:PMC3345104.
89. Cazander G, Jukema GN, Nibbering PH. Complement activation and inhibition in wound healing. Clin. Dev. Immunol. 2012;2012:534291. PMCID:PMC3546472.
90. Triantafilou K, Hughes TR, Triantafilou M, Morgan BP. The complement membrane attack complex triggers intracellular Ca2+ fluxes leading to NLRP3 inflammasome activation. J. Cell Sci. 2013 Apr 23.
91. Aleshin AE, Schraufstatter IU, Stec B, Bankston LA, Liddington RC, DiScipio RG. Structure of complement C6 suggests a mechanism for initiation and unidirectional, sequential assembly of membrane attack complex (MAC). J. Biol. Chem. 2012 Mar 23;287(13):10210–22. PMCID:PMC3323040.
92. Hadders MA, Bubeck D, Roversi P, Hakobyan S, Forneris F, Morgan BP, Pangburn MK, Llorca O, Lea SM, Gros P. Assembly and regulation of the membrane attack complex based on structures of C5b6 and sC5b9. Cell Rep. 2012 Mar 29;1(3):200–7. PMCID:PMC3314296.
93. Kinsella JA, Tobin WO, Hamilton G, McCabe DJ. Platelet activation, function, and reactivity in atherosclerotic carotid artery stenosis: a systematic review of the literature. Int. J.Stroke 2012 Sep 27.
94. Sheriff J, Soares JS, Xenos M, Jesty J, Bluestein D. Evaluation of shear-induced platelet activation models under constant and dynamic shear stress loading conditions relevant to devices. Ann. Biomed. Eng 2013 Jun;41(6):1279–96. PMCID:PMC3640664.
95. Woolley R, Prendergast U, Jose B, Kenny D, McDonagh C. A rapid, topographical platelet activation assay. Analyst 2013 Jun 11.
96. Kao WJ. Evaluation of protein-modulated macrophage behavior on biomaterials: designing biomimetic materials for cellular engineering. Biomaterials 1999 Dec;20(23–24):2213–21.
97. Reddi AH. Morphogenesis and tissue engineering of bone and cartilage: inductive signals, stem cells, and biomimetic biomaterials. Tissue Eng 2000 Aug;6(4):351–9.
98. Mardilovich A, Kokkoli E. Biomimetic peptide-amphiphiles for functional biomaterials: the role of GRGDSP and PHSRN. Biomacromolecules. 2004 May;5(3):950–7.
99. Prasad CK, Krishnan LK. Regulation of endothelial cell phenotype by biomimetic matrix coated on biomaterials for cardiovascular tissue engineering. Acta Biomater. 2008 Jan;4(1):182–91.
100. Ko YG, Ma PX. Surface-grafting of phosphates onto a polymer for potential biomimetic functionalization of biomaterials. J. Colloid Interface Sci. 2009 Feb 1;330(1):77–83. PMCID:PMC2645349.
101. Novak MT, Bryers JD, Reichert WM. Biomimetic strategies based on viruses and bacteria for the development of immune evasive biomaterials. Biomaterials 2009 Apr;30(11):1989–2005. PMCID:PMC2673477.
102. Luz GM, Mano JF. Biomimetic design of materials and biomaterials inspired by the structure of nacre. Philos. Trans. A Math. Phys.Eng Sci. 2009 Apr 28;367(1893):1587–605.
103. Nguyen EH, Schwartz MP, Murphy WL. Biomimetic approaches to control soluble concentration gradients in biomaterials. Macromol. Biosci. 2011 Apr 8;11(4):483–92.

REFERENCES

104. Francolini I, Crisante F, Martinelli A, D'Ilario L, Piozzi A. Synthesis of biomimetic segmented polyurethanes as antifouling biomaterials. Acta Biomater. 2012 Feb;8(2):549–58.
105. Vandecandelaere N, Rey C, Drouet C. Biomimetic apatite-based biomaterials: on the critical impact of synthesis and post-synthesis parameters. J. Mater. Sci. Mater. Med. 2012 Nov;23(11):2593–606.
106. Rahmany MB, Van DM. Biomimetic approaches to modulate cellular adhesion in biomaterials: A review. Acta Biomater. 2013 Mar;9(3):5431–7.
107. Weng Y, Chen J, Tu Q, Li Q, Maitz MF, Huang N. Biomimetic modification of metallic cardiovascular biomaterials: from function mimicking to endothelialization in vivo. Interface Focus. 2012 Jun 6;2(3):356–65. PMCID:PMC3363017.
108. Shin H, Jo S, Mikos AG. Biomimetic materials for tissue engineering. Biomaterials 2003 Nov;24(24):4353–64.
109. Hersel U, Dahmen C, Kessler H. RGD modified polymers: biomaterials for stimulated cell adhesion and beyond. Biomaterials 2003 Nov;24(24):4385–415.
110. Li J, Ding M, Fu Q, Tan H, Xie X, Zhong Y. A novel strategy to graft RGD peptide on biomaterials surfaces for endothelization of small-diamater vascular grafts and tissue engineering blood vessel. J. Mater. Sci. Mater. Med. 2008 Jul;19(7):2595–603.
111. Hennessy KM, Clem WC, Phipps MC, Sawyer AA, Shaikh FM, Bellis SL. The effect of RGD peptides on osseointegration of hydroxyapatite biomaterials. Biomaterials 2008 Jul;29(21):3075–83. PMCID:PMC2465812.
112. Bellis SL. Advantages of RGD peptides for directing cell association with biomaterials. Biomaterials 2011 Jun;32(18):4205–10. PMCID:PMC3091033.
113. Glass J, Blevitt J, Dickerson K, Pierschbacher M, Craig WS. Cell attachment and motility on materials modified by surface-active RGD-containing peptides. Ann. N.Y. Acad. Sci. 1994 Nov 30;745:177–86.
114. Engstrand T. Biomaterials and biologics in craniofacial reconstruction. J. Craniofac. Surg. 2012 Jan;23(1):239–42.
115. Keatch RP, Schor AM, Vorstius JB, Schor SL. Biomaterials in regenerative medicine: engineering to recapitulate the natural. Curr. Opin. Biotechnol. 2012 Aug;23(4):579–82.
116. Prewitz M, Seib FP, Pompe T, Werner C. Polymeric biomaterials for stem cell bioengineering. Macromol. Rapid Commun. 2012 Sep 14;33(17):1420–31.
117. Gilbert TW, Sellaro TL, Badylak SF. Decellularization of tissues and organs. Biomaterials 2006 Jul;27(19):3675–83.
118. Baptista PM, Orlando G, Mirmalek-Sani SH, Siddiqui M, Atala A, Soker S. Whole organ decellularization—a tool for bioscaffold fabrication and organ bioengineering. Conf. Proc. IEEE Eng Med. Biol. Soc. 2009; 2009:6526–9.
119. Crapo PM, Gilbert TW, Badylak SF. An overview of tissue and whole organ decellularization processes. Biomaterials 2011 Apr;32(12):3233–43. PMCID:PMC3084613.
120. Gilbert TW. Strategies for tissue and organ decellularization. J. Cell Biochem. 2012 Jul;113(7):2217–22.
121. Park KM, Woo HM. Systemic decellularization for multi-organ scaffolds in rats. Transplant. Proc. 2012 May;44(4):1151–4.
122. Arenas-Herrera JE, Ko IK, Atala A, Yoo JJ. Decellularization for whole organ bioengineering. Biomed. Mater. 2013 Feb;8(1):014106.

123. Park JB. The use of hydrogels in bone-tissue engineering. Med. Oral Patol. Oral Cir. Bucal. 2011 Jan;16(1):e115–e118.

124. Aurand ER, Lampe KJ, Bjugstad KB. Defining and designing polymers and hydrogels for neural tissue engineering. Neurosci. Res. 2012 Mar;72(3):199–213. PMCID:PMC3408056.

125. Vulic K, Shoichet MS. Tunable growth factor delivery from injectable hydrogels for tissue engineering. J. Am. Chem. Soc. 2012 Jan 18;134(2):882–5. PMCID:PMC3260740.

126. Hu J, Hou Y, Park H, Choi B, Hou S, Chung A, Lee M. Visible light crosslinkable chitosan hydrogels for tissue engineering. Acta Biomater. 2012 May;8(5):1730–8.

127. Ekenseair AK, Boere KW, Tzouanas SN, Vo TN, Kasper FK, Mikos AG. Synthesis and characterization of thermally and chemically gelling injectable hydrogels for tissue engineering. Biomacromolecules. 2012 Jun 11;13(6):1908–15. PMCID:PMC3372601.

128. Pandit V, Zuidema J, Venuto KN. Macione J, Dai G, Gilbert RJ, Kotha S. Evaluation of Multifunctional Polysaccharide Hydrogels with Varying Stiffness for Bone Tissue Engineering. Tissue Eng Part A 2013 Jun 2.

129. Jenkins AD, Kratochvfl P, Stepto RFT, U.W. Suter. Glossary of Basic Terms in Polymer Science. Pure and Applied Chemistry 2013;68(12):2287–311.

130. Saxena AK, Marler J, Benvenuto M, Willital GH, Vacanti JP. Skeletal muscle tissue engineering using isolated myoblasts on synthetic biodegradable polymers: preliminary studies. Tissue Eng 1999 Dec;5(6):525–32.

131. Stock UA, Mayer JE, Jr. Tissue engineering of cardiac valves on the basis of PGA/PLA Co-polymers. J. Long Term Eff. Med. Implants 2001;11(3–4):249–60.

132. Dang JM, Leong KW. Natural polymers for gene delivery and tissue engineering. Adv. Drug Deliv. Rev. 2006 Jul 7;58(4):487–99.

133. Kim HN, Kang DH, Kim MS, Jiao A, Kim DH, Suh KY. Patterning methods for polymers in cell and tissue engineering. Ann. Biomed. Eng 2012 Jun;40(6):1339–55.

134. Girones MJ, Mendez JA, San RJ. Bioresorbable and nonresorbable polymers for bone tissue engineering. Curr. Pharm. Des 2012;18(18):2536–57.

135. Wei C, Cai L, Sonawane B, Wang S, Dong J. High-precision flexible fabrication of tissue engineering scaffolds using distinct polymers. Biofabrication. 2012 May 25;4(2):025009.

136. Lalwani G, Henslee AM, Farshid B, Parmar P, Lin L, Qin YX. Kasper FK, Mikos AG, Sitharaman B. Tungsten disulfide nanotubes reinforced biodegradable polymers for bone tissue engineering. Acta Biomater. 2013 May 29.

137. Athanasiou KA, Eswaramoorthy R, Hadidi P, Hu JC. Self-Organization and the Self-Assembling Process in Tissue Engineering. Annu. Rev. Biomed. Eng 2013 May 20.

138. Schiffmann Y. Self-organization in biology and development. Prog. Biophys. Mol. Biol. 1997;68(2–3):145–205.

139. Coffey DS. Self-organization, complexity and chaos: the new biology for medicine. Nat. Med. 1998 Aug;4(8):882–5.

140. Karsenti E. Self-organization in cell biology: a brief history. Nat. Rev. Mol. Cell Biol. 2008 Mar;9(3):255–62.

141. Ricard J. Systems biology and the origins of life? part II. Are biochemical networks possible ancestors of living systems? networks of catalysed chemical reactions: non-equilibrium, self-organization and evolution. C R Biol. 2010 Nov;333(11–12):769–78.
142. Saetzler K, Sonnenschein C, Soto AM. Systems biology beyond networks: generating order from disorder through self-organization. Semin. Cancer Biol. 2011 Jun;21(3):165–74. PMCID:PMC3148307.
143. Fairman R, Akerfeldt KS. Peptides as novel smart materials. Curr. Opin. Struct. Biol. 2005 Aug;15(4):453–63.
144. Levi DS, Kusnezov N, Carman GP. Smart materials applications for pediatric cardiovascular devices. Pediatr. Res. 2008 May;63(5):552–8.
145. Bizdoaca N, Tarnita D, Tarnita DN. Modular adaptive implant based on smart materials. Rom. J. Morphol. Embryol. 2008;49(4):507–12.
146. Qian K, Wan J, Huang X, Yang P, Liu B, Yu C. A smart glycol-directed nanodevice from rationally designed macroporous materials. Chemistry 2010 Jan 18;16(3):822–8.
147. Lavalle P, Voegel JC, Vautier D, Senger B, Schaaf P, Ball V. Dynamic aspects of films prepared by a sequential deposition of species: perspectives for smart and responsive materials. Adv. Mater. 2011 Mar 11;23(10):1191–221.
148. Rodriguez-Cabello JC, Girotti A, Ribeiro A, Arias FJ. Synthesis of genetically engineered protein polymers (recombinamers) as an example of advanced self-assembled smart materials. Methods Mol. Biol. 2012;811:17–38.
149. Jochum FD, Theato P. Temperature- and light-responsive smart polymer materials. Chem. Soc.Rev. 2012 Aug 7.
150. Chrzanowski W, Khademhosseini A. Biologically inspired 'smart' materials. Adv. Drug Deliv. Rev. 2013 Apr;65(4):403–4.
151. Lee KY, Peters MC, Anderson KW, Mooney DJ. Controlled growth factor release from synthetic extracellular matrices. Nature 2000 Dec 21;408(6815):998–1000.
152. Bearinger JP, Terrettaz S, Michel R, Tirelli N, Vogel H, Textor M, Hubbell JA. Chemisorbed poly(propylene sulphide)-based copolymers resist biomolecular interactions. Nat. Mater. 2003 Apr;2(4):259–64.
153. Napoli A, Boerakker MJ, Tirelli N, Nolte RJ, Sommerdijk NA, Hubbell JA. Glucose-oxidase based self-destructing polymeric vesicles. Langmuir 2004 Apr 27;20(9):3487–91.
154. Napoli A, Valentini M, Tirelli N, Muller M, Hubbell JA. Oxidation-responsive polymeric vesicles. Nat. Mater. 2004 Mar;3(3):183–9.
155. Kusonwiriyawong C, van de Wetering P, Hubbell JA, Merkle HP, Walter E. Evaluation of pH-dependent membrane-disruptive properties of poly(acrylic acid) derived polymers. Eur. J. Pharm. Biopharm 2003 Sep;56(2):237–46.
156. Seliktar D, Zisch AH, Lutolf MP, Wrana JL, Hubbell JA. MMP-2 sensitive, VEGF-bearing bioactive hydrogels for promotion of vascular healing. J. Biomed. Mater. Res. A 2004 Mar 15;68(4):704–16.
157. Heinegard D. Extracellular matrix: pathobiology and signaling. Biol. Chem. 2013 Jun 1;394(6):805–6.
158. Clause KC, Barker TH. Extracellular matrix signaling in morphogenesis and repair. Curr. Opin.Biotechnol. 2013 May 28.

4

TISSUE FABRICATION TECHNOLOGY

Learning Objectives

After completing this chapter, students should be able to:

1. Describe self-organization technology for the fabrication of 3D artificial tissue.
2. Explain the process of cell sheet engineering for the fabrication of 3D artificial tissue.
3. Describe cell and organ printing as it applies to the tissue fabrication process.
4. Explain scaffold-based tissue engineering as a process of bioengineering 3D artificial tissue.
5. Discuss solid freeform fabrication of bioengineering 3D scaffolds.
6. Describe the concept of soft lithography as it relates to microfluidics and explain how this technology has been used to develop "organ-on-a-chip" models.
7. Describe technologies for cell patterning.
8. Design an idealized system to support fabrication of 3D artificial heart muscle.

Introduction to Tissue Engineering: Applications and Challenges, First Edition. Ravi Birla.
© 2014 The Institute of Electrical and Electronics Engineers, Inc. Published 2014 by John Wiley & Sons, Inc.

CHAPTER OVERVIEW

In this chapter, we will present techniques for the fabrication of 3D artificial tissue. We begin with a discussion of scaffold-free methods, which make use of the extracellular matrix that has been generated by the cells for artificial tissue fabrication. Cell-sheeting engineering is another scaffold-free tissue fabrication strategy; in this case, temperature- sensitive surfaces are used to fabricate 2D cells sheets that can be stacked together to form 3D artificial tissue. We will also discuss scaffold-based tissue engineering, including acellular grafts, polymeric scaffolds, and biodegradable hydrogels for tissue engineering. Scaffold-free technologies and scaffold-based strategies are the most commonly used methods of fabricating 3D artificial tissue. However, over the years, many new methods have been developed to support tissue fabrication, some of which are covered in this chapter. We discuss cell and organ printing as it relates to tissue engineering. Cell/organ printing makes use of 3D printing technology to fabricate 3D artificial tissue one layer at a time. This method has far-reaching applications in the tissue engineering space. A similar technique has been applied when fabricating 3D scaffolds and is known as solid freeform fabrication; in this case, complex 3D scaffolds are fabricated by printing one layer at a time. Cell patterning is another technique that has received a lot of attention and is based on techniques that regulate the spatial positioning of isolated cells. Cell patterning techniques are designed to replicate the complex organization of multiple cell types found in mammalian tissue. Microfluidic techniques are designed to fabricate microchannels, which are often used to simulate flow conditions within capillary networks and can also be used to regulate the spatial positioning of cells. Collectively, the strategies we present in this chapter provide researchers with a tool kit of fabrication techniques that can be applied toward the fabrication of 3D artificial tissue.

4.1 INTRODUCTION TO TISSUE FABRICATION TECHNOLOGIES

Tissue fabrication technologies have been developed to fabricate 3D tissue artificial tissue. As we have seen throughout the course of this book, 3D artificial tissue is fabricated by coupling isolated cells with biomaterials to support functional integration; tissue fabrication technologies have been developed to support this process. The objective of these technologies is to fabricate 3D artificial tissue that is similar in form and function to mammalian tissue. In nature, mammalian tissue is highly organized with multiple cell types and extracellular matrix components interacting in an orchestrated way to support formation and function of tissue. During early days of tissue engineering, 3D artificial tissue was fabricated by direct injection of isolated cells within 3D scaffolds. However, over the years significant advancements have been made in the development of novel tissue fabrication technologies to support fabrication of 3D artificial tissue. In this chapter, we will study many of these technologies.

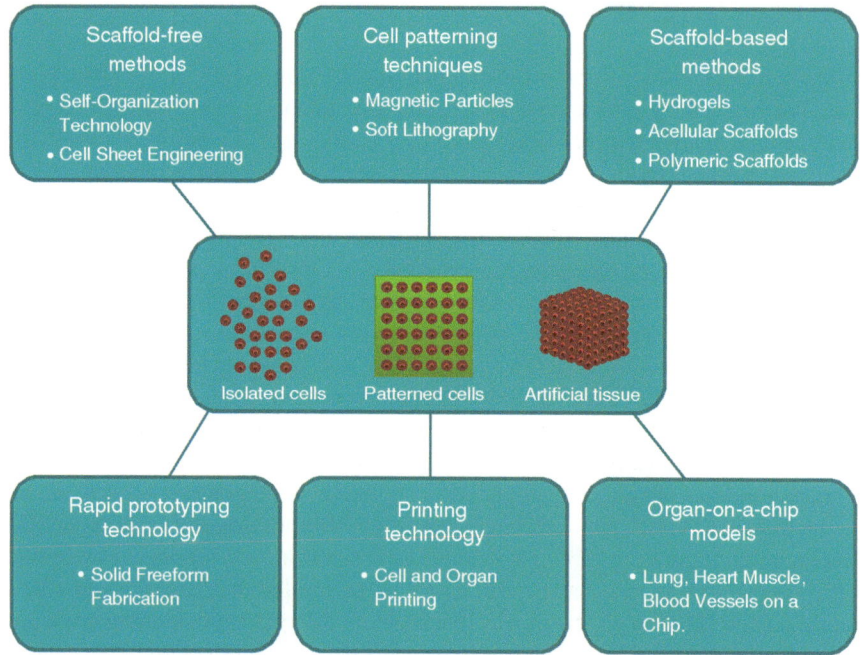

Figure 4.1 Overview of Tissue Fabrication Technologies—There are six categories of tissue fabrication technologies: scaffold-free methods, cell patterning techniques, scaffold-based methods, rapid prototyping technology, printing technology, and "organ-on-a-chip" model.

Tissue fabrication technologies can be classified into six categories, as shown in Figure 4.1, which include scaffold-free methods, cell patterning techniques, scaffold-based methods, rapid prototyping technologies, printing technology, and "organ-on-a-chip" models.

Scaffold-free methods have been developed to support tissue fabrication in the absence of any external scaffolding; rather, the extracellular matrix is generated by cells (1,2). These methods are based on the hypothesis that ECM fabricated by cells will prove to be superior to any external synthetic scaffolding material. Scaffold-free methods are based on the theory that extracellular matrix fabricated by cells will be superior in form and function to synthetic materials and therefore will provide a perfect scaffold to support cell-matrix interaction and formation of functional artificial tissue. Some examples of scaffold-free technologies that have been developed to support artificial tissue fabrication include self-organization strategies and cell sheet engineering.

Scaffold-based tissue fabrication models extensively used in the tissue engineering field (3). As we saw in Chapter 1, seminal work in the field of tissue engineering was based on this technology. The primary advantage of using scaffold-based methods is the ability to regulate material properties like 3D architecture, fiber geometry, and pore size, orientation, and alignment. This ability to control scaffold properties leads to process customization and the ability to bioengineer specific scaffolds for specific applications. Acellular scaffolds, polymer scaffolds, and biodegradable scaffolds are examples of scaffolds used in tissue engineering.

Rapid prototyping systems have recently come into play and represent a very interesting area of research. These systems are designed to fabricate scaffolds or entire artificial tissue by creating one layer of the tissue/scaffold at a time in the x-y-direction and then building additional layers in the z-direction (4,5). Using this strategy, the process for scaffold/tissue fabrication can be spatially controlled. Solid freeform fabrication has been used in the production of scaffolds, while cell and organ printing has been used for cell patterning and tissue fabrication.

Soft lithography (6,7) and microfluidics (8–10) have been used for the development of the "organ-on-a- chip" model. Soft lithography is used to create a specific pattern of microchannels on a culture substrate and can be been used to fabricate microvascular networks. This technology has been used for cell patterning, which allows spatial regulation in the placement of cells on a 2D culture surface. Recently, this technology has also been adapted for "organ-on-a-chip" models. "Organ-on-a-chip" models are used to understand organ functionality in an isolated *in vitro* system; the models are not developed for the fabrication of transplantable tissue.

Cell patterning technologies are designed to create 2D patterns of cells on a culture substrate, including specific positioning of cells to control alignment and orientation of cells (11–14). Several strategies have been developed to pattern cells on 2D culture surfaces, including cell printing, soft lithography, and use of magnetic nanoparticles. Cell patterning has been extended to fabricate 3D artificial tissue using printing technologies to control 3D positioning of cells; this is referred to as organ printing.

4.2 SELF-ORGANIZATION TECHNOLOGY

Introduction—Self-organization technology was pioneered in 2001 for the fabrication of artificial skeletal muscle and later adapted for the fabrication of 3D artificial heart muscle (1,2,15). Self-organization technology is based on the fabrication of extracellular matrix by cells that then use the newly formed ECM to support artificial tissue fabrication. This technology is an example of a scaffold-free tissue fabrication process and does not require external or synthetic scaffolding; rather, scaffolding is produced by cells. The hypothesis is that extracellular matrix produced by cells is superior in form and function to synthetic scaffolding and can support formation of 3D artificial tissue superior to that of synthetic scaffolding. The formation of 3D artificial tissue is by organization and alignment of cells relative to newly formed extracellular matrix. Artificial tissue formation is governed by self-organization of cells and ECM in the absence of any external signals or cues. There are two important concepts related to self-organization technology: 1) there are no synthetic scaffolds, and extracellular matrix is produced by cells, and 2) fabrication of 3D artificial tissue is by spontaneous remodeling of cells and not by external cues or based on the geometry of synthetic scaffolding.

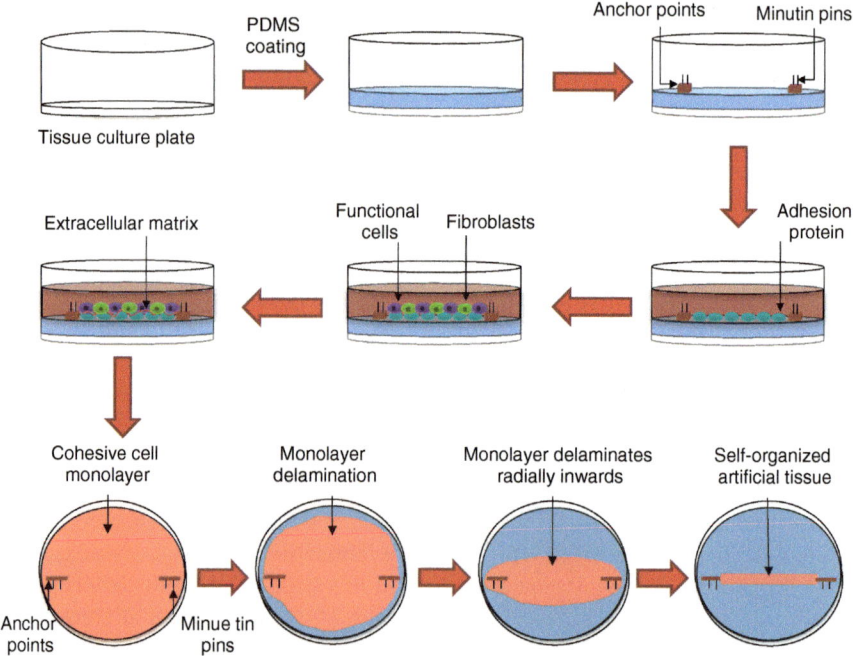

Figure 4.2 Self-Organization Technology for Fabrication of 3D Artificial Tissue—A tissue culture surface coated with PDMS and anchor points are positioned at the center of the culture surface. The culture surface is coated with an adhesion protein. Isolated cells are plated on the culture surface and attached to the adhesion protein. During culture, cells proliferate and form a cohesive cell monolayer; cells also generate extracellular matrix that serves to provide scaffolding during tissue remodeling and tissue fabrication. During time in culture, the extracellular matrix degrades with regular media changes. The cohesive cell monolayer delaminates, either due to the spontaneous contractions or due to the basal tension of cells. Monolayer delamination starts at the periphery of the culture surface and progresses radially inwards toward the center of the plate. At the center of the tissue culture plate, the delaminating monolayer attaches to anchor points that were pre-engineered onto the culture surface. Subsequent remodeling results in the fabrication of 3D artificial tissue.

Methodology—The methodology to support the fabrication of artificial 3D tissue using self-organization technology is presented in Figure 4.2.

The first step in the process is coating a culture surface with polydimethylsiloxane (PDMS), a culture surface that does not support cell adhesion very well (the function of the PDMS surface will become clear at the end of this discussion). Anchor points are positioned at the center of the tissue culture and consist of silk sutures that are secured in position using minutin pins (one of the functions of the PDMS is to allow placement of pins, as PDMS is a rubber like material). The culture surface is then coated with an adhesion protein; laminin has been used extensively to support artificial tissue fabrication, although other adhesion proteins like fibronectin and collagen have worked well in concert with self-organization technology. PDMS does not support cell adhesion, and therefore the function of the adhesion protein is to support attachment of cells to the culture surface. The specific adhesion protein used and the concentration at which it is used are experimentally determined variables and vary between tissue applications.

Once the culture surface has been coated with adhesion protein, cells are plated on the surface. It is important to plate a mixed cell population that consists of at least two cell types: the functional cell type, which would be cardiac myocytes for heart muscle fabrication, and fibroblasts, which produce extracellular matrix. The ratio of functional cell type to fibroblasts is an experimentally determined variable. Once attached, functional cells like cardiac myocytes begin the process of remodeling and formation of intercellular connectivity between neighboring cells. In addition, fibroblasts begin the process of extracellular matrix production. The cells functionally couple with the extracellular matrix to form a cohesive cell monolayer that covers the entire culture surface.

The formation of a cohesive cell monolayer is a critical prerequisite for the fabrication of artificial tissue and is supported by the adhesion protein, as cells remain anchored to the underlying protein. A few days after formation of the cohesive cell monolayer, the cell monolayer detaches from the underlying culture surface in a process known as monolayer delamination. This occurs due to the spontaneous contractions of the cells (if they are cardiac myocytes) or due to the basal tension created by other cell types. During the course of the culture period, adhesion protein dissolves, exposing cells to underlying PDMS surface; this process aids delamination of the cohesive cell monolayer. Over time, monolayer delamination continues, and the cohesive cell monolayer progresses toward the center of the culture plate. This process is spontaneous and does not require any external intervention like chemical or electrical stimulation. At the center of the plate, the delaminating cell monolayer attaches to anchor points, resulting in the formation of 3D artificial.

Discussion of Self-Organization Technology—The novelty of this technology is that 3D tissue formation is spontaneous and regulated by cells without any external stimuli. Chemical or electrical stimuli are not required for tissue formation, and even the delamination process is self-regulated. Delamination of the cohesive cell monolayer is regulated by spontaneous contractions of cardiac myocytes or basal tension of other cell types, while 3D tissue formation is a result of self-assembly of the cohesive cell monolayer. Extracellular matrix is produced by fibroblasts and has clear advantages over synthetic scaffolding; the composition is physiological, has attachment sites for cells and mechanical properties required for fabrication of artificial tissue, and is closer in form and function to mammalian tissue.

4.3 CELL SHEET ENGINEERING

Cell sheet engineering is another example of a scaffold-free technology and does not rely on external scaffolding to support fabrication of artificial tissue

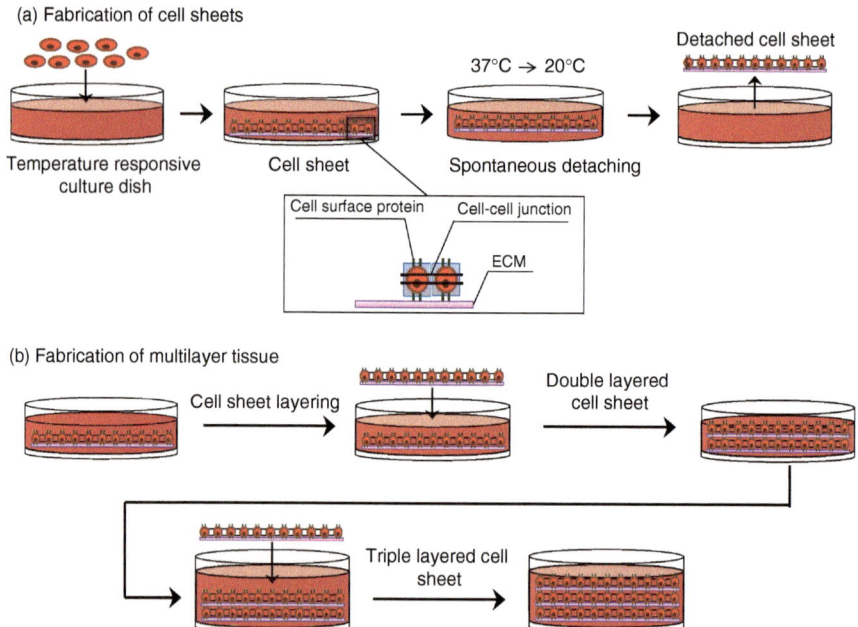

Figure 4.3 Cell Sheet Engineering for the Fabrication of Artificial Tissue—(a) Fabrication of Cell Sheets—Isolated cells are plated on the surface of a culture plate coated with a temperature-sensitive molecule. At 37°C, the temperature-responsive surface supports cell attachment. During the initial phases of cell culture, cells attach to the temperature-responsive surface, proliferate, and form a cohesive cell monolayer. Once a cohesive cell monolayer has been formed, the culture temperature is reduced to 20°C. At the lower temperature, the temperature-responsive surface does not support cell attachment. This in turn results in detachment of the cohesive cell monolayer, which at this point is known as a cell sheet. (b) Fabrication of Multilayer tissue—Individual cell sheets can be stacked together to form 3D artificial tissue, which consists of multiple layers of cell sheets stacked on top of each other.

(16–22). As with self-organization technology, cell sheet engineering relies upon extracellular matrix produced by cells; the newly formed extracellular matrix provides scaffolding during artificial tissue formation and remodeling. As we have seen before, extracellular matrix produced by cells has the right composition and distribution of proteins and other components required for cell attachment and 3D tissue fabrication.

The process of fabricating artificial tissue using cell sheet engineering is shown in Figure 4.3.

At the heart of the process is a temperature-sensitive molecule known as poly(N-isopoplyacrylaminde) (PIPAAm) (21). Under normal cell culture conditions at 37°C, PIPAAm can support cell attachment and proliferation. However, if the temperature is reduced to below 32°C, the PIPAAm surface properties change and can no longer support cell attachment; at lower temperatures, attached cells detach from the culture surface. In essence, the PIPAAm molecules change from a *"cell-friendly"* surface to a *"cell-unfriendly"* surface as the temperature is reduced from 37°C to 32°C; this process is at the heart of cell sheet engineering.

The first step in the tissue fabrication process using cell sheet engineering is to coat a standard polystyrene culture surface with PIPAAm. Isolated cells are then plated on the modified tissue culture surface. During initial stages of cell attachment and culture, the temperature is maintained at 37°C to support cell attachment on the PIPAAm surface; cells attach to the culture surface and proliferate to form a confluent monolayer (similar to the case with self-organization technology). The cells form a cohesive monolayer, supporting cell-cell interaction and cellular remodeling and begin the process of extracellular matrix production, which requires presence of fibroblasts or other cell types known to engage in generation of extracellular matrix.

After formation of a cohesive cell monolayer, the culture temperature is reduced from 37°C to a temperature below 32°C (usually 20°C), which changes the properties of PIPAAm from *"cell-friendly"* to *"cell-unfriendly."* This change results in detachment of the cell monolayer from the culture surface. The detached cell monolayer is referred to as a cell sheet since it is exactly that: a sheet of cells. The cell sheet remains intact after detachment from the culture surface, retaining intercellular connectivity. Once an individual sheet has detached from the culture surface, it can be physically transferred and placed on top of another cell sheet. This process results in the fabrication of 3D artificial tissue with a thickness of two cell monolayers. This process continues, and multiple cell sheets can be added on top of eachother, supporting fabrication of multilayer artificial tissue.

The process of cell sheeting engineering can be compared with self-organization technology. Let us look at the similarities first. Both processes are based on scaffold-free technology and depend on extracellular matrix produced by cells. In addition, formation of a cohesive cell monolayer is required for both processes, and the presence of extracellular matrix-producing cells is essential for both technologies. There are also significant differences between the two processes: in the case of self-organization technology, the process is regulated by adhesion proteins like laminin, while in the case of cell sheet engineering, the process is regulated by temperature responsive culture surfaces.

4.4 SCAFFOLD-BASED TISSUE FABRICATION

Scaffold-based tissue fabrication has been seminal in the development of the field of tissue engineering (23–25), as much of the early work was based on this technology. Indeed, one of the seminal papers in the field, which first demonstrated the feasibility of fabricating artificial liver tissue equivalents, was based on the use of this technology. Due to the prevalence of scaffold-based tissue engineering in the field, we described the process of tissue fabrication based on scaffolding technology during our discussion in Chapter 1. Scaffold-based models continue to receive

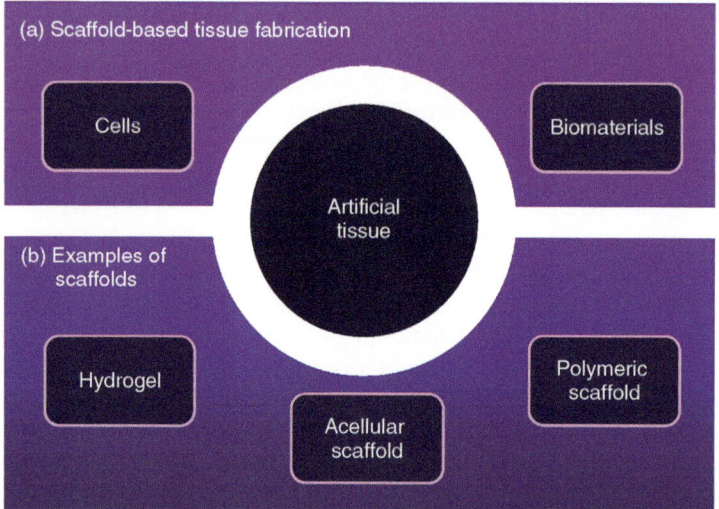

Figure 4.4 (a) **Scaffold-Based Tissue Engineering**—Technology is based on cellularization of 3D scaffolds with isolated cells. (b) **Examples of Scaffolds**—Acellular grafts, polymeric scaffolds, and hydrogels have been used to support the fabrication of 3D artificial tissue.

attention in recent tissue engineering studies, and this technology has significance in the field.

Scaffold-based tissue fabrication makes use of an external or synthetic 3D structure to simulate properties of the ECM (Figure 4.4).

Let us compare scaffold-based methods to self-organization strategies and cell sheet engineering. In the case of scaffold-free models, extracellular matrix is generated by cells during culture and utilized for artificial tissue fabrication. Compared to this approach, scaffold-based technologies rely on fabrication of a suitable scaffold to support formation of artificial tissue.

There are several advantages to the use of scaffold-based methods to support artificial tissue fabrication (23–25). The size, shape, and geometry of the scaffold can be controlled by changing processing variables, thereby allowing a greater degree of freedom in terms of determining scaffold geometry. Similarly, properties of the scaffold, including tensile properties, biomimetic properties, and biocompatibility, can also be modulated based on the specific tissue engineering application. Examples of variables that can be modulated include type and composition of individual monomer units and reaction conditions including temperature, pressure, and pH.

There are three strategies that have been used to fabricate artificial tissue using scaffold-based technology: acellular scaffolds, hydrogels, and polymer scaffolds (discussed in Chapter 3). Acellular scaffolds are obtained from tissue biopsies after complete removal of all cellular components, a process that leaves behind an intact extracellular matrix. As tissue specimens are obtained from mammalian tissue biopsies, the composition and distribution of the extracellular matrix is perfect for tissue fabrication. Hydrogels contain a high percentage of water and often contain naturally occurring molecules like collagen, fibrin, and alginate, which support cell-matrix interactions leading to the fabrication of functional tissue. Polymeric scaffolds are rigid structures fabricated from monomer units using tightly controlled reaction conditions.

The elements of scaffold-based tissue engineering have been described before in several chapters and will be summarized here. The strategy of fabricating artificial tissue using scaffold-based technology has been described in Chapter 1. Depending on the choice of scaffolding used (acellular scaffolds, hydrogels, polymeric scaffolds), material fabrication and characterization efforts will be implemented and optimized, which has been discussed in Chapter 3. Prior to fabrication of artificial tissue, cell sourcing is important and requires knowledge about cell biology, stem cell engineering, and cell culture, all of which have been covered in Chapter 2. The implementation of bioreactors for scaffold cellularization is also important and is described later in Chapter 6.

There are several variables that need to be optimized for scaffold-based tissue fabrication technology. The number of cells, purity of the cells, and proportion of different cell types are important determinants of tissue function and need to be optimized experimentally. This is particularly important with the use of primary cells, as tissue digestion results in a mixed cell population consisting of functional cells, supporting structural cells, vascular, and nerve cells. As one example, digestion of hearts will result in a mixed cell population that consists of cardiac myocytes, cardiac fibroblasts, and vascular cells including endothelial cells and smooth muscle cells. During tissue fabrication, the number of different cell types will need to be calculated and the relative proportion of each cell type will need to be optimized for scaffold cellularization.

Cell retention is an important variable when working with tissue fabrication technologies. A specific number of cells are added to the scaffold; however, only a certain percentage of cells are retained within the 3D scaffold. The remaining cells are washed out with regular media changes, particularly within the first 24 hours. Cell retention is defined as the number of cells within the scaffold at any given time, expressed as a percentage of the total number of cells used to populate the scaffold; cell retention needs to be experimentally determined and optimized. Cell retention is less of a challenge with scaffold-free technologies like self-organization strategies or cell sheet engineering, as a high percentage of cells are retained once integrated within the cohesive cell monolayer or cell sheet; cells are not lost with consecutive media changes.

Scaffold-based tissue engineering continues to be a preferred model for tissue fabrication and has found widespread applications in tissue engineering. Together with scaffold-free technologies, these methods continue to be explored for a variety of tissue applications and for the development of artificial tissue. Scaffold-based and scaffold-free technologies have relative advantages and disadvantages and are suited for different tissue fabrication applications.

4.5 CELL AND ORGAN PRINTING

We have looked at several strategies of fabricating 3D artificial tissue including scaffold-based and scaffold-free technologies like self-organization and cell sheet engineering. These approaches focus on building macroscopic artificial tissue starting with predefined structures: 3D scaffolds for scaffold-based methods, cohesive cell monolayers for self-organization, or cell sheets for cell sheet engineering. Cell and organ printing uses a different concept altogether and is based on a ground-up fabrication approach, starting with a single building block that consists of a few cells trapped within a hydrogel to fabricate or build artificial tissue by controlling spatial positioning of these building blocks (26–32). This approach may be compared to any fabrication methodology, including building a house, which requires placement of one brick at a time, or a building project with LEGOs, which again requires placement of individual blocks to fabricate complex 3D structures. The advantage of using cell and organ printing for tissue fabrication is the ability to control spatial placement of cells relative to other cells or ECM proteins; this in turn provides the ability to "build" 3D artificial tissue without any predefined geometrical constraints imposed by the scaffold, cell monolayer, or cell sheet.

Cell and organ printing may be compared to inkjet printing, in which individual droplets of ink are transferred through a nozzle and deposited on the surface of paper. In comparison, cell and organ printing technologies are based on transferring cells or groups of cells through a fluid nozzle onto a biologically adapted printing surface. While cell/organ printing is innovative, it is not without challenges, some of which include: what carrier will be used to deliver cells through a fluid nozzle? What will be done to ensure cells are not damaged in the transfer process? What will be used as the paper equivalent to inkjet printing—the surface on which the cells will be delivered? How will the spatial resolution be controlled? How will cells adhere to the printing surface on which they are delivered? How will individual cells or groups of cells bond with each other to form 3D structures?

Let us begin with some definitions related to this cell/organ printing (33,34):

- *Bioprinting*—use of computer-aided transfer processes for patterning and assembling living and non-living materials with prescribed 2D or 3D organization.
- *Bioink*—cell or biomaterial containing solution used in bioprinting technologies. For cell printing, cells are suspended in culture media and supplemented with a thickening agent that must be biocompatible (e.g. an alginate solution). Maintaining homogeneity of the cell suspension with time is crucial for reproducible bioprinting.
- *Biopaper*—surface onto which the bio-ink is deposited during bioprinting. This can be compared with traditional paper during inkjet printing.
- *Cell printing*—bioprinting process used for 2D cell patterning by depositing bio-ink on the surface of biopaper.
- *Organ printing*—bioprinting process used for fabrication of 3D tissue by depositing bio-ink on the surface of biopaper.

CELL AND ORGAN PRINTING

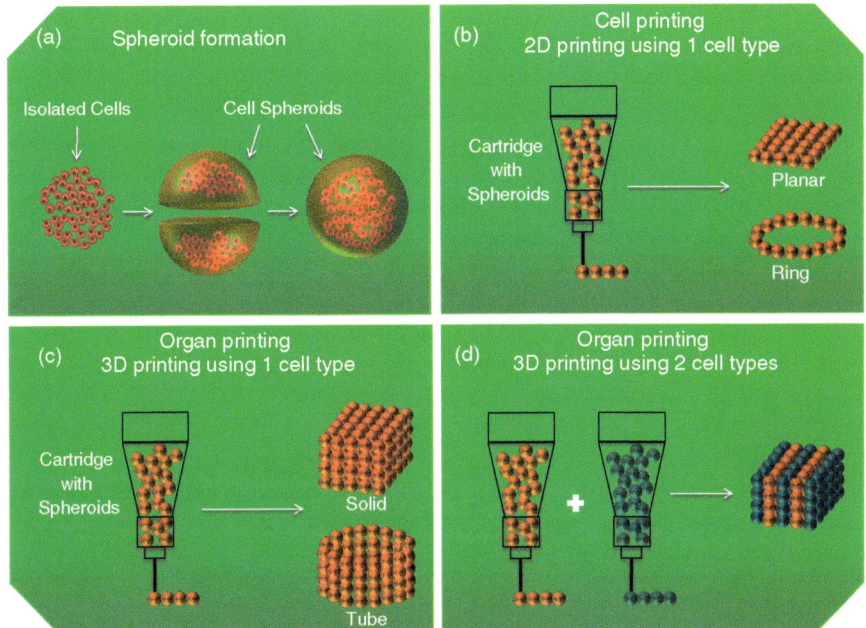

Figure 4.5 Cell and Organ Printing—(a) Spheroid Formation—Isolated cells are secured within a hydrogel to support formation of cell spheroids, which are building blocks for cell and organ printing technology. (b) Cell Printing–2D Printing using 1 cell type—Spheroids can be transferred from a nozzle to the surface to form several patterns, including planar, ring, and linear patterns. (c) Organ Printing— 3D Cell Printing using 1 Cell Type—A second layer of spheroids can be printed on the surface of the first spheroid to support formation of 3D artificial tissue. (d) Organ Printing—3D Cell Printing using 2 Cell Types—Complex organs can be fabricated using organ printing technologies with two or more different cell types.

We next provide a general description of the cell/organ printing process (Figure 4.5).

Cells are first suspended in a hydrogel solution (e.g. alginate or collagen) to form spheroids. Spheroids are the fundamental unit for cell and organ printing; isolated cells will be damaged during travel through the fluid nozzle in the bioprinter. The choice of hydrogel, concentration at which it is used, size of an individual spheroid, and cell density are experimental variables that need to be optimized based on the specific application. Once spheroids have been prepared, they are loaded into the bioprinter. A custom culture surface needs to be prepared for bioprinting of spheroids; hydrogels can be used as biopaper to receive and bind the cell-loaded spheroids. Other examples of culture surfaces include binding agents that secure the hydrogel upon contact or gelling agents, which polymerize upon contact and stabilize the hydrogel.

The bioprinter transfers spheroids from a hold reservoir through a fluid nozzle to the biopaper using robotic controls and software algorithms. Computer-aided design (CAD) is used to design 2D cell patterns or 3D artificial tissue, while computer aided machining (CAM) is used to control the bioprinting process. Spatial resolution of up to 1 μm can be achieved. Spheroids are positioned on the biopaper using CAD drawings, and adhesives are used to secure individual spheroids to one another. A single layer of spheroids is patterned and is followed by a second layer and then a third layer and so on. This process results in fabrication of multilayer 3D artificial tissue. Depending on the complexity of the bioprinter, the number of fluid nozzles can be increased to accommodate different cell types. A two-nozzle system is required for dispensing two cell types, and a three-nozzle system is required for three cell types. In addition to dispensing cells, hydrogels can also be transferred from the bioprinter to the biopaper, and complex patterns of cells and hydrogels can be formed. Planar, solid, and tubular structures can be generated using bioprinting technology and can be combined to fabricate vascularized tissue; tubular structures can be generated using endothelial cells, and complex patterns of capillary networks can be generated.

Cell and organ printing has tremendous potential to have a significant impact on tissue engineering. From a conceptual standpoint, the bottom-up strategy for tissue fabrication and spatial resolution offered by bioprinters make this technology very appealing. However, a word of caution is warranted, and it should be appreciated that the current state of the technology is at the feasibility stage. While proof of concept studies have been successful, fabrication of complex multicellular functional tissue has not been demonstrated. In addition, the capital cost of bioprinters is currently high and can limit widespread acceptability of this technology.

4.6 SOLID FREEFORM FABRICATION

Solid freeform fabrication (SFF) refers to a group of technologies that build 3D scaffolds using a layer-by-layer approach (35,36). Collectively, these technologies are known as rapid prototyping (RP) methods. Cell and organ printing also fall within the general classification of RP technologies. SFF methods are bottom-up strategies and are designed to fabricate scaffolds with precise control over 3D architecture and morphology. The detailed architecture of the scaffold is designed using computer-aided design (CAD), and the CAD drawings are used to drive prototyping systems to fabricate the 3D scaffold. Compared to traditional scaffold fabrication technologies, SFF offers tight control over 3D geometry of the scaffold, including porosity, fiber alignment and distribution, mechanical properties, and placement of different molecules relative to each other. While SSF technologies offer considerable advantages over other scaffold fabrication methods, they have not received widespread applicability across tissue engineering laboratories due to the high initial capital investment and the high degree of personnel training required for these methods.

SFF refers to a group of technologies used to fabricate 3D scaffolds including stereo lithography, selective laser sintering, and fusion deposition modeling (35,36). The principle is similar in all of these techniques; a polymer is transferred to a surface using a nozzle, and the x-y position of the polymer is controlled by robotic arms using pre-programmed algorithms. Using this strategy, a complex 3D pattern is created by carefully controlling the position of the polymer. Once the first layer has been completed, a second layer of polymer is placed in the z-direction, and this process continues until a complex 3D scaffold has been fabricated. This strategy of fabricating a complex 3D scaffold by carefully positioning the polymer in the x, y, and z directions provides control over the scaffold's architecture and properties.

In the case of selective layer sintering, a laser beam is used to heat the polymer above its glass transition temperature, after it has been positioned on the platform (37,38). As the polymer cools, it transitions from a rubbery to a glassy state, resulting in bonding of neighboring particles. This process leads to formation of solid structures. In the case of stereolithography, a photocurable polymer is used. This means that polymerization takes place in the presence of ultraviolet light. As in the case for selective layer sintering, the polymer is positioned in the x-y plane, and UV light is used to photopolymerize the polymer. This results in binding of the neighboring molecules and formation of planar structures. As the process continues, a second layer of polymer is added in the z-direction and again treated with UV light. This process continues until a complex 3D scaffold has been fabricated. In the case of fusion deposition modeling, temperatures beyond the glass transition temperatures are used to melt the polymer, and the molten polymer is extruded through a nozzle onto a fabrication platform in the x-y plane. As the temperature is reduced, the melted polymer solidifies, resulting in the fabrication of a planar scaffold. This process is repeated, and second and additional layers are added in the z-plane; temperature is used to control binding of the layers resulting in 3D scaffolds.

4.7 SOFT LITHOGRAPHY AND MICROFLUIDICS

Soft lithography is a microfabrication technology used to engineer microfluidic devices, particularly microvascular networks (6,7). There are many applications of microfluidic devices. Microfluidic channels have been engineered to simulate the flow of red blood cells through capillary networks (39–41), or to study intercellular connectivity between cardiac myocytes (42,43). Microfluidics has also been used to study the effect of fluid shear stress on endothelial cells (44–46), and to regulate differentiation of stem cells in response to predefined flow regimes (47–52). Some other applications include separation of cells or analytical assessment of metabolic activity of cells (53–55); these concepts have even been extended to "lab-on-a-chip" models—in which complex analytical analysis can be performed on microfabricated devices (56–59). Another interesting application has been in the fabrication of micro-organs using microfabricated devices with the heart being one example (60,61). The microfabricated heart consists of a pumping chamber and microchannels to simulate capillaries, all within a closed loop configuration (61).

The strength of microfluidics technology lies in the fabrication of microchannels in complex configurations on a very small scale. These devices can be used to perform complex operations on a very small scale, eliminating the need for the ancillary support apparatus required at larger scales. Microfabrication devices have not been used to support tissue fabrication, and it is difficult to envision a case where microfluidic devices will be used to bioengineer transplantable artificial tissue. Rather, microfluidics devices can be used as a tool to understand effects of the microenvironment on cellular interaction and function; this data can directly feed into the tissue fabrication process.

The process of microfabrication using soft lithography is shown in Figure 4.6.

This is a two-step process: the first step involves fabrication of a reusable stamp using replica molding, and the second step is microcontact printing in which a specific pattern is transferred to a surface using this stamp (62). The first step in the process is fabrication of a stamp using PDMS with a predefined pattern. A negative of the pattern is generated on a silicon wafer using photoresist (a chemical that responds to UV light) and exposure to ultraviolet light. The entire silicon wafer is coated with photoresist to form a uniform layer. The photoresist surface is then selectively exposed to UV light; photoresist regions exposed to UV light break and are washed away in a solvent. This process leaves a patterned surface of photoresist on the silicon wafer. PDMS is then poured on the silicon wafer, allowed to cure, and then removed to form a negative of the pattern created by the photoresist. At this stage, the PDMS surface is known as the "stamp". This stamp is

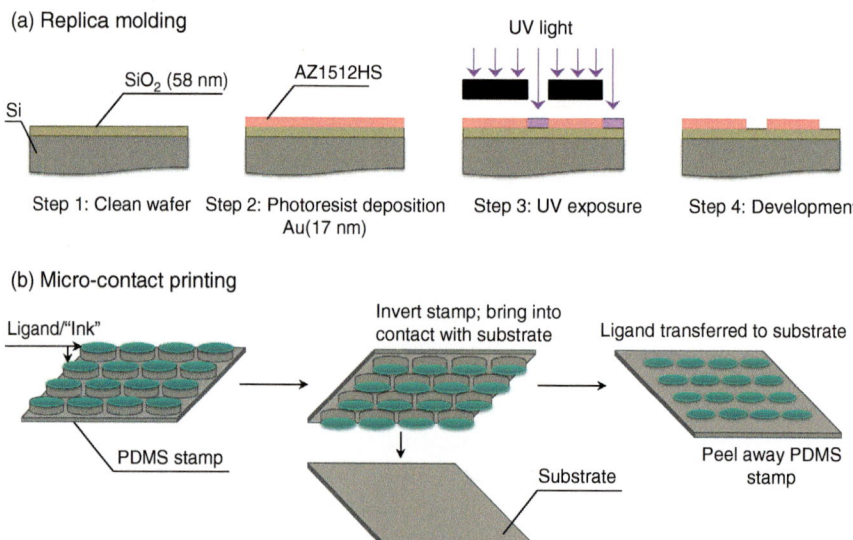

Figure 4.6 Soft Lithography—(a) Replica Molding—A PDMS surface is coated with photoresist and selectively exposed to UV light. Regions of photoresist that are not exposed to UV light are retained on the PDMS surface, resulting in formation of a specific pattern. This process leads to the formation of a stamp. (b) Microcontact Printing—The stamp that was fabricated in the first step of the process is coated with an ink and is transferred to a culture surface by direct contact with the surface.

used to create patterned surfaces over time and can be used over an extended time period.

The PDMS stamp fabricated using soft lithography is used in a process known as microsurface printing (62). The stamp is inked with a specific molecule and transferred to a substrate. This process is carried out by physical contact between the stamp and the substrate and requires a very short time—not more than seconds. Once the molecule has been transferred to the substrate, the void spaces can be filled by a second molecule of interest.

There are several interesting applications of microfabrication technologies in tissue engineering, particularly in the area of "organ-on-a-chip". The concept has been to recreate organ level functionality within a microfluidics device. These systems are not geared toward fabrication of transplantable tissue, but rather as model systems to understand organ-level function within an isolated microfluidics system. An example has been the fabrication of "organ-on-a-chip" models for lungs (63). During normal mammalian function, contraction of the diaphragm causes a reduction in intrapleural pressure and expansion of the alveoli, leading to influx of air in contact with the epithelium layer. Oxygen transfer takes place from the epithelium layer of the alveoli and to the endothelium layer of adjacent capillaries (63). This process has been recreated by fabricating microchannels, layering epithelium cells on one side and endothelial cells on the other, and fabricating side chambers maintained under a vacuum. As the air pressure increases in the microchamber from the side with epithelial cells (which represent alveoli function), the microchamber expands, resulting in flow of oxygen to the endothelium side, which represents capillary function (63).

Another interesting application of microfluidics technology has been in the fabrication of artificial heart muscle (60). In this case, a patterned surface was created by fabrication of microchannels on a substrate, and cardiac myocytes were cultured within these channels (60). The confined space of the microchannels supported intercellular connectivity between adjacent cardiac myocytes, while geometry of the channels resulted in alignment of newly formed heart muscle tissue. Microfluidic channels have also been used to fabricate microvascular networks by generating organized arrays of microchannels connected by fluid inlet and outlet channels, thereby creating a closed-loop circulatory system (64).

4.8 CELL PATTERNING

The process by which the spatial placement of cells is controlled to create an organized pattern of cell monolayers or 3D tissue is known as cell patterning (28–32). The rationale for development of cell patterning technologies is to control organization of different cell types relative to one and other and to the extracellular matrix, just as is found in mammalian tissue. Fabrication of 3D artificial tissue requires a high level of spatial resolution in order to bioengineer tissue that is similar in form and function to mammalian tissue. In an earlier section, we looked at some examples of technologies used for cell patterning, including cell/organ

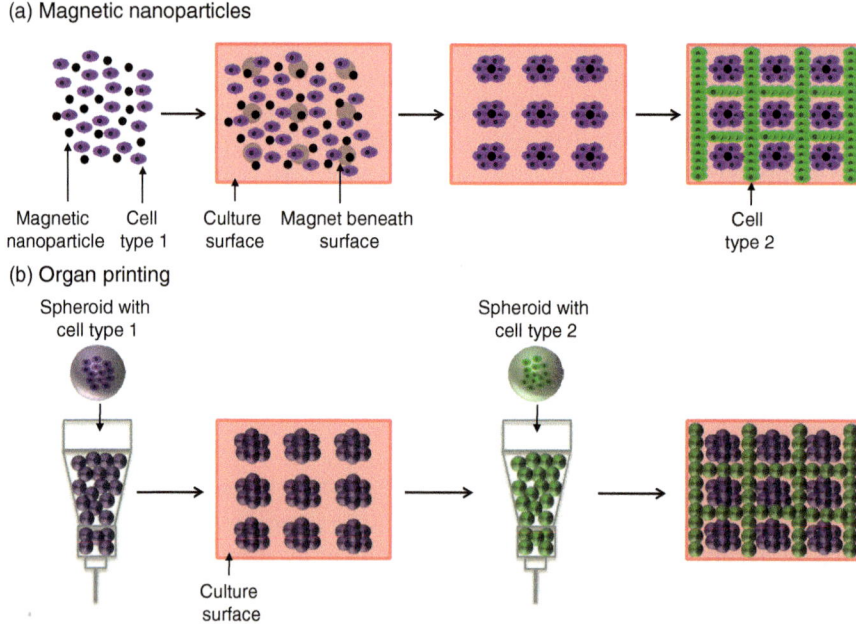

Figure 4.7 Cell Patterning. (a) Nanoparticles for Cell Patterning—Isolated cells are mixed with magnetically charged nanoparticles and plated on the culture substrate. A magnetic field is applied beneath the culture surface, and cells organize around this magnetic field. Cells form a pattern that correlates to the magnetic field. A second cell type is added directly to the culture surface without any nanoparticles, and it fills void spaces. (b) Organ Printing for Cell Patterning—Spheroids are formed using two different cell types and are loaded in cartridges. Spheroids are transferred on a culture surface, and their cell pattern is determined based on a predefined algorithm.

printing, soft lithography, and microfluidic channels. Here, we look at examples of these and other technologies as they are applied for cell patterning (Figures 4.7, 4.8 and 4.9).

One interesting approach to cell patterning has been the use of nanoparticles in conjunction with controlled magnetic fields (65). Nanoparticles are mixed with a cell suspension and plated on a culture surface. Application of a controlled magnetic field is used to guide placement of these cells at specific locations on the culture surface. Using this technology, isolated cells can be patterned into many different configurations, including small cell islands, during monolayer culture. In addition, nanoparticles can be used to preferentially regulate spatial placement of two different cell types. For example, the first cell type can be mixed with nanoparticles and patterned to any configuration using magnetic fields. The void spaces on the culture surface can be later filled by plating the second cell type without any nanoparticles. The second cell type will attach to the culture surface in the void spaces, leading to a controlled pattern of two different cell types.

CELL PATTERNING

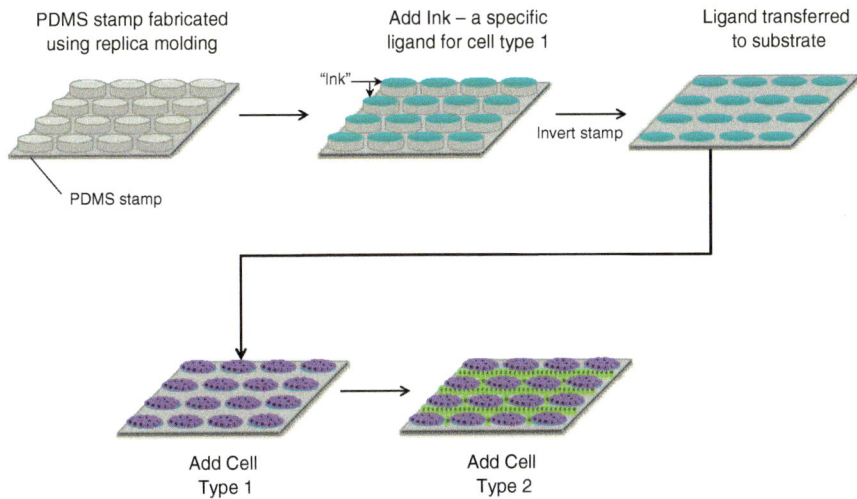

Figure 4.8 Soft Lithography for Cell Patterning—A PDMS stamp is fabricated using replica molding and coated with an adhesion protein. Adhesion protein is transferred to the culture surface by direct contact between the PDMS stamp and the culture surface. Cells are plated on the culture surface and attach to form a specific cell pattern that correlates with the pattern of the adhesion protein. A second cell type is plated on the culture surface and fills void spaces.

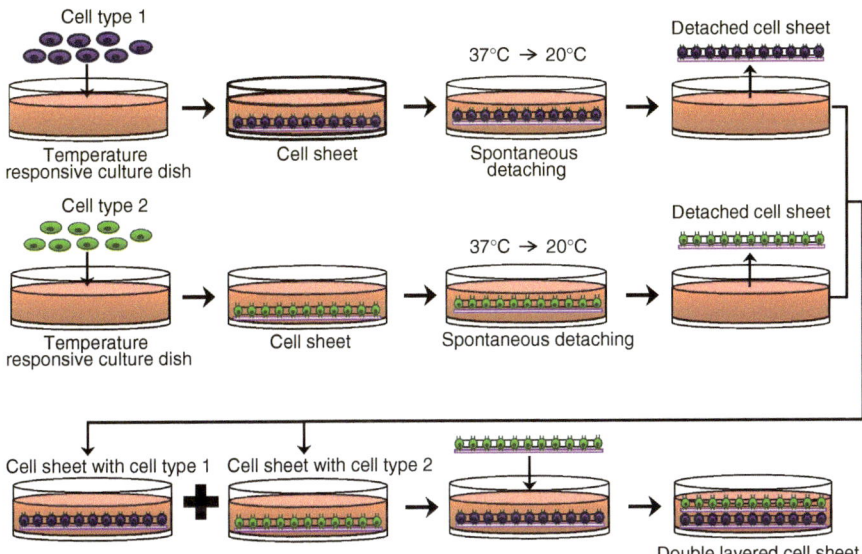

Figure 4.9 Cell Sheet Engineering for Cell Patterning—Cell sheets are fabricated using two different cell types by plating the cells on a culture surface that has been coated with temperature-responsive molecules. Individual sheets are fabricated using different cell types that can be layered on top of one another to form 3D tissue with a specific pattern.

Soft lithography has been extensively used for cell patterning (66). The first step in the process involves fabrication of a stamp using replica molding; this stamp can be designed in a specific configuration and can be used as a template to transfer cells or adhesion proteins on a culture surface. Once fabricated, cells or adhesion proteins can be coated on the stamp and directly transferred to the culture surface, thereby creating a pattern of cells/proteins that replicate the pattern of the stamp. Alternatively, an adhesion protein can be transferred to a culture surface and used as a binding site for cells. In both cases, the technology can be adapted to pattern two different cell types: the first cell type is coated on the culture surface using the stamp, and the second cell type is plated to fill the void spaces.

Using cell/organ printing, isolated cells are suspended within a spheroid and used as building blocks for cell/tissue patterning (67). Spheroids are transferred through a robotically controlled nozzle that regulates spatial placement. Using this technology, a single cell type can be suspended in a spheroid and patterned on the culture surface by controlling spatial placement of spheroids. This process can be adapted for two different cell types—a second nozzle can be used to deliver spheroids with a second cell type.

Cell sheet engineering can be used for cell patterning to fabricate multilayer tissue (68). As may be recalled, cell sheet engineering is based on detachment of a cohesive cell monolayer from an underlying substrate coated with a temperature-sensitive molecule. This process leads to formation of cell sheets that can be stacked together to form multilayer tissue. This process can be adapted to fabricate specific patterns of 3D tissue by fabricating a cell sheet using a single cell type and then layering it on the surface of the cell sheet fabricated with a second cell type. Using this method, 3D tissue can be fabricated using multiple cell sheets with different cell types.

4.9 IDEALIZED SYSTEM TO SUPPORT TISSUE FABRICATION

An idealized system to support fabrication of 3D artificial tissue is shown in Figure 4.10.

The system consists of modules for different cell types and for scaffolds that allow control over placement of scaffolds and cells in order to support 3D tissue formation. In this system, the scaffold provides temporary support during the initial stages of artificial tissue fabrication, degrades over time in a controlled manner, and is replaced by extracellular matrix produced by cells. This process leads to formation of scaffold-free artificial tissue, which has a physiological composition of proteins and other ECM compounds.

As shown in Figure 4.10, the idealized system is designed to support fabrication of artificial heart muscle. Isolated cardiac myocytes are suspended within spheroids and transferred to a nozzle; these spheroids are patterned on a culture substrate to support linear alignment of cells which supports cell-cell interaction. This process is carried out until a cohesive monolayer of cardiac myocytes has been formed, followed by layering of a sheet of polymeric molecules that serve to

SUMMARY 149

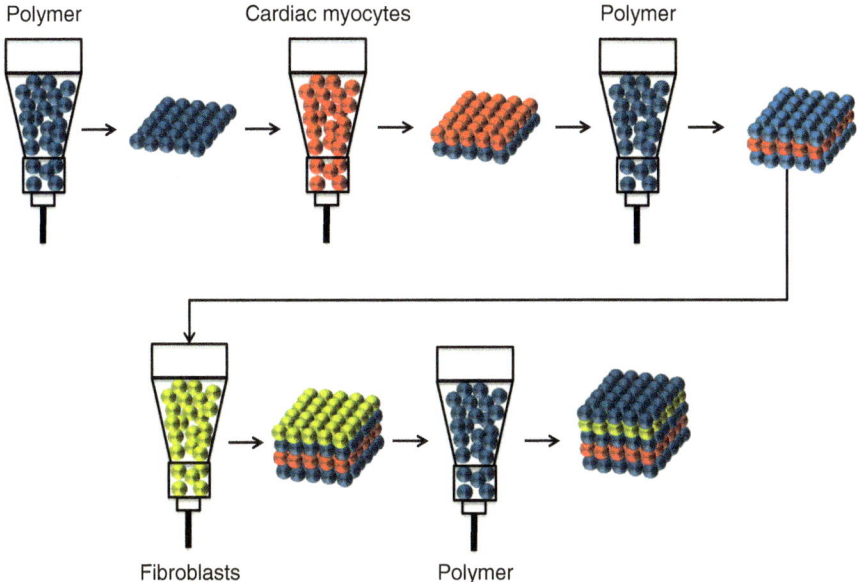

Figure 4.10 Idealized System for Heart Muscle Fabrication—Organ printing is used to fabricate a multilayer system of cardiac myocytes, fibroblasts and a polymer. Over time, the polymer degrades and is replaced by ECM produced by fibroblasts. This process leads to the fabrication of scaffold-free artificial heart muscle fabricated by organ printing.

provide structural support during tissue fabrication. After fabrication of these two layers (cardiac myocytes and polymeric molecules), the third layer is generated by layering fibroblast cells on top of the polymer sheet. The fourth layer is again fabricated using polymers to stabilize the multilayer heart muscle. The next step is formation of a capillary network by controlled patterning of endothelial cells on top of the polymer layer. This is followed by patterning another polymer layer, followed by fibroblast cells, polymers, and cardiac myocytes, leading to fabrication of complex multi-cellular 3D tissue. The polymer is designed to serve as a temporary scaffold and is replaced by ECM components produced by fibroblasts, thereby leading to 3D scaffold-free heart muscle tissue.

SUMMARY

Current State of the Art—Over the years, there have been several strategies that have been developed to support fabrication of 3D artificial tissue. In this chapter, we have looked at several such strategies, including scaffold-free methods, scaffold-based methods, cell and organ printing, solid freeform fabrication, soft lithography, and cell patterning. These techniques provide researchers with a vast tool kit to use when fabricating 3D artificial tissue. The objective for all of these strategies is the same: to replicate the complex 3D architecture of mammalian tissue.

Thoughts for the Future—The current generation of tissue fabrication technologies has been developed for the development of one aspect of artificial tissue. For example, cell and organ printing techniques are designed to support the fabrication of 3D tissue using cells, while solid freeform fabrication strategies are focused on the scaffold. In addition, cell patterning techniques are focused on the spatial alignment of cells, while soft lithography and microfluidics are geared toward the fabrication of microvasculature and more toward understanding cellular behavior. As tissue fabrication technologies are being developed, there is a growing need for a new generation of strategies that can incorporate multidimensional mammalian tissue. A new generation of tissue fabrication technologies needs to be developed that incorporates cellular and extracellular components, including vasculature and innervation. These tissue fabrication technologies will form a critical component of the tissue fabrication process, and these technologies are essential for the fabrication of 3D artificial tissue that mimics the complex architecture of mammalian tissue.

PRACTICE QUESTIONS

1. This chapter is about tissue fabrication technology. Provide a general discussion about why tissue fabrication is important for the development of artificial tissue and organs. What do you think is involved in the fabrication of artificial tissue and organs?

2. In this chapter we described the fabrication of artificial tissue using scaffold-free technology. We discussed self-organization strategies and cell sheet engineering. Describe what the term "scaffold-free technology" means. What are the relative advantages and disadvantages of using scaffold-free technology for the fabrication of artificial tissue? Describe self-organization strategies and cell sheet engineering.

3. Describe the process of self-organization for the fabrication of 3D artificial tissue. What variables affect tissue formation and function?

4. Describe the process of cell sheet engineering for the fabrication of 3D artificial tissue. What variables affect tissue formation and function?

5. We studied scaffold-based tissue engineering and looked at acellular scaffolds, polymers, and hydrogels. Discuss each of these as they relate to biomaterial synthesis and tissue fabrication. What are the relative advantages and disadvantages of acellular scaffolds, polymers, and hydrogels as applied to tissue engineering and tissue fabrication?

6. Scaffold-based and scaffold-free methods have been used extensively in tissue engineering. Describe the relative advantages and disadvantages of these two strategies.

7. Pick any tissue or organ fabrication application and develop a strategy for bioengineering your selected tissue or organ using self-organization strategies. What specific method will you choose and why?
8. Describe the process of cell and organ printing. Explain the process and describe how complex 3D structures can be formed using multiple cell types. What are some of the scientific and technological hurdles that need to be overcome in order to improve this technology?
9. Describe how you would use organ printing to fabricate functional 3D artificial heart muscle with embedded vasculature.
10. Explain how soft lithography is used to fabricate microfluidics channels.
11. Why are microfluidic channels important in tissue engineering? How can microfluidics channels be used to support tissue fabrication?
12. We looked at several models for tissue and "organ-on-a-chip": heart muscle, lungs, and blood vessels. Develop a model for a total bioartificial heart- on-a- chip.
13. What is cell patterning and why is it important for tissue engineering? During the course of this chapter, we looked at several strategies for cell patterning. Pick any one and describe how your chosen strategies can be used for cell patterning.
14. During the course of this chapter, we looked at several tissue fabrication technologies. What are some scientific and technological challenges associated with tissue fabrication technologies? What can be done to overcome these scientific and technological challenges?
15. We have studied several tissue fabrication technologies used in tissue engineering. However, there are many different fabrication technologies that are used in different fields, particularly in engineering. Based on your experiences with other fabrication technologies, develop a new method that can be used to support the fabrication of 3D artificial tissue.

REFERENCES

1. Baar K, Birla R, Boluyt MO, Borschel GH, Arruda EM, Dennis RG. Self-organization of rat cardiac cells into contractile 3-D cardiac tissue. FASEB J. 2005 Feb;19(2):275–7.
2. Khait L, Hodonsky CJ, Birla RK. Variable optimization for the formation of three-dimensional self-organized heart muscle. In Vitro Cell Dev. Biol. Anim 2009 Dec;45(10):592–601.
3. Blan NR, Birla RK. Design and fabrication of heart muscle using scaffold-based tissue engineering. J. Biomed. Mater. Res. A 2008 Jul;86(1):195–208.
4. Huang Y, He K, Wang X. Rapid prototyping of a hybrid hierarchical polyurethane-cell/hydrogel construct for regenerative medicine. Mater. Sci. Eng C Mater. Biol. Appl. 2013 Aug 1;33(6):3220–9.

5. Schrank E, Hitch L, Wallace K, Moore R, Stanhope S. Assessment of a Virtual Functional Prototyping Process for the Rapid Manufacture of Passive-Dynamic Ankle-Foot Orthoses. J. Biomech. Eng. 2013 Jun 1.
6. Wang Y, Balowski J, Phillips C, Phillips R, Sims CE, Allbritton NL. Benchtop micromolding of polystyrene by soft lithography. Lab Chip. 2011 Sep 21;11(18):3089–97. PMCID:PMC3454527.
7. Pan J, Yung CS, Common JE, Amini S, Miserez A, Birgitte LE, Kang L. Fabrication of a 3D hair follicle-like hydrogel by soft lithography. J. Biomed. Mater. Res. A 2013 Mar 30.
8. Silpe JE, Nunes JK, Poortinga AT, Stone HA. Generation of antibubbles from core-shell double emulsion templates produced using microfluidics. Langmuir 2013 Jun 12.
9. Zhao CX. Multiphase flow microfluidics for the production of single or multiple emulsions for drug delivery. Adv. Drug Deliv. Rev. 2013 Jun 12.
10. Baker BM, Trappmann B, Stapleton SC, Toro E, Chen CS. Microfluidics embedded within extracellular matrix to define vascular architectures and pattern diffusive gradients. Lab Chip. 2013 Jun 20.
11. Frampton JP, White JB, Abraham AT, Takayama S. Cell co-culture patterning using aqueous two-phase systems. J. Vis. Exp. 2013;(73).
12. Cosson S, Allazetta S, Lutolf MP. Patterning of cell-instructive hydrogels by hydrodynamic flow focusing. Lab Chip. 2013 Jun 7;13(11):2099–105.
13. Whatley BR, Li X, Zhang N, Wen X. Magnetic-directed patterning of cell spheroids. J. Biomed. Mater. Res. A 2013 May 13.
14. Ho CT, Lin RZ, Chen RJ, Chin CK, Gong SE, Chang HY, Peng HL, Hsu L, Yew TR, Chang SF, et al. Liver-cell patterning Lab Chip: mimicking the morphology of liver lobule tissue. Lab Chip. 2013 Jun 6.
15. Kosnik PE, Faulkner JA, Dennis RG. Functional development of engineered skeletal muscle from adult and neonatal rats. Tissue Eng. 2001 Oct;7(5):573–84.
16. Shimizu T, Yamato M, Kikuchi A, Okano T. Cell sheet engineering for myocardial tissue reconstruction. Biomaterials 2003 Jun;24(13):2309–16.
17. Yang J, Yamato M, Kohno C, Nishimoto A, Sekine H, Fukai F, Okano T. Cell sheet engineering: recreating tissues without biodegradable scaffolds. Biomater. 2005 Nov;26(33):6415–22.
18. Yang J, Yamato M, Shimizu T, Sekine H, Ohashi K, Kanzaki M, Ohki T, Nishida K, Okano T. Reconstruction of functional tissues with cell sheet engineering. Biomater. 2007 Dec;28(34):5033–43.
19. Lee JI, Nishimura R, Sakai H, Sasaki N, Kenmochi T. A newly developed immunoisolated bioartificial pancreas with cell sheet engineering. Cell Transplant 2008;17(1–2):51–9.
20. Wu KH, Mo XM, Liu YL. Cell sheet engineering for the injured heart. Med. Hypotheses 2008 Nov;71(5):700–2.
21. Nagase K, Kobayashi J, Okano T. Temperature-responsive intelligent interfaces for biomolecular separation and cell sheet engineering. J. R Soc. Interface 2009 Jun 6; 6 Suppl 3:S293-S309. PMCID:PMC2690096.
22. Elloumi-Hannachi I, Yamato M, Okano T. Cell sheet engineering: a unique nanotechnology for scaffold-free tissue reconstruction with clinical applications in regenerative medicine. J. Intern. Med. 2010 Jan;267(1):54–70.

23. Hutmacher DW, Sittinger M, Risbud MV. Scaffold-based tissue engineering: rationale for computer-aided design and solid free-form fabrication systems. Trends Biotechnol. 2004 Jul;22(7):354–62.
24. Hutmacher DW, Cool S. Concepts of scaffold-based tissue engineering–the rationale to use solid free-form fabrication techniques. J. Cell Mol. Med. 2007 Jul;11(4):654–69.
25. Sundelacruz S, Kaplan DL. Stem cell- and scaffold-based tissue engineering approaches to osteochondral regenerative medicine. Semin. Cell Dev. Biol. 2009 Aug;20(6):646–55. PMCID:PMC2737137.
26. Varghese D, Deshpande M, Xu T, Kesari P, Ohri S, Boland T. Advances in tissue engineering: cell printing. J. Thorac. Cardiovasc. Surg. 2005 Feb;129(2):470–2.
27. Ilkhanizadeh S, Teixeira AI, Hermanson O. Inkjet printing of macromolecules on hydrogels to steer neural stem cell differentiation. Biomater. 2007 Sep;28(27):3936–43.
28. Liberski A, Zhang R, Bradley M. Laser printing mediated cell patterning. Chem. Commun. (Camb.) 2009 Dec 28;(48):7509–11.
29. Koch L, Deiwick A, Schlie S, Michael S, Gruene M, Coger V, Zychlinski D, Schambach A, Reimers K, Vogt PM, et al. Skin tissue generation by laser cell printing. Biotechnol. Bioeng. 2012 Jul;109(7):1855–63.
30. Yamaguchi S, Ueno A, Akiyama Y, Morishima K. Cell patterning through inkjet printing of one cell per droplet. Biofabrication. 2012; Dec 4(4):045005.
31. Matsusaki M, Sakaue K, Kadowaki K, Akashi M. Three-dimensional human tissue chips fabricated by rapid and automatic inkjet cell printing. Adv. Healthc. Mater. 2013 Apr;2(4):534–9.
32. Koch L, Gruene M, Unger C, Chichkov B. Laser assisted cell printing. Curr. Pharm. Biotechnol 2013 Jan;14(1):91–7.
33. Ozbolat IT, Yu Y. Bioprinting toward organ fabrication: challenges and future trends. IEEE Trans. Biomed. Eng 2013 Mar;60(3):691–9.
34. Guillemot F, Mironov V, Nakamura M. Bioprinting is coming of age: Report from the International Conference on Bioprinting and Biofabrication in Bordeaux (3B'09). Biofabrication. 2010 Mar;2(1):010201.
35. Leong KF, Cheah CM, Chua CK. Solid freeform fabrication of three-dimensional scaffolds for engineering replacement tissues and organs. Biomaterials 2003 Jun;24(13):2363–78.
36. Suri S, Han LH, Zhang W, Singh A, Chen S, Schmidt CE. Solid freeform fabrication of designer scaffolds of hyaluronic acid for nerve tissue engineering. Biomed. Microdevices 2011 Dec;13(6):983–93.
37. Gu YW, Khor KA, Cheang P. Bone-like apatite layer formation on hydroxyapatite prepared by spark plasma sintering (SPS). Biomater. 2004 Aug;25(18):4127–34.
38. Shishkovskiy IV, Morozov YG, Kuznetsov MV, Parkin IP. Electromotive force measurements in the combustion wave front during layer-by-layer surface laser sintering of exothermic powder compositions. Phys. Chem. Chem. Phys. 2009 May 14;11(18):3503–8.
39. Ramser K, Enger J, Goksor M, Hanstorp D, Logg K, Kall M. A microfluidic system enabling Raman measurements of the oxygenation cycle in single optically trapped red blood cells. Lab Chip. 2005 Apr;5(4):431–6.
40. Zhao C, Cheng X. Microfluidic separation of viruses from blood cells based on intrinsic transport processes. Biomicrofluidics. 2011 Sep;5(3):32004–3200410. PMCID:PMC3194787.

41. Kwan JM, Guo Q, Kyluik-Price DL, Ma H, Scott MD. Microfluidic analysis of cellular deformability of normal and oxidatively-damaged red blood cells. Am. J. Hematol. 2013 May 15.
42. Klauke N, Smith GL, Cooper JM. Microfluidic partitioning of the extracellular space around single cardiac myocytes. Anal. Chem. 2007 Feb 1;79(3):1205–12.
43. Klauke N, Smith G, Cooper JM. Microfluidic systems to examine intercellular coupling of pairs of cardiac myocytes. Lab Chip. 2007 Jun;7(6):731–9.
44. Kaji H, Yokoi T, Kawashima T, Nishizawa M. Directing the flow of medium in controlled cocultures of HeLa cells and human umbilical vein endothelial cells with a microfluidic device. Lab Chip. 2010 Sep 21;10(18):2374–9.
45. Chen KC, Lee TP, Pan YC, Chiang CL, Chen CL, Yang YH, Chiang BL, Lee H, Wo AM. Detection of circulating endothelial cells via a microfluidic disk. Clin. Chem. 2011 Apr;57(4):586–92.
46. Liu MC, Shih HC, Wu JG, Weng TW, Wu CY, Lu JC, Tung YC. Electrofluidic pressure sensor embedded microfluidic device: a study of endothelial cells under hydrostatic pressure and shear stress combinations. Lab Chip. 2013 May 7;13(9):1743–53.
47. Wu HW, Lin XZ, Hwang SM, Lee GB. The culture and differentiation of amniotic stem cells using a microfluidic system. Biomed. Microdevices. 2009 Aug;11(4):869–81.
48. Wan CR, Chung S, Kamm RD. Differentiation of embryonic stem cells into cardiomyocytes in a compliant microfluidic system. Ann. Biomed. Eng 2011 Jun;39(6):1840–7.
49. Schirhagl R, Fuereder I, Hall EW, Medeiros BC, Zare RN. Microfluidic purification and analysis of hematopoietic stem cells from bone marrow. Lab Chip 2011 Sep 21;11(18):3130–5.
50. Blagovic K, Kim LY, Voldman J. Microfluidic perfusion for regulating diffusible signaling in stem cells. PLoS One 2011;6(8):e22892. PMCID:PMC3150375.
51. Wadhawan N, Kalkat H, Natarajan K, Ma X, Gajjeraman S, Nandagopal S, Hao N, Li J, Zhang M, Deng J, et al. Growth and positioning of adipose-derived stem cells in microfluidic devices. Lab Chip. 2012 Nov 21;12(22):4829–34.
52. Lesher-Perez SC, Frampton JP, Takayama S. Microfluidic systems: a new toolbox for pluripotent stem cells. Biotechnol. J. 2013 Feb;8(2):180–91.
53. Kraly JR, Holcomb RE, Guan Q, Henry CS. Review: Microfluidic applications in metabolomics and metabolic profiling. Anal. Chim. Acta 2009 Oct 19;653(1):23–35. PMCID:PMC2791705.
54. Cheng W, Klauke N, Smith G, Cooper JM. Microfluidic cell arrays for metabolic monitoring of stimulated cardiomyocytes. Electrophoresis 2010 Apr;31(8):1405–13.
55. Legendre A, Baudoin R, Alberto G, Paullier P, Naudot M, Bricks T, Brocheton J, Jacques S, Cotton J, Leclerc E. Metabolic characterization of primary rat hepatocytes cultivated in parallel microfluidic biochips. J. Pharm. Sci. 2013 Feb 19.
56. Kim KH. Lab-on-a-chip for Urology. Int. Neurourol. J. 2013 Mar;17(1):1. PMCID:PMC3627991.
57. Nguyen NT, Shaegh SA, Kashaninejad N, Phan DT. Design, fabrication and characterization of drug delivery systems based on lab-on-a-chip technology. Adv. Drug Deliv. Rev. 2013 May 29.
58. Krishna KS, Li Y, Li S, Kumar CS. Lab-on-a-chip synthesis of inorganic nanomaterials and quantum dots for biomedical applications. Adv. Drug Deliv. Rev. 2013 May 29.

REFERENCES

59. Yu L, Ng SR, Xu Y, Dong H, Wang YJ, Li CM. Advances of lab-on-a-chip in isolation, detection and post-processing of circulating tumour cells. Lab Chip. 2013 Jun 17.
60. Grosberg A, Alford PW, McCain ML, Parker KK. Ensembles of engineered cardiac tissues for physiological and pharmacological study: heart on a chip. Lab Chip. 2011 Dec 21;11(24):4165–73.
61. Tanaka Y, Sato K, Shimizu T, Yamato M, Okano T, Kitamori T. A micro-spherical heart pump powered by cultured cardiomyocytes. Lab Chip. 2007 Feb;7(2):207–12.
62. Whitesides GM, Ostuni E, Takayama S, Jiang X, Ingber DE. Soft lithography in biology and biochemistry. Annu. Rev. Biomed. Eng. 2001;3:335–73.
63. Huh D, Matthews BD, Mammoto A, Montoya-Zavala M, Hsin HY, Ingber DE. Reconstituting organ-level lung functions on a chip. Science 2010 Jun 25;328(5986):1662–8.
64. Franco C, Gerhardt H. Tissue engineering: Blood vessels on a chip. Nature 2012 Aug 23;488(7412):465–6.
65. Tseng P, Di CD, Judy JW. Rapid and dynamic intracellular patterning of cell-internalized magnetic fluorescent nanoparticles. Nano. Lett. 2009 Aug;9(8):3053–9.
66. Kane RS, Takayama S, Ostuni E, Ingber DE, Whitesides GM. Patterning proteins and cells using soft lithography. Biomaterials 1999 Dec;20(23–24):2363–76.
67. Gaebel R, Ma N, Liu J, Guan J, Koch L, Klopsch C, Gruene M, Toelk A, Wang W, Mark P, et al. Patterning human stem cells and endothelial cells with laser printing for cardiac regeneration. Biomaterials 2011 Dec;32(35):9218–30.
68. Hannachi IE, Yamato M, Okano T. Cell sheet technology and cell patterning for biofabrication. Biofabrication. 2009 Jun;1(2):022002.

5

VASCULARIZATION OF ARTIFICIAL TISSUE

Learning Objectives

After completing this chapter, students should be able to:

1. Explain the need for vascularization during tissue fabrication and development.
2. Describe and discuss seminal publications in angiogenesis research by Dr. Judah Folkman.
3. Define vasculogenesis, angiogenesis, and arteriogenesis and understand differences between these three processes.
4. Identify triggers that drive vasculogenesis, angiogenesis and arteriogenesis.
5. Describe molecular mechanisms of vasculogenesis, angiogenesis, and arteriogenesis.
6. Discuss the process of therapeutic angiogenesis as it relates to regenerative medicine.
7. Explain why vascularization is important during tissue fabrication.
8. Describe strategies that have been used to incorporate vasculature within artificial tissue.
9. Explain differences between biologically replicated, biologically mediated, and biologically inspired strategies for vascularization of 3D artificial tissue.

Introduction to Tissue Engineering: Applications and Challenges, First Edition. Ravi Birla.
© 2014 The Institute of Electrical and Electronics Engineers, Inc. Published 2014 by John Wiley & Sons, Inc.

INTRODUCTION

10. Give specific examples of vascularization strategies based on biologically replicated, biologically mediated, and biologically inspired models.
11. Design a process to engineer vasculature within artificial tissue.

CHAPTER OVERVIEW

We begin this chapter with a discussion centered on the need for neovascularization during 3D tissue fabrication. We next move on to describe seminal work in the field of angiogenesis by Dr. Judah Folkman and how this work relates to tissue engineering. During human development and growth, there are three distinct mechanisms that give rise to blood vessels: vasculogenesis, angiogenesis, and arteriogenesis. During the course of this chapter, we describe these three processes and the molecular mechanisms that lead to blood vessel formation. We also describe how these three processes can be used to support vascularization of 3D artificial tissue. Next, we look at one specific application of angiogenesis, therapeutic angiogenesis, which is a clinical strategy designed to support revascularization of diseased or injured avascular tissue. Using this framework of blood vessel formation during human development and growth, we discuss ways in which this information can be relayed back to the tissue fabrication process. We next compare *in vitro* models of vascularization with *in vivo* models and provide a discussion on the relative advantages and disadvantages of each strategy. After this, we present an idealized case for inducing vascularization within 3D artificial tissue. We next provide a framework for vascularization strategies in tissue engineering and present a flow chart to aid in the decision making process. Next, we describe vascularization strategies in tissue engineering, which fall into one of three categories: biologically replicated, biologically mediated, and biologically inspired. For each of the three vascularization strategies, we describe the underlying principles and provide examples from the literature. Finally, we conclude this chapter with a discussion of scientific and technological challenges that need to be overcome to develop efficient vascularization strategies to support the fabrication of 3D artificial tissue.

5.1 INTRODUCTION

In the human body, the vasculature serves as a distribution network to deliver oxygen, glucose, and other nutrients to various tissue and cells and remove waste products like carbon dioxide and lactic acid. The smallest blood vessels are known as capillaries and consist of endothelial cells supported by a basement membrane. These capillaries are found in close proximity to cells, with no cells being more than 100–200 µm away from a capillary. Cells can be supported via diffusion as a delivery mechanism for oxygen and other nutrients over short distances (~100–200 µm); vascularization is required beyond this critical range. Capillaries are a part of the circulatory system, which consists of arterioles feeding into the capillaries and venules receiving throughput from capillaries. The capillary network couples with larger arteries and veins that then feed into the heart and complete the circulatory system.

Cells need to be in close proximity to a vascular network to support viability, and this scenario applies to tissue engineering as well. The process of tissue fabrication begins with a few cells cultured in an appropriate 3D environment. During culture, these cells proliferate and support tissue growth and maturation, providing impetus for tissue development and growth. During early stages of tissue fabrication, cell viability can be supported by diffusion (Figure 5.1). However, as the tissue matures

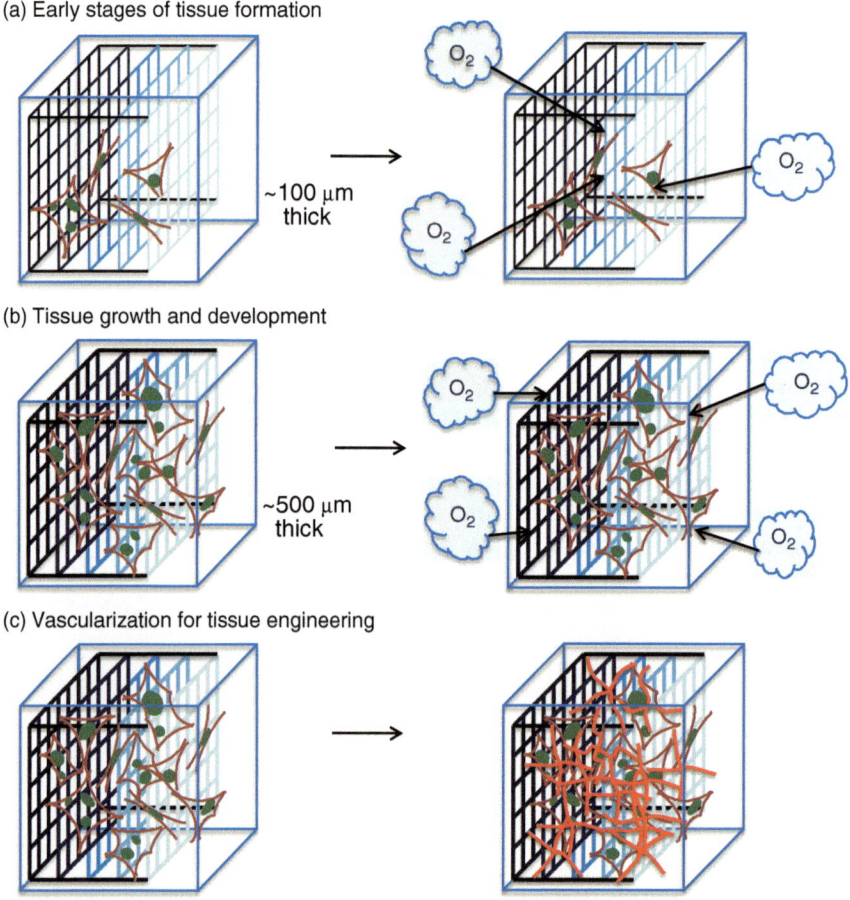

Figure 5.1 Vascularization for Tissue Engineering—(a) Early Stages of Tissue Development—Diffusion of oxygen is sufficient to support metabolic activity of cells. **(b) Tissue Growth and Development**—As the thickness of the artificial tissue increases, diffusion of oxygen is not able to support the metabolic activity of cells, particularly those within the inner core of the artificial tissue. **(c) Vascularization for Tissue Engineering**—As the thickness of the artificial tissue increases, vascularization is important to support the metabolic activity of cells within 3D artificial tissue.

and increases in thickness, diffusion is no longer sufficient to support the metabolic requirements of the cells. This is particularly the case when the thickness of the tissue-engineered constructs increases beyond 200 µm; at this stage, it is critical to engineer a vasculature within 3D artificial tissue.

5.2 SEMINAL PUBLICATIONS IN ANGIOGENESIS RESEARCH

There were a series of seminal studies in the early 1970s by Dr. Judah Folkman at Harvard Medical School that have paved the field of angiogenesis (1–11). In a 1971 publication, Dr. Folkman isolated and characterized tumor angiogenesis factor (TAF) and showed that TAF could support endothelial cell proliferation and formation of new capillaries (8). Based on his findings, Dr. Folkman proposed the following working hypothesis (8):

> "*It appears that most solid tumors, whether they originate from a cell transformed by virus or carcinogen, or whether they begin as a metastatic implant, must exist early as a small population of cells dependent upon nutrients which diffuse from the extravascular space. Such a pinpoint colony eventually expands to a size where simple diffusion of nutrients (and wastes) is insufficient. New capillaries are elicited and the tumor then enters a phase in which nutrients arrive by perfusion. It is possible that TAF is responsible for this final stage. It is tempting to suggest that tumor growth might be arrested at a very small size if the angiogenesis activity of this factor could be blocked.*"

The working hypothesis put forth by Dr. Folkman was that cancerous cells are proliferative and that during initial stages of growth, nutrient delivery takes place by diffusion. However, as the cell mass grows, diffusion is no longer able to support cell viability, and vascularization is important to support viability of the tissue mass. Dr. Folkman went on to suggest that blocking formation of new blood vessels could restrict the proliferation of cancerous cells and could be used for therapeutic purposes; Dr. Folkman demonstrated the feasibility of this approach in a later publication (10).

The publications by Dr. Folkman were seminal in the field of angiogenesis, as they provided a clear link between angiogenesis and cancer formation and proposed blocking angiogenesis as a therapeutic strategy for the treatment of cancer. This initial work has led to significant advancements in our understanding of our molecular mechanisms, angiogenesis, and the development of novel anti-angiogenic strategies. This work also has clear implications for tissue engineering. As cancer cells proliferate, vascularization is required for nutrient delivery to support tissue growth and development. Similarly, during tissue fabrication, cell proliferation is initially supported by nutrient diffusion until reaching a certain critical mass, after which vascularization is needed. In the case of cancerous cells, the objective is to block angiogenesis, whereas in the case of tissue fabrication, the objective is to induce angiogenesis.

5.3 VASCULARIZATION DEFINED

In this section, we provide the reader with important terminology associated with blood vessel formation during embryogenesis, human development, and in response to injury and other stimuli. The development and growth of vasculature is a very tightly regulated process and takes place in response to specific stimuli. In addition, different stimuli like hypoxia and fluid shear stress, initiate different signaling pathways, which in turn result in different mechanisms of blood vessel growth and remodeling. In this section, we provide an overview of these distinct processes for blood vessel development, and in subsequent sections, we provide a discussion of molecular mechanisms involved. We introduce the following concepts in this section: vasculogenesis, angiogenesis, and arteriogenesis (Figure 5.2).

During embryogenesis, fertilization of an oocyte by a sperm leads to zygote formation, which refers to the fertilized oocyte. A single cell then divides progressively to give rise to two cells and then four, eight, and so on until formation of an embryo occurs. During these early stages of embryogenesis, vascularization is absent, and nutrients are delivered to all cells via diffusion. However, as the embryo grows and increases in mass, diffusion is no longer able to supply nutrients to all of its cells, and vascularization becomes important. At critical time points during embryogenesis, mesodermal stem cells differentiate to form angioblasts and then endothelial cells, which in turn support capillary formation.

Vasculogenesis refers to initial events in vascular growth in which endothelial cell precursors (angioblasts) migrate to discrete locations, differentiate in situ, and assemble into solid endothelial cords, later forming a plexus with endothelial tubes (12–16). Vasculogenesis takes place very early during embryonesis and gives rise to the entire circulatory system by expansion and development of the vasculature.

	Vasculogenesis	Angiogenesis	Arteriologenesis
Definition	Blood vessel formation during embryogenesis	Formation of capillaries from existing ones	Maturation and development of blood vessels from existing ones
Starting from	Isolated stem cells	Capillaries	Mature vessels
Ending with	Capillaries	Capillaries	Mature vessels
When	Embryogenesis	Embryogenesis and adult life	Adult life
Trigger	VEGF expression	Hypoxia	Fluid shear stress

Figure 5.2 Vasculogenesis, Angiogenesis, and Arteriogenesis—The terms vasculogenseis, angiogenesis, and arteriogenesis have distinct meanings relating to blood vessel formation. Vasculogenesis refers to the process of blood vessel formation during embryogenesis. Angiogenesis refers to the formation of new blood vessels from existing ones. Arteriogenesis refers to the maturation and development of existing blood vessels.

In a sense, vasculogenesis can be viewed as embryogenesis for the vasculature—a very accurate description of the process. Figure 5.2 shows some characteristics of the process and highlights differences between vasculogenesis, angiogenesis, and arteriogenesis.

The process of vasculogenesis starts with very early stem cells: the mesodermal cells and hemangoblasts. The end results of vasculogenesis are early capillaries that do not have smooth muscle cells or vaso-contraction and vaso-relaxation properties. Since vasculogenesis is associated with embryogenesis and early human development, the triggers that initiate capillary formation are based on changes in the genetic profile of early embryonic stem cells. In particular, vascular endothelial growth factor (VEGF) has been associated with vasculogenesis and is the earliest known marker of vascular lineage. Commitment of mesodermal stem cells to a vascular lineage is defined on the basis of VEGF expression.

Angiogenesis refers to the growth, expansion, and remodeling of primitive blood vessels formed during vasculogenesis to form a mature vascular network (17,18). An important distinction can be made between angiogenesis and vasculogenesis: the former requires pre-existing vessels for formation of new ones, and the latter does not have this requirement. Another important distinction is based on time frame, as vasculogenesis occurs during embryogenesis, and angiogenesis occurs later on during development. In fact, angiogenesis succeeds vasculogenesis.

Sprouting angiogenesis is the process by which an existing blood vessel gives rise to another blood vessel by sprouting endothelial cells toward the angiogenic stimuli. This process increases the blood vessel count in tissue that is otherwise lacking vasculature. A second process by which angiogenesis occurs is known as intussusceptive angiogenesis, in which an existing blood vessel splits to form two blood vessels. This process is faster than sprouting angiogenesis as it does not require proliferation of endothelial cells and depends on reorganization of existing cells. Intussusceptive angiogenesis has only been recently discovered in 1986, and considerably less is known about it compared to sprouting angiogenesis, which has been known and studied for decades.

We have seen that vasculogenesis and angiogenesis are two steps in the process of blood vessel formation and organization. The third stage in this process is arteriogenesis, a process that results in an increase in the diameter of blood vessels, along with other functional modifications. Arteriogenesis is the process by which blood vessels increase in diameter to form muscular arteries and incorporate smooth muscle cells and vaso-contraction and vaso-relaxation properties (19–22). The primary stimulus for arteriogenesis is fluid stress, while the primary stimulus for angiogenesis is ischemia or the lack of oxygen.

5.4 MOLECULAR MECHANISM OF VASCULOGENESIS

During embryogenesis, there is a rapid expansion in cell mass from the single cell oocyte to two-, four-, and eight-cell structures. During this early expansion in cell mass, cell viability can be supported by diffusion of oxygen and other nutrients to all cells. However, as cell masses increase beyond a certain critical limit, diffusion is no longer sufficient to support the metabolic activity of all cells. Neovascularization then becomes important, and it comes as no surprise that the circulatory system is the first system to be developed during embryogenesis, as early as embryonic day (E) 6.5–7. In this section, we present the molecular mechanisms of vasculogenesis in three steps, starting with a description of stem cells involved in the process, followed by morphological changes that take place, and finally, the molecular signals that regulate the vasculogenesis process.

We begin our discussion of vasculogenesis by looking at key stem cell sources (23,24). Early during embryogenesis, three germ layers that give rise to all cells in the human body are formed. These are known as the endoderm, mesoderm, and the ectoderm; they represent the inner, middle, and outer layers of the embryoblast (early cell mass), respectively, The mesodermal stem cells give rise to all cells within the circulatory system and can be considered as initial precursor cells for vasculogenesis. Mesodermal stem cells differentiate to form hemangioblasts, which are early progenitor stem cells that give rise to components of blood and the vascular system during early embryogenesis. Hemangioblasts have been defined as mesodermal progenitor cells committed to the generation of endothelial cells and blood cells, sometimes and perhaps always via a hematogenous endothelium intermediate. Hemangioblasts are multipotent and can differentiate to form endothelial cells or hematopoietic stem cells; hematopoietic stem cells give rise to other blood cells including monocytes, macrophages, neutrophils, basophils eosinophils, erythrocytes, megakaryocytes, platelets, T-cells, and B-cells. Hemangioblasts give rise to angioblasts, which are intermediate stem cells committed to endothelial cell differentiation. The differentiation pathway for the formation of vascular cells progresses from multipotent mesodermal stem cells to multipotent hemangioblasts, which later give rise to angioblasts, cells committed to forming endothelial cells, and multipotent hematopoietic stem cells.

We next explore the steps in capillary formation during vasculogenesis. The formation of blood vessels via vasculogenesis during embryogenesis can be divided into 5 stages (25):

> ***Step One*** Differentiation of Mesodermal cells to Hemangioblasts—Very early during vasculogenesis, mesodermal cells differentiate to form hemangioblasts, early precursor cells that have the potential to be differentiated to all cells in the vasculature. The term hemangioblast is given to the common blood island precursor cell that eventually gives rise to both endothelial and hematopoietic cells.
>
> ***Step Two*** Formation of Blood Islands—Hemangioblasts form clusters known as blood islands, which are the first semblance of capillaries.
>
> ***Step Three*** Differentiation of Hemangioblasts—Once blood islands have formed, hemangioblasts differentiate to form angioblasts, precursor cells that give rise to endothelial cells. Angioblasts are organized on the outer side of blood islands. Hemangioblasts, positioned on the inside of blood islands, give rise to hematopoietic stem cells.

Step Four Blood Vessel Formation—Proliferation, migration, and association of angioblasts gives rise to the very first primitive blood vessels.

Step Five Lumenization—Lumenization is the process by which primitive vessels form organized capillaries. Angioblasts differentiate to form endothelial cells, while hematopoietic stem cells differentiate to more specialized functions. Tight junctions form between endothelial cells, basement membrane is deposited, and pericyte recruitment takes place.

The triggers that result in vasculogenesis are very different from the triggers that initiate angiogenesis and arteriogenesis. As we discuss in the next two sections, formation of new capillaries by angiogenesis is in response to hypoxia, while the formation of muscular blood vessels by arteriogenesis is in response to fluid shear stress. In both of these cases, a very specific external signal triggers new blood vessel formation. However, vasculogenesis is a developmentally regulated process and triggers are molecular and genetic, as opposed to environmental. Fibroblast growth factors, the hedgehog family of morphogens, vascular endothelial growth factors and their receptors, and transforming growth factors and their receptors have been indicated as important modulators of vasculogenesis (25).

5.5 MOLECULAR MECHANISM OF ANGIOGENESIS

Angiogenesis is the process of new capillary formation from existing capillaries in response to hypoxic conditions. Angiogenesis can occur during embryogenesis and adulthood. While the process is very specific, the word "angiogenesis" has often been used fairly loosely with reference to new blood vessel formation, particularly when applied to new blood vessel formation within 3D tissue constructs.

In this section, we will study the mechanism of new blood vessel formation via sprouting angiogenesis, a process in which new blood vessels are formed from existing ones. There are two important concepts that need to be defined here. First, endothelial cells are important for normal blood vessel function, as they line the luminal surface of all vessels, providing a non-thrombogenic surface during blood flow; endothelial cells are also important during angiogenesis. However, all endothelial cells that line the luminal surface of vessels are not the same, and at least three categories have been identified based on distinct cellular specifications: 1) tip, 2) stalk, and 3) phalanx cells (26,27). The second concept is the stimuli required to induce angiogenesis. Under normal physiological conditions, endothelial cells are not activated and do not participate in angiogenesis. The need for angiogenesis arises under hypoxic conditions, where tissue is deficient in oxygen and other nutrients. *How do endothelial cells know there is hypoxia tissue nearby and that they need to begin the process of angiogenesis?* Cells within hypoxic tissue release a protein known as VEGF, and the concentration of this protein is proportional to the degree of hypoxia. Endothelial cells have surface receptors for VEGF and can sense hypoxia tissue and initiate a cascade of intracellular signaling events that leads to the formation of new blood vessels (26,27). This very elaborate signaling mechanism provides a basis for communication between hypoxic tissue and endothelial cells, which then respond by initiating a sequence of molecular changes leading to the formation of new blood vessels.

How exactly is a new blood vessel formed? As we have stated before, all endothelial cells are not the same. When an angiogenic signal like hypoxia is introduced via VEGF signaling, endothelial cells that are exposed to the highest concentration of VEGF are selected to become tip cells (27). Tip cells act at the forefront and lead the formation of new blood vessels in the direction of the angiogenic signal. In each sprout, a single tip cell determines the vessel. Tip cells are highly polarized and use filopodia to guide a sprouting vessel toward an angiogenic signal. The tip cells are non-proliferating. Endothelial stalk cells follow behind the tip cells, proliferate to form elongated stalks, and create a lumen. Further away from the tip cells are endothelial phalanx cells that are lumenized, non-proliferating cells that sense and regulate perfusion in the persistent sprout (27). As sprouting continues to be led by tip cells, two ends of sprouting vessels connect to form a luminal blood vessel. This results in initiation of blood flow through the newly formed blood vessel, thereby reducing the concentration of the angiogenic signal VEGF and causing a reduction in angiogenic sprouting. Pericyte recruitment is required for stabilization of the nascent blood vessel, and this stabilization is followed by extracellular matrix production and deposition (27).

5.6 MOLECULAR MECHANISM OF ARTERIOGENESIS

Arteriogenesis is the formation of new blood vessels from existing ones in response to specific physiological stimuli like changes in fluid shear stresses or pathological conditions like stenosis or blood vessel occlusion (28–33) (in cases of plaque formation occurs during atherosclerosis). Earlier in the chapter, we looked at vascuogenesis as a mechanism of blood vessel formation during embryogenesis, and we looked at angiogenesis as a mechanism of blood vessel formation from existing vessels. Compared to these mechanisms, arteriogenesis is different based on its initiation trigger, which is often a change in the fluid stress environment within existing blood vessels. There are similarities between arteriogenesis and angiogenesis. New blood vessels are formed by growth from existing ones; however, the triggers are different, with fluid shear stress for arteriogenesis and hypoxia for angiogenesis. Vasculogenesis is a different process altogether; it involves growth and development of blood vessels from early precursor stem cells during embryogenesis.

During arteriogenesis, blood vessel growth takes place in response to fluid shear stress or changes in the stress environment caused by pathological conditions like vessel occlusion (28–33). Upon blockage of a certain portion of an artery, redistribution of fluid stresses results in an increase of shear stresses within neighboring parts of the occluded vessel. In response to changes in the fluid shear stress, endothelial cells release growth factors such as TGF-β. This release in turn correlates to an increase in the rate of proliferation of other endothelial cells and smooth muscle cells, a necessary prerequisite for arteriogenesis (increase in the rate of proliferation of endothelial cells (ECs) and smooth muscle cells (SMCs) is necessary for new blood vessel formation). Smooth muscle cell proliferation and remodeling is a very important component of arteriogenesis, as an increase in SMC number is a prerequisite for new blood vessel formation. There can be greater than a 20-fold increase in SMC numbers during arteriogenesis in humans. Another important component in the process of arteriogenesis is an increase in the rate of MMP synthesis, which is responsible for the degradation of ECM components, thereby "loosening" existing vessels to support growth and expansion, particularly for SMC proliferation and remodeling. An increase in activity of MMP-2 and MMP-9 has been observed during arteriogenesis along with an increase in degradation of elastin and other extracellular matrix components. Although exact cellular and molecular mechanisms for arteriogenesis have not been fully elucidated, the current knowledge base alludes to the following seven steps (28,33):

Step One Changes in the fluid shear stress environment due to occlusion or some other factor serve as the trigger for arteriogenesis.

Step Two Endothelial cells on the luminal surface of blood vessels sense these changes in shear stress environment using biological sensors, which may be cell surface integrins.

Step Three Endothelial cells respond to changes in fluid shear stress by an increase in the expression of adhesion molecules like monocyte chemoattractant protein-1 (MCP-1), intercellular adhesion molecule-1 (ICAM-1), and vascular cell adhesion molecule-1 (VCAM-1).

Step Four Increase in the expression of adhesion molecules by endothelial cells results in recruitment of circulating monocytes at the site of arteriogenesis; these monocytes are anchored to the adhesion molecules.

Step Five Once recruited to the site of arteriogenesis, monocytes produce proteases such as matrix metalloproteinase and uPA, which act to degrade extracellular matrix components. Degradation of ECM is important to "loosen" the tissue in order to support smooth muscle cell proliferation and migration.

Step Six The proteins released by monocytes serve to degrade vascular extracellular matrix components.

Step Seven Degradation products of the vascular extracellular matrix, particularly elastin, act as a stimulant for smooth muscle cell proliferation and migration toward the site of arteriogenesis.

While the detailed cellular and molecular mechanisms have not been fully elucidated for arteriogenesis, the seven steps outlined above provide an outline of the general outline that occurs in response to fluid stresses leading to the formation of muscular blood vessels.

5.7 THERAPEUTIC ANGIOGENESIS

Ischemia during Myocardial Infarction—Ischemic tissue is prevalent in the left ventricle after a myocardial infarction, and is due to plaque formation in the coronary artery that occludes the vessels and limits blood supply to heart muscle. This has detrimental consequences and leads to cardiac myocyte cell death and loss of left ventricular function. An experimental strategy to counter ischemia of heart muscle is therapeutic angiogenesis, designed to promote reperfusion (re-establishment of blood flow) through angiogenesis in ischemic tissue, thereby reducing cell death and restoring lost ventricular function (34–37).

What exactly is Therapeutic Angiogenesis? Vasculogenesis, angiogenesis, and arteriogenesis induce the formation of new blood vessels during embryogenesis and normal human development. The process of vascularization can also be stimulated in diseased or ischemic tissue as a therapeutic strategy. During myocardial infarction, blood supply to the left ventricle is compromised due to plaque formation in the coronary artery; this in turn causes cell death and loss of heart function. In theory, if we can develop a therapeutic strategy to increase blood supply to the left ventricle, this will reduce cardiac myocyte cell death and restore lost myocardial function. This strategy is known as therapeutic angiogenesis, which refers to the stimulation of angiogenesis for therapeutic purposes (34–37). New blood vessel formation is due to outgrowth from existing vessels. Hence, the process of angiogenesis occurs rather than vasculogenesis. Therapeutic angiogenesis has been defined as the use of angiogenic factors to induce formation of a collateral blood supply, effectively bypassing an occluded diseased blood vessel in patients with damaged coronary or peripheral myocardial tissue, or other types of damaged tissue.

Agents for Therapeutic Angiogenesis—Any agent that promotes vascularization during human development can be used for therapeutic angiogenesis (38–43). Molecular and developmental biology have provided many such potential candidates; a subset of these agents has been selected for experimental testing and validation. Commonly used agents for therapeutic angiogenesis include proteins like vascular endothelial growth factor (VEGF) (44–47) and fibroblast growth factor (FGF) (45,48). These angiogenic factors have been tested by direct delivery of the protein to the site of myocardial infarction or by delivering the gene that encodes the protein. In other words, protein and gene therapy have been used as modes of action to support therapeutic angiogenesis. Unsurprisingly, cell transplantation has also been used, particularly with endothelial progenitor cells (EPCs) (49–53) and bone marrow-derived mesenchymal cells (BMCs) (54–57).

Modes of Delivery—Intracoronary delivery or intramyocardial injection are two strategies for therapeutic agents (58–62). Intracoronary delivery involves the use of catheters to deliver therapeutic agents directly to the coronary artery, which then transports these agents to the heart muscle. In the case of intramyocardial delivery, the therapeutic agent is directly injected to the heart muscle using catheter-based methods.

Therapeutic Angiogenesis and Regenerative Medicine—In Chapter 1, we studied the field of regenerative medicine in terms of any stimulus that initiates the self-healing process in the human body. After studying therapeutic angiogenesis, it should be clear that therapeutic angiogenesis is one form of regenerative medicine. The therapeutic agent (protein, gene, or cell) delivered to the site of injury is designed to simulate a regenerative response in the host.

Mode of Action—Therapeutic angiogenesis involves delivery of genes, proteins, or cells to the site of injury to revascularize the ischemic tissue. *What exactly is the mechanism responsible for vascularization and improvement in functional outcome?* There are at least three potential mechanisms that have been proposed:

1. *Direct Angiogenesis*—When the therapeutic agent is an angiogenic factor like VEGF or FGF, the mode of action is direct angiogenesis, as these proteins act on existing endothelial cells to promote proliferation and capillary formation (48). Similarly, when the therapeutic agent is EPCs, it is thought that cells directly integrate within existing vasculature, supporting repair and/or new blood vessel formation.
2. *Paracrine Signaling*—If the therapeutic agent does not have a direct effect on angiogenesis (as in the case of bone marrow-derived mesenchymal stem cells), it is thought that these cells release soluble factor into the host environment (35,38–39). The soluble factor acts as a paracrine agent and communicates with endothelial cells to promote new blood vessel formation.
3. *Recruiting of Circulating EPCs*—The third hypothesis is also indirect. It suggests that transplanted agents, particularly non-angiogenic cells, recruit EPCs from the circulation, causing EPCs to home to the site of injury (49). Once circulating EPCs are recruited to the site of injury, they have a direct effect on angiogenesis by supporting formation of new blood vessels.

Assessment of Reperfusion—The success of therapeutic angiogenesis can be measured by many different metrics, though the most direct measure is the number and function of new blood vessels. This can be measured directly using histological techniques, which require tissue specimens stained with antibodies that bind to specific endothelial cell markers, like von Willibrand factor (vWF). These sections are used to obtain capillary counts and allow comparison of samples that have been treated with an angiogenic agent versus controls. This capillary count and comparison provides a very direct metric to measure reperfusion of ischemic tissue. A second approach is angiography, which involves delivery of a contract agent to the tissue, followed by x-ray imaging (63–65). Angiography can also provide a direct measure to blood vessels, in addition to assessment of flow.

5.8 TISSUE ENGINEERING AND VASCULARIZATION

We have looked at vasculogenesis, angiogenesis and related fields, and blood vessel development during embryonesis and human development. The critical question to

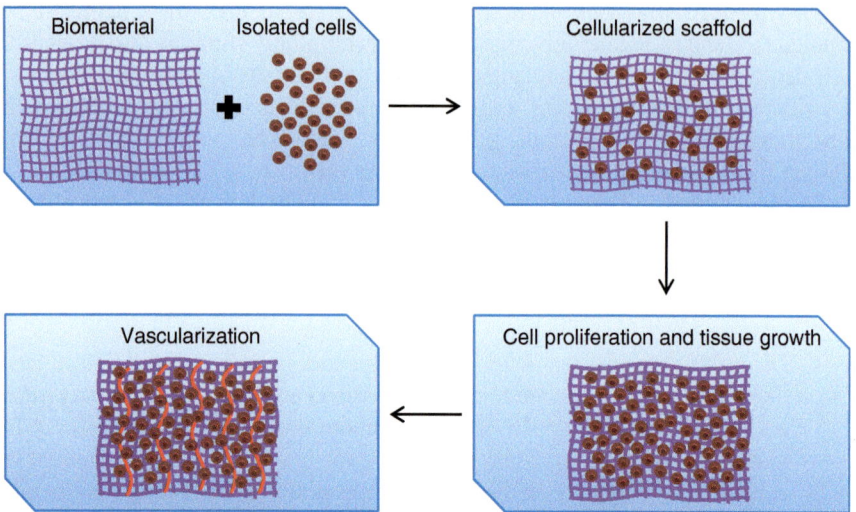

Figure 5.3 Vascularization During 3D Artificial Tissue Formation—During the tissue fabrication process, isolated cells are coupled with biomaterials to support fabrication of 3D artificial tissue. Cell proliferation and subsequent remodeling leads to artificial tissue development and maturation. During this process, induction of vascularization is a critical step in the development of 3D artificial tissue.

address is: *how is this related to tissue engineering?* We can answer this question by discussing the end point objective in tissue engineering, which is 3D tissue and organ fabrication. During 3D tissue development, we begin with isolated cells and develop strategies to fabricate artificial tissue and organs from these isolated cells. At some point during 3D tissue development and maturation, blood vessels are required to support the metabolic activity of artificial tissue, Figure 5.3.

This process is analogous to embryogenesis, in which fertilization leads to oocyte formation and subsequent cell division leads to an increase in cell mass. At some point along this pathway, vasculogenesis is initiated and is followed by angiogenesis and arteriogenesis. As we study the process of blood vessel formation from early embryogenesis through human development, we can apply this knowledge to tissue engineering. Important questions that need to be answered are:

- *During early embryogenesis, the cell mass is avascular and lacks any blood vessels. At some point during embryogenesis, vasculogenesis starts. At what point does vasculogenesis begin, and what are the signals that determine capillary formation?*
- *During tissue engineering, isolated cells come together to form 3D tissue. During early stages of 3D tissue development, there is no vasculature. What is the critical size of the 3D tissue that can remain viable in the absence of vasculature?*

- As 3D tissue grows, the process of blood vessel development takes place. What are the critical signals that need to be created during 3D tissue development in order to support vasculature formation?

Studying these questions will help understand the relationship between angiogenesis and tissue engineering and the interplay between the two disciplines. The objective is to understand the signals that determine blood vessel development during embryogenesis and human development, and translate this understanding toward the fabrication of 3D artificial tissue and organs.

5.9 CONCEPTUAL FRAMEWORK FOR VASCULARIZATION DURING ARTIFICIAL TISSUE FORMATION

In this section, we present a general overview of vascularization for tissue engineering. At some point during tissue fabrication, development, or maturation, we need to incorporate a vasculature. Based on what we have learned about vascularization and the specific molecular mechanisms involved, we need to ask the question: *what strategy can we adopt to engineer a vasculature within artificial tissue?*

We begin this section by addressing vascularization strategies on a conceptual and theoretical basis and then move on to look at specific methodologies that have been used to achieve these strategies. Before diving into the current methods of inducing vascularization, we need to take a step back and start from the beginning and ask one question: *based on our understanding of vascularization, if we had to develop a strategy to incorporate blood vessels within artificial tissue, where would we start?* Our discussion is centered on the premise that we have not reviewed the literature; therefore, our objective is to start with a fresh mind and an empty plate and explore the possibilities. *Where and how do we begin?*

A general scheme of vascularization for bioengineered tissue is present in Figure 5.4.

In this scheme, we present three options for the vascularization of 3D artificial tissue using strategies based on vasculogenesis, angiogenesis, and arteriogenesis. Our first case is based on vasculogenesis, and the general scheme begins with early progenitor stem cells like mesodermal stem cells (Figure 5.4a). The objective is to control external cues to promote differentiation of mesodermal cells to hemangioblasts and then to other cell types required for blood vessel formation (angioblasts, hematopoietic stem cells, and endothelial cells). Conceptually, this scheme is aligned with the objectives of tissue engineering—tissue fabrication being able to recreate embryogenesis coupled with vasculogenesis. Some challenges associated with this strategy are:

- *What will be the source of stem cells? Will the mesodermal cells be obtained by differentiation of human embryonic stem cells, and if so, are there ethical issues involved?*

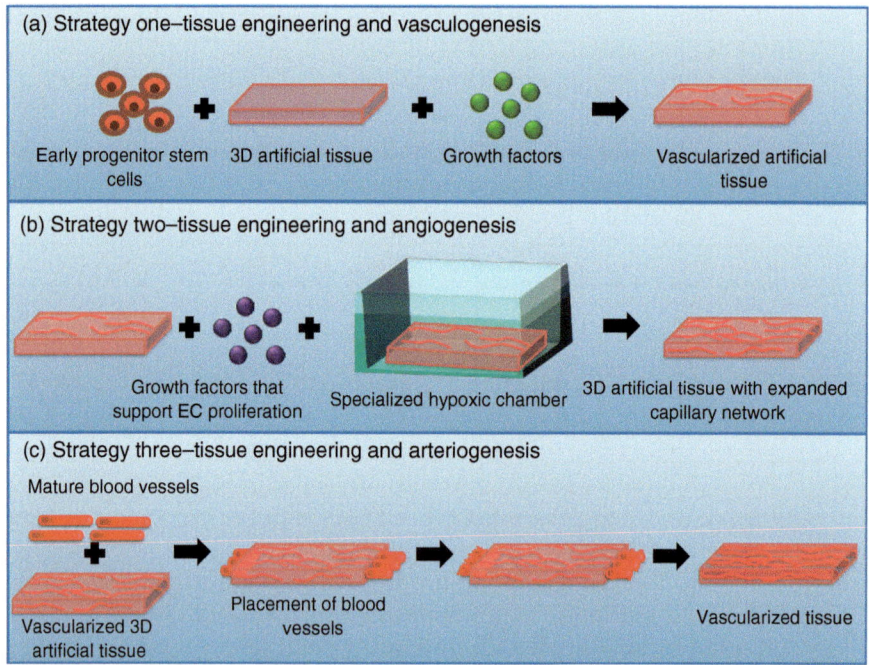

Figure 5.4 Tissue Engineering and Vascularization—(**a**) **Tissue Engineering and Vasculogenesis**—Early progenitor stem cells can be used to support the formation of capillaries during artificial tissue fabrication. (**b**) **Tissue Engineering and Angiogenesis**—Starting with vascularized artificial tissue, angiogenesis can be induced to promote formation of new blood vessels from existing ones. In this example, growth factors and hypoxia are used as drivers of angiogenesis. (**c**) **Tissue Engineering and Arteriogenesis**—Arteriogenesis is the formation and/or maturation of blood vessels from existing ones. In this example, blood vessels are positioned at specific locations in proximity to artificial tissue and are stimulated to support formation of new blood vessels. New blood vessels invade the artificial tissue and support vascularization of 3D artificial tissue.

- *What are the signals that drive the differentiation fate of early progenitor stem cells toward a vascular lineage, and how do we create these signals in vitro?*
- *If we are able to control the differentiation fate of progenitor cells toward a vascular lineage, how do we support the formation of capillaries and capillary networks? What if the cells stay in an isolated state and do not remodel to form complex vascular networks?*

The second strategy is to incorporate vasculature using methodology based on angiogenesis, which refers to the development of new capillaries from existing ones. As we have seen before, capillary formation during angiogenesis requires the presence of existing capillaries and requires a hypoxic culture environment. Therefore, from a tissue engineering standpoint, one can engineer capillaries using vasculogenesis as described before and then subjecting the newly formed capillaries to hypoxic conditions in order to induce angiogenesis, as shown in Figure 5.4b. This process can, in theory, lead to the formation of an entire vascular network within 3D artificial tissue. There are many scientific and technological challenges in building an entire vasculature by recreating molecular mechanisms found during embryogenesis. Identification of a suitable stem cell source and defining signals that drive the differentiation fate of these stem cells toward a vascular lineage remain challenging. In addition, the following can be added to the list:

- *How do we create a hypoxic environment in vitro that replicates properties of an in vivo hypoxic microenvironment?*
- *What mechanisms will be implemented to support long term perfusion of the newly formed capillary network?*
- *What will be the orientation and alignment of newly formed capillaries relative to cells within the artificial tissue?*

We now explore the use of arteriogenesis to induce vasculature within artificial tissue. The starting point for arteriogenesis is the presence of muscular blood vessels, something which is clearly lacking in 3D artificial tissue (Figure 5.4c). In order to engineer a vasculature within artificial tissue using arteriogenesis, the starting substrate has to be muscular blood vessels. We can isolate and maintain blood vessels in culture using a closed loop perfusion system and anchor the vessels to both ends of 3D artificial tissue. We can then create occlusions by reducing the flow rate at specific points on the blood vessel; this reduction of flow rate will lead to changes in the fluid shear stress. This process will initiate arteriogenesis and can lead to the formation of a new vascular network within 3D artificial tissue; we then connect the vascular networks from both blood vessels to create a continuous fluid flow loop. Some of the scientific and technological challenges associated with this strategy include:

- *Long-term culture of muscular blood vessels in vitro is difficult, and the ability to engineer a perfusion loop and create an occlusion that represents pathological conditions is not trivial.*
- *Arteriogenesis requires participation of circulating monocytes, which are absent during in vitro culture.*
- *Interconnectivity of the two vascular beds that originate from the two source vessels is not a spontaneous event and requires complex interventional strategies.*

In this section, we have provided an interface between *in vivo* vascularization (what happens in nature) and *in vitro* vascularization strategies (what we want to accomplish in tissue engineering) for 3D artificial tissue. We presented three conceptual examples to translate our understanding of vasculogenesis, angiogenesis, and arteriogenesis toward the development of *in vitro* vascularization strategies. Our objective in presenting these examples is to demonstrate the relationship between vascularization in nature and in tissue engineering. In addition, we provided some of the scientific and technological challenges in developing vasculature for 3D artificial tissue. This background is necessary as an entry point into the journey into neovascularization strategies for 3D artificial tissue that is described in the next few sections.

5.10 IN VIVO MODELS OF VASCULARIZATION

In the previous section, we looked at several *in vitro* approaches to engineer vasculature within artificial tissue. Our objective in all of the models was to develop *in vitro* conditions that replicate the process of vascularization as it occurs *in vivo*. Stated another way, the objective was to recapitulate *in vivo* conditions to support *in vitro* vascularization of 3D artificial tissue. This has been a very aggressive area of research with many different strategies being evaluated. However, if we look at this problem from a totally different perspective, we can envision a completely different strategy. Let us begin by re-evaluating our objective, which is to develop a vasculature within artificial tissue to recreate *in vivo* conditions under controlled *in vitro* conditions. *However, instead of recreating physiological conditions in vitro, what if we designed a system to engineer vasculature in vivo?* Such a strategy would require creating a custom chamber to house artificial tissue followed by implantation of this chamber *in vivo*; which will result in exposure to controlled physiological culture environment. This in turn will serve as a platform for vascularization of 3D artificial tissue. Since this is a fairly new concept and not as obvious as *in vitro* strategies, let us take a moment to explore this further by discussing the concept of *in vivo* vascularization, some of the advantages of this strategy, and the challenges of developing and implementing these strategies.

What exactly is in vivo vascularization?—The concept of *in vivo* vascularization revolves around culturing bioengineered tissue within specialized chambers that can be implanted to support the formation of new blood vessels within 3D artificial tissue (66–68) (Figure 5.5).

Once implanted, artificial tissue remains *in vivo* for a culture period of approximately 3–4 weeks, during which time the neovascularization of implanted tissue occurs due to host response. After implantation, artificial tissue is removed and separated from the chamber, leading to vascularization of implanted 3D artificial tissue.

Why would in vivo angiogenesis work? This strategy is based on the hypothesis that the *in vivo* culture environment has the right physiological cues for formation, development, and maturation of blood vessels. This involves presence of progenitor stem cells and the other cell types required for vascularization. In addition to cells, *in vivo* physiological conditions also have the right stimuli for blood vessel development, like fluid shear stress for arteriogenesis and hypoxia for angiogenesis, along with an abundance of growth factors like VEGF and FGF to support blood vessel formation. The microenvironment is also ideal for vascularization and consists of the right temperature, pH, ion concentration, oxygen saturation, and presence of

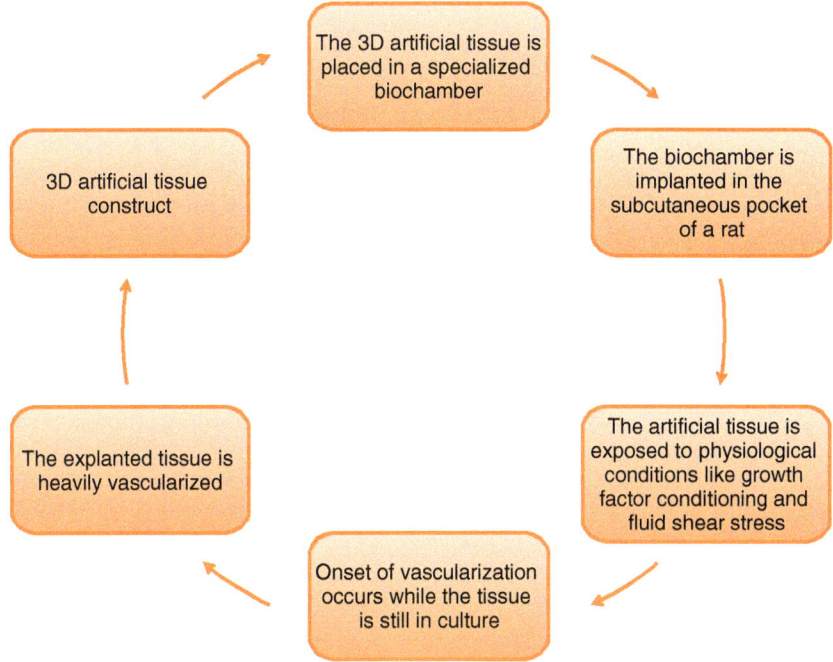

Figure 5.5 *In Vivo* **Models of Vascularization**—Artificial tissue can be secured within a custom biochamber and implanted *in vivo* to support vascularization of the 3D artificial tissue.

other components. Collectively, presence of all the required raw materials and presence of physiological stimulation cues, along with an optimized microenvironment, provides a strong rationale for blood vessel formation upon implantation.

What are some of the challenges with in vivo vascularization strategies? While *in vivo* vascularization strategies have many advantages, they are not without limitations. Development of a culture biochamber to support viability of artificial tissue during *in vivo* culture is challenging. The design of the biochamber has to support the culture and viability of 3D artificial tissue while protecting implanted tissue from host immune response and supporting physiological interaction with the host culture environment. In addition, the *in vivo* culture environment is harsh and can lead to damage of implanted artificial tissue. For example, fluid shear stresses resulting from arterial blood flow can be high. Artificial tissue may not be able to withstand high pressures, which can lead to physical damage of the tissue, cell death, and other adverse effects. Another challenge is the inability to regulate stimulation conditions during implantation; artificial tissue is exposed to the physiological culture environment without the ability to change variables. For example, if the arterial blood pressure is too high for artificial tissue, there is no way to bring it down to a lower level; this level of control can only be accomplished during *in vitro* culture. Finally, like with any other *in vivo* studies, there is significant variability between experimental groups due to the inability to exhibit any control over physiological stimulation parameters; this reduces uniformity across experimental groups.

How does in vivo vascularization of artificial tissue compare with in vitro strategies? In vitro approaches are designed to replicate many of the conditions present during *in vivo* culture. At times, *in vitro* strategies are able to replicate some, though not all, of these conditions due to the complexity of the *in vivo* culture environment. This problem is solved by designing and implementing *in vivo* strategies for vascularization of 3D artificial tissue; these strategies have the right stimulation environment in place. *In vivo* strategies, however, do limit the amount of user control over the process of vascularization, which reduces the extent of customization of vascular networks within 3D artificial tissue.

What are some examples of potential implantation sites to support vascularization of artificial tissue? In theory, any implantation site that has been used for devices in the past can be used as a potential implantation site for vascularization. For proof of concept studies and experimental validation, subcutaneous implantation of artificial tissue has been used extensively (69). In such a case, the tissue is implanted in a subcutaneous pocket to support vascularization and explanted after a period of 3–4 weeks with a significant amount of neovascularization within the artificial tissue.

5.11 IDEALIZED VASCULARIZATION STRATEGY FOR TISSUE ENGINEERING

We started this chapter by looking at vasculature formation during normal human development and physiological conditions and studied molecular mechanisms responsible for vasculogenesis, angiogenesis, and arteriogenesis. We then looked at two general strategies for fabricating vasculature within 3D artificial tissue, including *in vitro* and *in vivo* methods. After exploring vascularization strategies in nature and in tissue engineering, we need to ask ourselves one question: *what is the best way to vascularize artificial tissue in an idealized system, and what are the scientific and technological challenges in order to address this?*

Our strategy is based on a biologically inspired approach designed to recreate blood vessel formation during human development and growth within 3D artificial tissue using early stem cells. Our idealized process is presented in Figure 5.6 and consists of four distinct phases: 1) formation of nascent capillary network, 2) expansion of capillary network, 3) growth of artificial tissue, and 4) perfusion of newly formed vascular network.

The first step in the process is to support formation of capillaries using early mesodermal stem cells. This requires isolation and expansion of mesodermal stem cells. This is followed by conditioning using a cocktail of growth factors to stimulate differentiation of mesodermal stem cells to endothelial cells and support organization of ECs for capillary formation. The composition and concentration of growth factors needs to be optimized to regulate differentiation of mesodermal stem cells; this part of the process is designed to recapitulate vasculogenesis during human

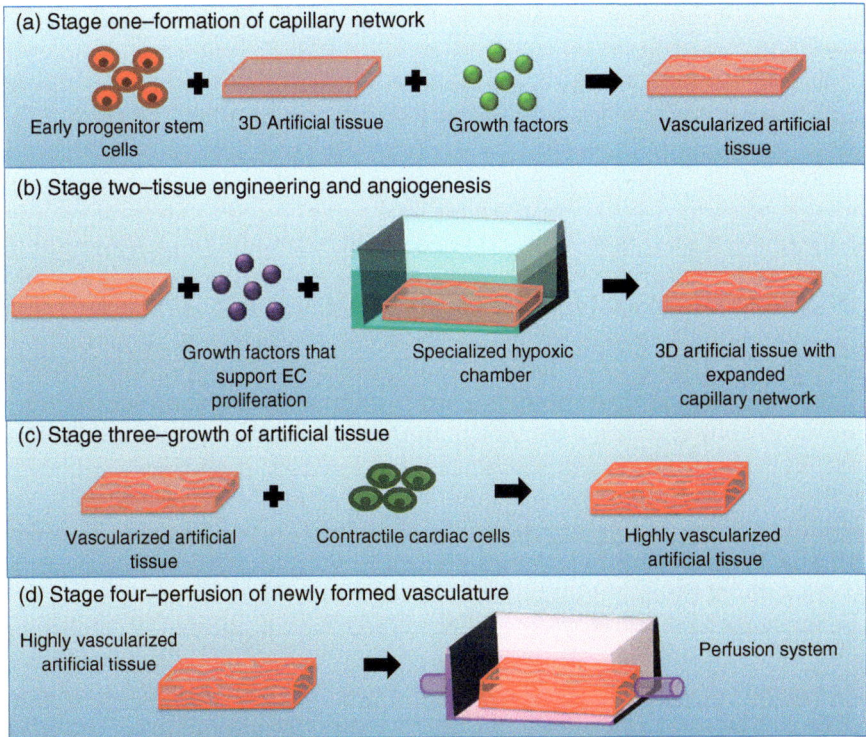

Figure 5.6 **Idealized Strategy for Vascularization of 3D Artificial Tissue**—(a) **Formation of Capillary Network**—The formation of new capillaries is promoted by the process of vasculogenesis. Early mesodermal stem cells are added to artificial tissue and stimulated with growth factors to support capillary formation. (b) **Expansion of Capillary Network**—The process of angiogenesis is used to expand the newly formed vasculature within 3D artificial tissue. Angiogenic factors and hypoxia are used as drivers of angiogenesis to support expansion of vasculature. (c) **Growth of Artificial Tissue**—Once a vascular network is in place, conditions are optimized to promote growth of 3D artificial tissue construct. (d) **Perfusion of Newly Formed Vasculature**—A perfusion system is developed to perfuse the newly formed vasculature.

development. This process will result in vascularized artificial tissue, though the capillary network will be rudimentary at this stage.

The second stage of the process is designed to expand the vascular network by creating *in vitro* conditions that replicate the angiogenic response during human development. Endothelial cells that have been obtained by differentiation of mesodermal stem cells will be used to support angiogenesis; these endothelial cells will be further stimulated by angiogenic factors like VEGF and FGF. The vascularized artificial tissue obtained from the studies of the first stage of the process will be coupled with angiogenic growth factors and incubated in a specialized hypoxic chamber designed to induce angiogenesis. After successfully completing stage two of the process, the end product will be artificial tissue with an integrated vasculature.

During stages one and two of our idealized process, we have implemented strategies to support the formation of a capillary network within 3D artificial tissue. This provides a platform to support the growth of the tissue that has been limited in thickness due to lack of a vasculature. Once a capillary bed has been engineered, we can increase the thickness of the artificial tissue by adding layers of cells and extracellular matrix, as the newly formed capillary network can now support the increase in metabolic activity resulting from an increase in tissue mass. Stage three of our idealized process is designed to accomplish this: increase thickness of the 3D artificial tissue making use of the newly formed vasculature to support tissue viability.

The fourth stage of our process consists of fabricating a perfusion system to deliver continuous media flow through the newly formed vascular network. Continuous media perfusion is designed to support metabolic requirements of artificial tissue, apply intraluminal pressure to the vascular network to prevent the blood vessels from collapsing, and support stability of the newly formed vessels. Continuous pulsatile media flow will also have a positive effect on the functional performance of 3D artificial tissue.

The model presented is oversimplified and is laden with scientific and technological challenges, some of which include the ability to control the differentiation pathway of mesodermal stem cells to form endothelial cells, and the ability to induce an angiogenic response using a hypoxic trigger. In addition, engineering challenges of fabricating multi-layer tissue constructs and complex perfusion systems are no less intimidating. However, the idealized process does provide a strategy that has the potential to create vascularized artificial tissue and identifies some of the challenges in doing so.

5.12 FLOW CHART AND DECISION MAKING

Earlier in this book, we defined tissue engineering in terms of tissue fabrication. We also emphasized that tissue fabrication should be viewed in terms of process flow charts, decision making, and process optimization, like any other engineering problem. We have maintained this theme throughout this book and continue here with vascularization. In this section, we present a process flow chart for vascularization and identify critical decision points. In this chapter, we have presented the molecular mechanisms of vascularization during embryogenesis and human development and used this as a platform to develop vascularization strategies for artificial tissue. We have observed that tissue engineering strategies often draw inspiration from these vascularization methods found in nature; other times, vascularization strategies are based on novel innovative methods outside of those found in nature. In this section, we bring this information together and present a process flow chart that can be used as a decision making tool for the development of vascularization strategies in tissue engineering. Our scheme is presented in Figure 5.7. The process

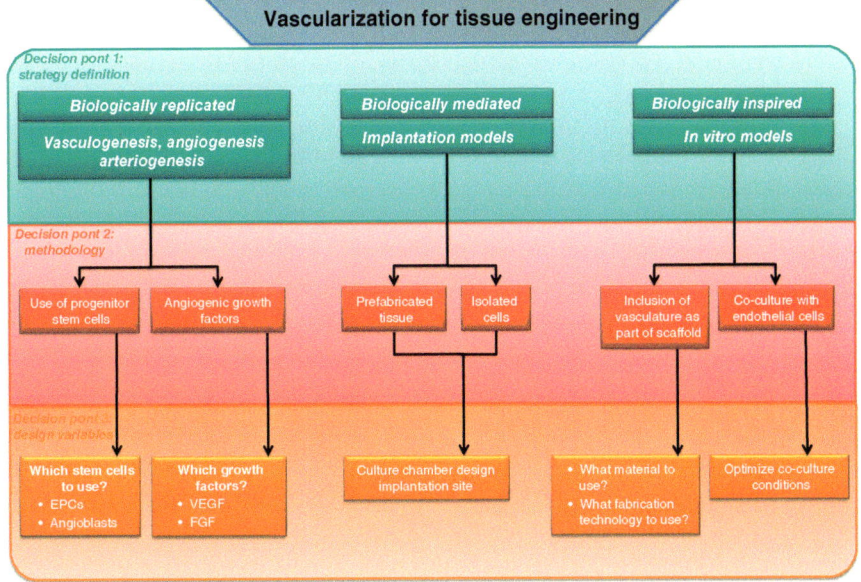

Figure 5.7 Process Flow Chart for Vascularization in Tissue Engineering—Three critical decision points are defined. The first step in the process is defining the vascularization strategy, which can be biologically replicated, biologically mediated, or biologically inspired. Once the vascularization strategy has been defined, the next two steps are focused on defining the specific methodology and statement of design variables.

flow chart is based on critical decision making points at three hierarchical levels, starting with defining a broad strategy and moving toward a specific methodology.

When presented with the challenge of fabricating vasculature within 3D artificial tissue, defining a starting point is vague, and the most difficult question to answer is—*where do I start?* The first decision point in our process flow chart is focused on addressing this issue by providing options for defining a broad vascularization strategy. The process flow chart for vascularization identifies three potential starting points, which are identified in the first level of decision making hierarchy. Based on current technology, there are three broad strategies for vascularization: biologically replicated, biologically mediated, and biologically inspired. In the first case, biologically replicated strategies are designed to replicate *in vivo* processes of vascularization *in vitro*. Biologically replicated processes are influenced by molecular biology, with the objective being understanding biological phenomena and defining controlled laboratory conditions to replicate these processes. These strategies are focused on defining *in vitro* conditions used to drive vasculogenesis, angiogenesis, and arteriogenesis; examples include the use of early progenitor stem cells and angiogenic growth factors to support capillary formation.

The second strategy is referred to as biologically mediated and includes vascularization models that are based on implantation of 3D artificial tissue. The term "biologically mediated" refers to the notion that successful implementation of these strategies requires intervention and mediation from recipient of the implanted tissue. Mediation of the vascularization process is a result of implantation of cells or artificial tissue.

The third strategy is referred to as "biologically inspired" and in this case, inspiration is drawn from biological process with an objective to replicate these processes using innovative *in vitro* strategies. The goal is not to replicate the biological process, but replicate functionality.

Once we have identified a broad strategy for the specific application, the next step is to identify a specific methodology. Specific examples of methodologies for each of the three strategies are provided in the process flow chart. For example, if we select "biologically replicated", we will develop a method that replicates vascularization as it occurs during human development. This can be done in one of two ways: use of early progenitor stem cells or use of angiogenic growth factors.

If we select "biologically mediated" as our vascularization strategy, our objective is to develop methods that support neovascularization of 3D artificial tissue by the host upon tissue implantation. This can be done one of two ways. In the first case, artificial tissue is fabricated under controlled *in vitro* laboratory conditions and then prefabricated tissue is implanted to support vascularization; tissue fabrication and vascularization are sequential processes. In the second case, isolated cells are placed within a culture chamber and then the culture chamber is implanted; fabrication of artificial tissue and vascularization occur simultaneously.

The third strategy is "biologically inspired", and in this case, we draw inspiration from nature and biological processes, and then develop strategies to replicate these biological processes. One example of this is the design of novel scaffold with pre-engineered vasculature; to achieve this objective, we will use prototyping techniques to fabricate blood vessels as a core component of the 3D scaffold, which is then populated with cells to support formation of vascularized 3D artificial tissue. A second example is the development of novel co-culture systems that recapitulate blood vessel formation with multiple cell types in culture.

The third and final step in the process is to define specific test variables and experimental conditions to evaluate these variables; this step involves specific experimentation and validation stage of the vascularization process. For example, if we decide to use a biologically replicated process based on progenitor stem cells, our experimentation strategy will need to focus on identifying stem cell type and source along with differentiation strategies to drive the phenotype of selected cells to a vascular lineage. Similarly, if our biologically replicated process is based on angiogenic growth factors, we need to identify specific growth factors, characterize the dose-dependent and time-course relationship, and develop strategies for temporal and spatial variations in growth factor delivery.

If we decide to go with a biologically mediated process based on implantation models, using either cells or artificial tissue, we need to select the implantation model, including the implantation site, *in vivo* culture time, and design of the implantation chamber. We could also go with a biologically inspired strategy for vascularization based on prevascularization of scaffolds. Specific experimental variables that need to be optimized would include the type of polymer and processing conditions and specific fabrication technologies for engineering vascular networks.

In summary, the process flow chart presented in Figure 5.7 provides a tool to assist in the decision making process for vascularization strategies in tissue engineering. There are several scientific and technological challenges that need to be addressed at every stage of the vascularization process. With this background, we proceed to give specific examples in tissue engineering in which many of these strategies have been implemented to fabricate vascularized 3D artificial tissue.

5.13 BIOLOGICALLY REPLICATED VASCULARIZATION STRATEGIES

As we have seen before, vascularization during embryogenesis and human development is through three distinct mechanisms: vasculogenesis, angiogenesis, and arteriogenesis. Tissue engineering strategies have been developed to replicate these vascularization strategies *in vitro* by culturing artificial tissue in controlled conditions that replicate these processes and promote vascularization. These strategies aim to replicate *in vivo* conditions *in vitro* and are therefore termed "biologically replicated vascularization strategies." There have been two areas of research that have been aggressively pursued. The first is focused on replicating vasculogenesis *in vitro*, starting with progenitor stem cells and culturing these cells under controlled conditions that trigger capillary formation. The second strategy is focused on recapitulating angiogenesis, starting with endothelial cells and using growth factors to promote capillary formation. In the following sections, we provide three examples of biologically replicated vasculature strategies for tissue engineering (Figure 5.8).

The first strategy is based on scaffold-guided differentiation of endothelial progenitor stem cells (70). In this method, endothelial progenitor cells were used and the trigger for capillary formation was provided by a custom fabricated scaffold. The steps in this process are outlined below (70):

Step One Custom scaffolds were fabricated using three extracellular matrix proteins—fibrinogen, fibronectin, and laminin at concentrations of 50 mg/ml, 5 µg/ml and 5 µg/ml respectively.

Step Two Primary mesenchymal stem cells were obtained from the bone marrow of rats and maintained in tissue culture flasks. The MSCs were then selected for EPCs, which were enriched and selectively grown from bulk culture of the MSCs. Using this process, a large number of EPCs were readily available for studies.

Step Three EPCs were seeded on the scaffold and cultured under controlled *in vitro* conditions for 21 days. At various time intervals, EPCs were analyzed to assess the differentiation fate of the EPCs to an endothelial lineage.

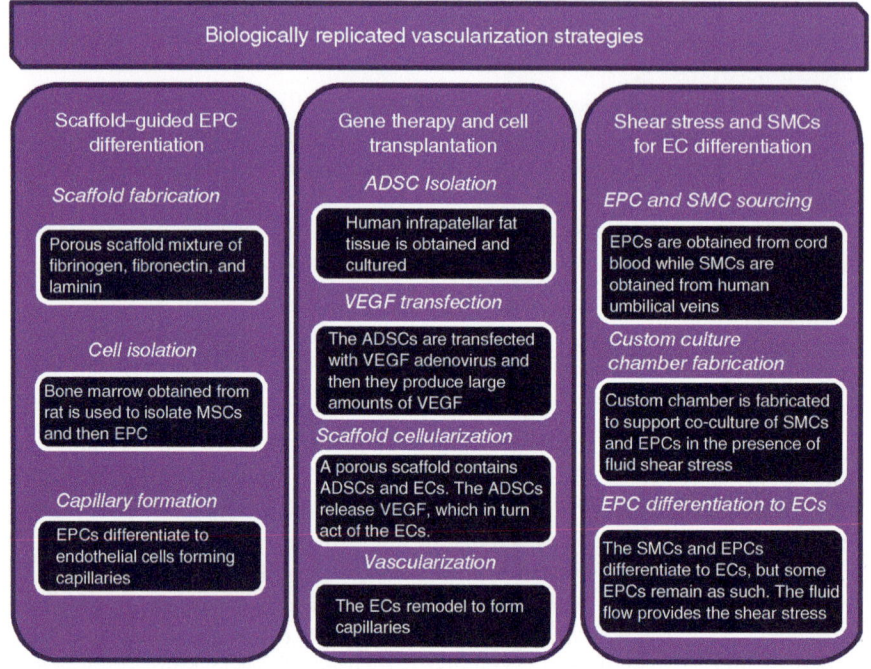

Figure 5.8 Biologically Replicated Vascularization Strategies for Tissue Engineering—Three examples of biologically replicated vascularization strategies: (a) Scaffold-Guided EPC Differentiation, (b) Gene therapy to deliver VEGF to drive the differentiation of adipose-derived stromal cells to form endothelial cells, and (c) Shear stress and contact with SMCs used for EPC differentiation to form endothelial cells.

The results of this study demonstrated that the EPCs were differentiating to form endothelial cells, and newly formed endothelial cells remodeled to form capillaries, resulting in vascularization of the scaffold. The trigger was engineered into the scaffold by using extracellular matrix components, which supported EPC differentiation to endothelial cells. This process was designed to recapitulate vasculogenesis by differentiation of early progenitor stem cells to a vascular lineage and use the differentiated cells to support capillary formation.

In the second approach, endothelial cells were stimulated using vascular endothelial growth factor to support capillary formation (71). The novelty of this strategy was that VEGF was released into the culture environment by adipose-derived stromal cells (ADSCs) using a genetic engineering approach (71). The details of this process are outlined below (71):

> ***Step One*** Adipose tissue was obtained from human donors from infrapatellar tissue and was subjected to an enzymatic digestion process to isolate ADSCs, which were then cultured and expanded *in vitro*.

Step Two ADSCs were transfected with VEGF-containing adenovirus constructed by using cotransfection of 293 cells. This process is an example of genetic engineering and is designed to increase the rate of production and release of VEGF in ADSCs.

Step Three Transfected ADSCs were co-cultured with endothelial cells in custom poly(lactide-co-glycolide) (PLAGA) scaffolds. The ADSCs served as a source of VEGF to stimulate capillary formation by endothelial cells.

Step Four In response to VEGF stimulation, endothelial cells remodeled to form capillaries, resulting in vascularization of the 3D scaffold.

The strategy was designed to promote vascularization of 3D artificial scaffolds based on remodeling of endothelial cells in response to VEGF as the trigger, a strategy aimed to replicate vasculogenesis. The use of genetically engineered ADSCs as the source of VEGF added to the novelty of this strategy.

A third strategy was based on the differentiation of EPCs to endothelial cells, which can then be used to vascularize artificial tissue (72). The trigger for EPC differentiation was fluid shear stress and co-culture with SMCs, a process aimed at replicating arteriogenesis to guide EPC differentiation (72). The steps in the process were (72):

Step One EPCs were obtained from a human source and isolated from cord blood while SMCs were obtained from human umbilical veins.

Step Two A custom culture chamber was fabricated to support co-culture of SMCs and EPCs in the presence of fluid shear stress. The SMCs and the EPCs were in direct contact with each other, and EPCs were also exposed to fluid shear stress. The SMCs were not exposed to fluid shear stresses.

Step Three EPCs differentiated to form ECs in response to fluid shear stress and direct cellular contact with SMCs.

In this model, the objective was to replicate the process of arteriogenesis. The study was focused on differentiation strategies required to drive EPC differentiation toward vascular lineage; differentiated cells were not used to support vascular formation. It was demonstrated that direct contact of EPCs with SMCs or exposure to fluid shear stress resulted in differentiation to ECs; furthermore, the coupled effect of fluid shear stress and contact of EPCs with SMCs increased the differentiation efficiency of EPCs to form endothelial cells.

5.14 BIOLOGICALLY MEDIATED VASCULARIZATION STRATEGIES

Biologically mediated vascularization strategies are based on implantation models and depend on the host to promote and support vascularization. This means that

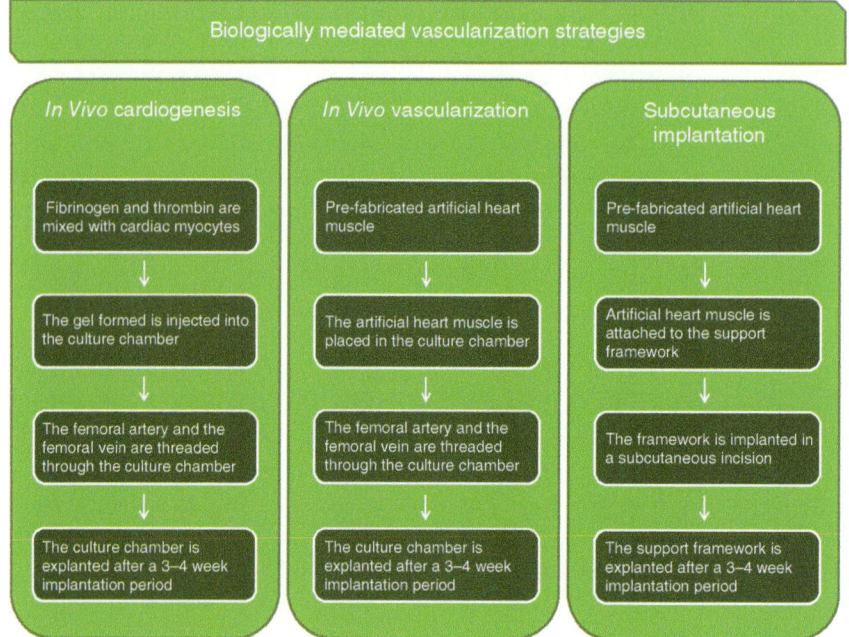

Figure 5.9 Biologically Mediated Vascularization Strategies for Tissue Engineering—Biologically mediated vascularization strategies rely upon implantation models for vascularization of 3D artificial tissue. Examples include *in vivo* cardiogenesis, *in vivo* vascularization, and subcutaneous implantation.

vascularization is mediated or facilitated by the host. While there have been several models published in the recent literature that showcase this methodology, we present three specific examples for illustrative purposes (Figure 5.9).

One approach has been to promote artificial tissue fabrication while vascularization takes place in a controlled culture chamber, all *in vivo* (73). This process has been used to engineer vascularized heart muscle using the following four step process (73):

Step One Primary cardiac myocytes were isolated and suspended in a complex 3D fibrin gel to provide support for artificial tissue formation.

Step Two The fibrin gel, with primary cardiac myocytes, was secured within a custom culture vessel designed to support artificial heart muscle formation during *in vivo* culture.

Step Three The culture chamber was implanted in close proximity to the femoral artery and femoral vein in recipient rats. The vascular pedicle serves as a source of nutrients for the cells as they remodel to form 3D tissue while providing pulsatile conditioning to enhance function of the cardiac myocytes. In addition, the vascular pedicle serves as a site for angiogenic sprouting, and the newly formed blood vessels are incorporated within the artificial heart muscle.

Step Four After a 3–4 week implantation, the culture chamber was explanted from the recipient animal. The culture chamber was separated from the fibrin gel, which, by the end of the implantation period, had remodeled to form vascularized 3D artificial heart muscle.

The four-step process leads to the formation of highly vascularized artificial heart muscle. During *in vivo* culture in the chamber, isolated cells remodel to form artificial heart muscle while the femoral artery and femoral vein serve as sites for vascular sprouting; both of these processes occur simultaneously. The femoral artery and femoral vein are retained as a part of the tissue construct. The novelty of this model lies in the fact that 3D tissue fabrication and vascularization happen at the same time and cells remodel around the newly formed vasculature. This process does not aim at replicating any *in vivo* vascularization strategy, but rather to develop a novel tissue engineering technology around angiogenic sprouting.

A second strategy, somewhat related to the previous one, was to fabricate artificial heart muscle *in vitro* and then implant the tissue *in vivo* to support vascularization. In this case, complete fabrication of 3D artificial tissue was conducted *in vitro* and then implanted *in vivo* to support vascularization. In this approach, vascularization takes place after tissue fabrication in contrast to the previous approach in which both processes occurred in tandem. This is a four-step process:

Step One Artificial heart muscle is fabricated *in vitro* using a published model to support the self-organization of primary cardiac myocytes to form scaffold-free tissue, known as cardioids.

Step Two Cardioids are secured within a custom culture chamber that has been designed to house 3D artificial tissue during implantation.

Step Three The culture chamber is implanted in close proximity to the femoral artery and femoral vein in recipient rats. The vascular pedicle serves the same functions described in the previous case, which include nutrient delivery to the cardioids, pulsatile conditioning to enhance cardioid function, and as a source for angiogenic sprouting to support vascularization.

Step Four After a 3–4 week implantation period, the culture chamber is explanted from the recipient animal and, as in the previous case, separated from cardioids. Implantation resulted in extensive vascularization of cardioids.

We have presented two models for vascularization. While the two strategies have similarities, there are significant differences between the two. In the first case, artificial tissue formation takes place in the culture chamber implanted *in vivo* while vascularization occurs simultaneously; tissue fabrication is in response to *in vivo* conditions. In the second case, artificial heart muscle is fabricated *in vitro* under controlled laboratory conditions; artificial tissue is transferred to a culture chamber and implanted *in vivo*, where vascularization takes place.

The third model presented for biologically mediated vascularization is based on the subcutaneous implantation of cardioids that have been fabricated under controlled *in vitro* conditions (69). In this case, artificial tissue is implanted in a subcutaneous pocket (69). The steps in this process are:

Step One Cardioids were fabricated under controlled *in vitro* conditions, as described for the previous model.

Step Two A custom support framework was fabricated to anchor cardioids during subcutaneous implantation. This support framework was different from the culture chamber described for the first two models. The function of the support framework was to provide attachment points for cardioids, which otherwise form a tissue mass due to spontaneous contractions.

Step Three The support framework with cardioids was implanted in a subcutaneous pocket in recipient animals.

Step Four After a 3–4 week implantation period, the support framework was removed from the site of implantation and the cardioids were separated; at the time of implantation, cardioids were extensively vascularized.

The third model is a variant of the first two, as it does not depend on direct interaction with a vascular pedicle for vascularization. Instead, the host response promotes vascularization of implanted artificial heart muscle.

5.15 BIOLOGICALLY INSPIRED VASCULARIZATION STRATEGIES

Thus far, we have looked at methods that are designed to replicate normal biological processes of vascularization *in vitro* or rely upon implantation models to induce vascularization. The third category is focused on developing technology inspired from nature; these processes draw inspiration from biological processes and use this inspiration to design novel strategies to induce vascularization (Figure 5.10). We will present three examples in the following sections.

The first strategy is based on technology to support scaffold fabrication with specific characteristics to support vascularization. In this approach, scaffolds were fabricated using a sugar leaching process, resulting in an average pore diameter of 100 μm (74). The pores served as sites for vascularization to support culture and proliferation of endothelial cells (74). The specific steps in the process are (74):

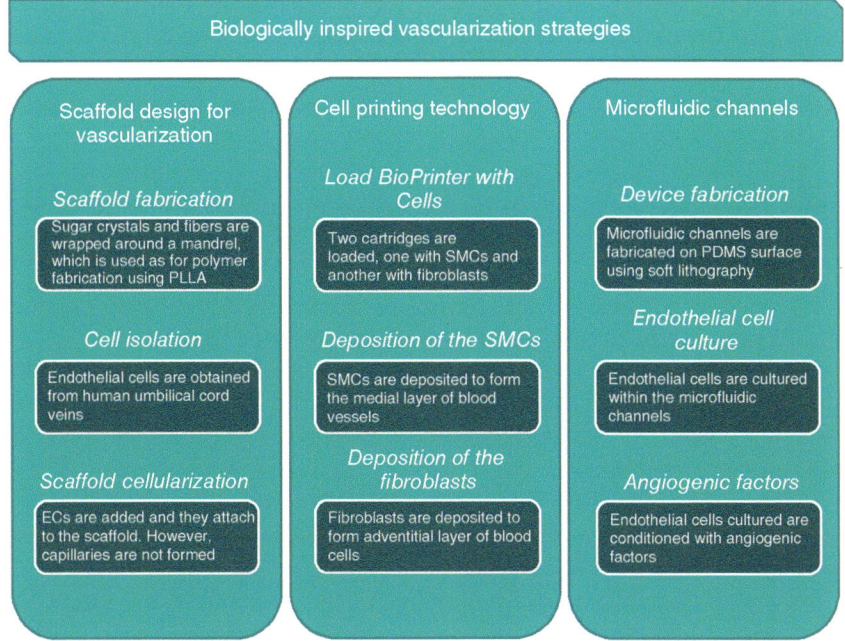

Figure 5.10 Biologically Inspired Vascularization Strategies for Tissue Engineering—These strategies draw inspiration from vascularization in nature; some examples are presented in the figure.

Step One Sugar crystals were solubilized and then heated and fibers wrapped concentrically around a mandrel. The mandrel was used as a framework for polymer fabrication using PLLA. PLLA polymer formed a 3D structure around the fibers and the 3D structure contained sugar crystals, which were solubilized to leave behind a highly porous polymer scaffold. These newly formed pores provided microchannels to support vascularization.

Step Two ECs were obtained from human umbilical veins and used to populate the porous scaffold.

Step Three The study demonstrated viability, attachment, and proliferation of the ECs within the pores of the scaffold. However, the study did not demonstrate capillary formation, which would require concentric organization of the ECs on the luminal surface of the pores within the scaffold.

A second strategy for engineering vascularization made use of cell printing techniques, which deposit droplets of cells to fabricate layers of tissue; individual layers can be combined to form complex artificial tissue (75). This strategy overcomes the need for synthetic scaffolds, as the cells produce their own components. This strategy has been used to fabricate 3D blood vessels consisting of the medial and advential layers, using steps described here (75):

Step One Cartridges were loaded with SMCs and fibroblasts.
Step Two SMCs were deposited to form the medial layer of blood vessels.
Step Three Fibroblasts were deposited to form the adventitial layer of the blood vessel.

This process resulted in fabrication of bi-layer blood vessels with expansion capability to add ECs to form and complete tri-layer blood vessels.

The third process was based on microfluidic channels fabricated using PDMS surfaces. These surfaces can be designed to form complex vascular networks and can also be seeded with angiogenic factors to support capillary formation (76). The steps in the process are (76):

Step One Microfluidic channels were fabricated on PDMS surface using soft lithography.
Step Two Endothelial cells were cultured within the microfluidic channels.
Step Three Endothelial cells that were cultured within the microfluidic channels were conditioned with angiogenic factors.

The steps described above were shown to support culture and proliferation of ECs, while angiogenic factors were shown to increase the rate of proliferation of ECs.

SUMMARY

Current State of the Art—During early stages of tissue fabrication, nutrient delivery to cells is supported by diffusion. As the cell mass increases, nutrient delivery by diffusion is not adequate to support cell viability, and vascularization is required. Seminal work in the field of angiogenesis research was conducted by Dr. Judah Folkman and lead to significant advancements in the field of angiogenesis. While vascularization is a generic term that refers to blood vessel formation, vasculogenesis, angiogenesis, and arteriogenesis refer to specific processes during human development and growth. Understanding molecular mechanisms of vascularization has allowed researchers to develop novel strategies to integrate blood vessels within the 3D architecture of artificial tissue. There are three categories of vascularization strategies for tissue engineering: biologically replicated, biologically mediated, and biologically inspired.

Thoughts for the Future—While there have been several interesting approaches to support vascularization of 3D artificial tissue, the ability to fabricate vascularized 3D tissue remains challenging and requires significant attention.

The most important area that needs to be developed is to carefully understand the triggers that drive vascularization during normal human development and growth (vasculogenesis, angiogenesis, and arteriogenesis) and use these triggers in tissue engineering. As we gain a better understanding of these triggers, this information can be translated to fabricate vascularized 3D artificial tissue.

PRACTICE QUESTIONS

1. Why is vascularization important for the development of artificial tissue?
2. Discuss the seminal work by Dr. Folkman in the area of angiogenesis.
3. Describe the terms vasculogenesis, angiogenesis, and arteriogenesis. What is the difference between these three processes?
4. Vasculogenesis, angiogenesis, and arteriogenesis are distinct processes for vascularization. Compare the relative advantages and disadvantages of each of these processes for vascularization of 3D artificial tissue.
5. Describe the molecular mechanism of vasculogenesis. What are the participating stem cells, processes for capillary formation, and molecular signals that guide the process of vasculogenesis?
6. Describe the molecular mechanism for angiogenesis. What are the participating stem cells, processes of capillary formation, and molecular signals that guide the process of angiogenesis?
7. Describe the molecular mechanism for arteriogenesis. What are the participating stem cells, processes of capillary formation, and molecular signals that guide the process of arteriogenesis?
8. Explain the concept of therapeutic angiogenesis. What are some potential therapeutic agents that can be used for therapeutic angiogenesis? What are the proposed mechanisms by which therapeutic angiogenesis provides a functional benefit to injured tissue?
9. Identify any clinical condition and explain how therapeutic angiogenesis may be used as a potential therapeutic strategy.
10. What is the difference between *in vivo* and *in vitro* vascularization models for artificial tissue? Compare the relative advantages and disadvantages of *in vivo* and *in vitro* vascularization models for artificial tissue.
11. What are some of the scientific and technological challenges associated with *in vivo* strategies for vascularization of 3D artificial tissue?
12. Develop an *in vivo* vascularization strategy for 3D artificial heart muscle.
13. Select any tissue engineering application and explain why *in vivo* or *in vitro* vascularization strategies would be better suited for your selected application.

14. Describe the idealized process for vascularization of 3D artificial tissue, as described in the chapter.
15. For any given tissue engineering application, develop your own idealized process for vascularization of 3D artificial tissue.
16. Describe the process flow chart used for development of vascularization strategies for artificial tissue.
17. Use the process flow chart for any selected tissue engineering application.
18. Explain the following terms: biologically replicated vascularization strategies, biologically mediated vascularization strategies, and biologically inspired vascularization strategies. What are the relative advantages and disadvantages of these strategies for vascularization of 3D artificial tissue?
19. Pick any tissue fabrication application. Develop three vascularization strategies for your selected application using biologically replicated, mediated, and inspired vascularization strategies.
20. Identify three critical challenges in the field of vascularization as it applies to vascularization for tissue engineering. What can be done to overcome these challenges?

REFERENCES

1. Folkman J. Tumor angiogenesis: a possible control point in tumor growth. Ann. Intern. Med. 1975 Jan;82(1):96–100.
2. Folkman J. Tumor angiogensis: role in regulation of tumor growth. Symp. Soc. Dev. Biol. 1974;30(0):43–52.
3. Folkman J. Tumor angiogenesis. Adv. Cancer Res. 1974;19(0):331–58.
4. Gimbrone MA, Jr., Cotran RS, Leapman SB, Folkman J. Tumor growth and neovascularization: an experimental model using the rabbit cornea. J. Natl. Cancer Inst. 1974 Feb;52(2):413–27.
5. Folkman J. Proceedings: Tumor angiogenesis factor. Cancer Res. 1974 Aug;34(8): 2109–13.
6. Folkman J. Anti-angiogenesis: new concept for therapy of solid tumors. Ann. Surg. 1972 Mar;175(3):409–16. PMCID:PMC1355186.
7. Folkman J. Angiogenesis in psoriasis: therapeutic implications. J. Invest Dermatol. 1972 Jul;59(1):40–3.
8. Folkman J, Merler E, Abernathy C, Williams G. Isolation of a tumor factor responsible for angiogenesis. J. Exp. Med. 1971 Feb 1;133(2):275–88. PMCID:PMC2138906.
9. Folkman J. Tumor angiogenesis: therapeutic implications. N. Engl. J. Med. 1971 Nov 18;285(21):1182–6.
10. Gimbrone MA, Jr., Leapman SB, Cotran RS, Folkman J. Tumor dormancy in vivo by prevention of neovascularization. J. Exp. Med. 1972 Aug 1;136(2):261–76. PMCID:PMC2139203.

REFERENCES

11. Folkman J, Hochberg M. Self-regulation of growth in three dimensions. J. Exp. Med. 1973 Oct 1;138(4):745–53. PMCID:PMC2180571.
12. Schmidt A, Brixius K, Bloch W. Endothelial precursor cell migration during vasculogenesis. Circ. Res. 2007 Jul 20;101(2):125–36.
13. Semenza GL. Vasculogenesis, angiogenesis, and arteriogenesis: mechanisms of blood vessel formation and remodeling. J. Cell Biochem. 2007 Nov 1;102(4):840–7.
14. Czirok A, Zamir EA, Szabo A, Little CD. Multicellular sprouting during vasculogenesis. Curr. Top. Dev. Biol. 2008;81:269–89. PMCID:PMC3025701.
15. Laschke MW, Giebels C, Menger MD. Vasculogenesis: a new piece of the endometriosis puzzle. Hum. Reprod. Update. 2011 Sep;17(5):628–36.
16. Czirok A, Little CD. Pattern formation during vasculogenesis. Birth Defects Res. C Embryo. Today 2012 Jun;96(2):153–62. PMCID:PMC3465733.
17. Giordano FJ. Angiogenesis: mechanisms, modulation, and targeted imaging. J. Nucl. Cardiol. 1999 Nov;6(6):664–71.
18. Harper J, Moses MA. Molecular regulation of tumor angiogenesis: mechanisms and therapeutic implications. EXS 2006;96:223–68.
19. Heil M, Schaper W. Cellular mechanisms of arteriogenesis. EXS 2005;94:181–91.
20. Deindl E, Schaper W. The art of arteriogenesis. Cell Biochem. Biophys. 2005;43(1):1–15.
21. Heil M, Schaper W. Insights into pathways of arteriogenesis. Curr. Pharm. Biotechnol. 2007 Feb;8(1):35–42.
22. Cai W, Schaper W. Mechanisms of arteriogenesis. Acta Biochim. Biophys. Sin. (Shanghai) 2008 Aug;40(8):681–92.
23. Kassmeyer S, Plendl J, Custodis P, Bahramsoltani M. New insights in vascular development: vasculogenesis and endothelial progenitor cells. Anat. Histol. Embryol. 2009 Feb;38(1):1–11.
24. Jin SW, Patterson C. The opening act: vasculogenesis and the origins of circulation. Arterioscler. Thromb. Vasc. Biol. 2009 May;29(5):623–9. PMCID:PMC3432309.
25. Patel-Hett S, D'Amore PA. Signal transduction in vasculogenesis and developmental angiogenesis. Int. J. Dev. Biol. 2011;55(4–5):353–63.
26. De SF, Segura I, De BK, Hohensinner PJ, Carmeliet P. Mechanisms of vessel branching: filopodia on endothelial tip cells lead the way. Arterioscler. Thromb. Vasc. Biol. 2009 May;29(5):639–49.
27. Carmeliet P, De SF, Loges S, Mazzone M. Branching morphogenesis and antiangiogenesis candidates: tip cells lead the way. Nat. Rev. Clin. Oncol. 2009 Jun;6(6):315–26.
28. van RN, Piek JJ, Schaper W, Bode C, Buschmann I. Arteriogenesis: mechanisms and modulation of collateral artery development. J. Nucl. Cardiol. 2001 Nov;8(6):687–93.
29. Hoefer IE, van RN, Rectenwald JE, Deindl E, Hua J, Jost M, Grundmann S, Voskuil M, Ozaki CK, Piek JJ, et al. Arteriogenesis proceeds via ICAM-1/Mac-1- mediated mechanisms. Circ. Res. 2004 May 14;94(9):1179–85.
30. Heil M, Schaper W. Cellular mechanisms of arteriogenesis. EXS 2005;94:181–91.
31. Grundmann S, Piek JJ, Pasterkamp G, Hoefer IE. Arteriogenesis: basic mechanisms and therapeutic stimulation. Eur. J. Clin. Invest. 2007 Oct;37(10):755–66.
32. van Oostrom MC, van OO, Quax PH, Verhaar MC, Hoefer IE. Insights into mechanisms behind arteriogenesis: what does the future hold? J. Leukoc. Biol. 2008 Dec;84(6):1379–91.

33. Cai W, Schaper W. Mechanisms of arteriogenesis. Acta Biochim. Biophys. Sin. (Shanghai) 2008 Aug;40(8):681–92.
34. Chu H, Wang Y. Therapeutic angiogenesis: controlled delivery of angiogenic factors. Ther. Deliv. 2012 Jun;3(6):693–714. PMCID:PMC3564557.
35. Deveza L, Choi J, Yang F. Therapeutic angiogenesis for treating cardiovascular diseases. Theranostics. 2012;2(8):801–14. PMCID:PMC3425124.
36. Ouma GO, Zafrir B, Mohler ER, III, Flugelman MY. Therapeutic Angiogenesis in Critical Limb Ischemia. Angiology 2012 Nov 4.
37. Said SS, Pickering JG, Mequanint K. Advances in growth factor delivery for therapeutic angiogenesis. J. Vasc. Res. 2013;50(1):35–51.
38. Colville-Nash PR, Willoughby DA. Growth factors in angiogenesis: current interest and therapeutic potential. Mol. Med. Today 1997 Jan;3(1):14–23.
39. Yoon YS, Johnson IA, Park JS, Diaz L, Losordo DW. Therapeutic myocardial angiogenesis with vascular endothelial growth factors. Mol. Cell Biochem. 2004 Sep;264(1–2):63–74.
40. Kontos CD, Annex BH. Engineered transcription factors for therapeutic angiogenesis. Curr. Opin. Mol. Ther. 2007 Apr;9(2):145–52.
41. Jones WS, Annex BH. Growth factors for therapeutic angiogenesis in peripheral arterial disease. Curr. Opin. Cardiol. 2007 Sep;22(5):458–63.
42. Nomi M, Miyake H, Sugita Y, Fujisawa M, Soker S. Role of growth factors and endothelial cells in therapeutic angiogenesis and tissue engineering. Curr. Stem Cell Res. Ther. 2006 Sep;1(3):333–43.
43. Sun Q, Silva EA, Wang A, Fritton JC, Mooney DJ, Schaffler MB, Grossman PM, Rajagopalan S. Sustained release of multiple growth factors from injectable polymeric system as a novel therapeutic approach towards angiogenesis. Pharm. Res. 2010 Feb;27(2):264–71. PMCID:PMC2812420.
44. Siemeister G, Martiny-Baron G, Marme D. The pivotal role of VEGF in tumor angiogenesis: molecular facts and therapeutic opportunities. Cancer Metastasis Rev. 1998 Jun;17(2):241–8.
45. Cross MJ, Claesson-Welsh L. FGF and VEGF function in angiogenesis: signalling pathways, biological responses and therapeutic inhibition. Trends Pharmacol. Sci. 2001 Apr;22(4):201–7.
46. Nowak DG, Amin EM, Rennel ES, Hoareau-Aveilla C, Gammons M, Damodoran G, Hagiwara M, Harper SJ, Woolard J, Ladomery MR, et al. Regulation of vascular endothelial growth factor (VEGF) splicing from pro-angiogenic to anti-angiogenic isoforms: a novel therapeutic strategy for angiogenesis. J. Biol. Chem. 2010 Feb 19;285(8):5532–40. PMCID:PMC2820781.
47. Sun Z, Huang P, Tong G, Lin J, Jin A, Rong P, Zhu L, Nie L, Niu G, Cao F, et al. VEGF-loaded graphene oxide as theranostics for multi-modality imaging-monitored targeting therapeutic angiogenesis of ischemic muscle. Nanoscale. 2013 Jun 17.
48. Aviles RJ, Annex BH, Lederman RJ. Testing clinical therapeutic angiogenesis using basic fibroblast growth factor (FGF-2). Br. J. Pharmacol. 2003 Oct;140(4):637–46. PMCID:PMC1350957.
49. Kudo FA, Nishibe T, Nishibe M, Yasuda K. Autologous transplantation of peripheral blood endothelial progenitor cells (CD34+) for therapeutic angiogenesis in patients with critical limb ischemia. Int. Angiol. 2003 Dec;22(4):344–8.

REFERENCES

50. Choi JH, Hur J, Yoon CH, Kim JH, Lee CS, Youn SW, Oh IY, Skurk C, Murohara T, Park YB, et al. Augmentation of therapeutic angiogenesis using genetically modified human endothelial progenitor cells with altered glycogen synthase kinase-3beta activity. J. Biol. Chem. 2004 Nov 19;279(47):49430–8.
51. Zou GM, Karikari C, Kabe Y, Handa H, Anders RA, Maitra A. The Ape-1/Ref-1 redox antagonist E3330 inhibits the growth of tumor endothelium and endothelial progenitor cells: therapeutic implications in tumor angiogenesis. J. Cell Physiol. 2009 Apr;219(1):209–18.
52. Rufaihah AJ, Haider HK, Heng BC, Ye L, Tan RS, Toh WS, Tian XF, Sim EK, Cao T. Therapeutic angiogenesis by transplantation of human embryonic stem cell-derived CD133+ endothelial progenitor cells for cardiac repair. Regen. Med. 2010 Mar;5(2):231–44.
53. Li JY, Su CH, Wu YJ, Tien TY, Hsieh CL, Chen CH, Tseng YM, Shi GY, Wu HL, Tsai CH, et al. Therapeutic angiogenesis of human early endothelial progenitor cells is enhanced by thrombomodulin. Arterioscler. Thromb. Vasc. Biol. 2011 Nov;31(11):2518–25.
54. Ishikane S, Ohnishi S, Yamahara K, Sada M, Harada K, Mishima K, Iwasaki K, Fujiwara M, Kitamura S, Nagaya N, et al. Allogeneic injection of fetal membrane-derived mesenchymal stem cells induces therapeutic angiogenesis in a rat model of hind limb ischemia. Stem Cells 2008 Oct;26(10):2625–33.
55. Liao W, Zhong J, Yu J, Xie J, Liu Y, Du L, Yang S, Liu P, Xu J, Wang J, et al. Therapeutic benefit of human umbilical cord derived mesenchymal stromal cells in intracerebral hemorrhage rat: implications of anti-inflammation and angiogenesis. Cell Physiol. Biochem. 2009;24(3–4):307–16.
56. Zhang Y, Zhang R, Li Y, He G, Zhang D, Zhang F. Simvastatin augments the efficacy of therapeutic angiogenesis induced by bone marrow-derived mesenchymal stem cells in a murine model of hindlimb ischemia. Mol. Biol. Rep. 2012 Jan;39(1):285–93.
57. Lee EJ, Park HW, Jeon HJ, Kim HS, Chang MS. Potentiated therapeutic angiogenesis by primed human mesenchymal stem cells in a mouse model of hindlimb ischemia. Regen. Med. 2013 May;8(3):283–93.
58. Rosengart TK, Patel SR, Crystal RG. Therapeutic angiogenesis: protein and gene therapy delivery strategies. J. Cardiovasc. Risk. 1999 Feb;6(1):29–40.
59. Kornowski R, Fuchs S, Leon MB, Epstein SE. Delivery strategies to achieve therapeutic myocardial angiogenesis. Circulation 2000 Feb 1;101(4):454–8.
60. Bhise NS, Shmueli RB, Sunshine JC, Tzeng SY, Green JJ. Drug delivery strategies for therapeutic angiogenesis and antiangiogenesis. Expert. Opin. Drug Deliv. 2011 Apr;8(4):485–504.
61. Ouma GO, Jonas RA, Usman MH, Mohler ER, III. Targets and delivery methods for therapeutic angiogenesis in peripheral artery disease. Vasc. Med. 2012 Jun;17(3):174–92.
62. Said SS, Pickering JG, Mequanint K. Advances in growth factor delivery for therapeutic angiogenesis. J. Vasc. Res. 2013;50(1):35–51.
63. Grist TM, Mistretta CA, Strother CM, Turski PA. Time-resolved angiography: Past, present, and future. J. Magn Reson. Imaging. 2012 Dec;36(6):1273–86.
64. Liotta R, Chughtai A, Agarwal PP. Computed tomography angiography of thoracic aortic aneurysms. Semin. Ultrasound CT MR 2012 Jun;33(3):235–46.

65. Nakazono T, Suzuki M, White CS. Computed tomography angiography of coronary artery bypass graft grafts. Semin. Roentgenol. 2012 Jul;47(3):240–52.
66. Elcin YM, Dixit V, Gitnick G. Extensive in vivo angiogenesis following controlled release of human vascular endothelial cell growth factor: implications for tissue engineering and wound healing. Artif. Organs. 2001 Jul;25(7):558–65.
67. Matsuda K, Falkenberg KJ, Woods AA, Choi YS, Morrison WA, Dilley RJ. Adipose-derived stem cells promote angiogenesis and tissue formation for in vivo tissue engineering. Tissue Eng Part A. 2013 Jun;19(11–12):1327–35. PMCID:PMC3638514.
68. Schumann P, Lindhorst D, von SC, Menzel N, Kampmann A, Tavassol F, Kokemuller H, Rana M, Gellrich NC, Rucker M. Accelerating the early angiogenesis of tissue engineering constructs in vivo by the use of stem cells cultured in matrigel. J. Biomed. Mater. Res. A. 2013 Jun 14.
69. Birla RK, Borschel GH, Dennis RG. In vivo conditioning of tissue-engineered heart muscle improves contractile performance. Artif. Organs. 2005 Nov;29(11):866–75.
70. Bu X, Yan Y, Zhang Z, Gu X, Wang M, Gong A, Sun X, Cui Y, Zeng Y. Properties of extracellular matrix-like scaffolds for the growth and differentiation of endothelial progenitor cells. J. Surg. Res. 2010 Nov;164(1):50–7.
71. Jabbarzadeh E, Starnes T, Khan YM, Jiang T, Wirtel AJ, Deng M, Lv Q, Nair LS, Doty SB, Laurencin CT. Induction of angiogenesis in tissue-engineered scaffolds designed for bone repair: a combined gene therapy-cell transplantation approach. Proc. Natl. Acad. Sci. U.S.A 2008 Aug 12;105(32):11099–104. PMCID:PMC2516212.
72. Ye C, Bai L, Yan ZQ, Wang YH, Jiang ZL. Shear stress and vascular smooth muscle cells promote endothelial differentiation of endothelial progenitor cells via activation of Akt. Clin. Biomech. (Bristol., Avon.) 2008;23 Suppl 1:S118–S124.
73. Birla RK, Borschel GH, Dennis RG, Brown DL. Myocardial engineering in vivo: formation and characterization of contractile, vascularized three-dimensional cardiac tissue. Tissue Eng. 2005 May;11(5–6):803–13.
74. Sun J, Wang Y, Qian Z, Hu C. An approach to architecture 3D scaffold with interconnective microchannel networks inducing angiogenesis for tissue engineering. J. Mater. Sci. Mater. Med. 2011 Nov;22(11):2565–71.
75. Norotte C, Marga FS, Niklason LE, Forgacs G. Scaffold-free vascular tissue engineering using bioprinting. Biomaterials 2009 Oct;30(30):5910–7. PMCID:PMC2748110.
76. Xiaozhen D, Shaoxi C, Qunfang Y, Jiahuan J, Xiaoqing Y, Xin X, Qifeng J, Albert Chih-Lueh W, Yi T. A novel in vitro angiogenesis model based on a microfluidic device. Chin Sci. Bull. 2011 Nov 1;56(31):3301–9. PMCID:PMC3254117.

6

BIOREACTORS FOR TISSUE ENGINEERING

Learning Objectives

After completing this chapter, students should be able to:

1. Provide examples of the role of biomechanical forces in nature.
2. Provide a definition for bioreactor.
3. Explain the difference between enabling and supporting technology for tissue engineering.
4. Describe the classification scheme for bioreactors and distinguish between bioreactors for cell culture, scaffold fabrication, tissue fabrication, and physiological conditioning.
5. Discuss design considerations for bioreactors.
6. Provide an example of an idealized bioreactor system.
7. Describe integration between tissue engineering and bioreactor technology and explain how bioreactors can be used at different stages of tissue fabrication.
8. Discuss the role of bioreactors in supporting and/or enabling fabrication of 3D artificial tissue.
9. Explain how bioreactors are used for mammalian cell culture.
10. Discuss the use of bioreactors for scaffold fabrication.

Introduction to Tissue Engineering: Applications and Challenges, First Edition. Ravi Birla.
© 2014 The Institute of Electrical and Electronics Engineers, Inc. Published 2014 by John Wiley & Sons, Inc.

11. Provide examples of bioreactors used for scaffold cellularization.
12. Explain the importance of perfusion in tissue engineering.
13. Explain the importance of stretch in tissue engineering, and describe the effect of stretch on smooth muscle cells, endothelial cells, and cardiac myocytes.
14. Explain the importance of electrical stimulation in tissue engineering.
15. Provide examples of bioreactors for electrical stimulation in tissue engineering.

CHAPTER OVERVIEW

In this chapter, we will look at the critical role of bioreactors in the fabrication of artificial tissue, and integration between the fields of bioreactor technology and tissue engineering. This chapter will start with examples of biomechanical forces in nature to illustrate the importance of these forces on tissue formation and function. This will be followed with a working definition of bioreactors, along with a classification scheme for bioreactors. We will present design considerations for bioreactors and follow this up with a description of an idealized bioreactor system. The idealized system is based on our wish list for functional capabilities for bioreactor technology. We then present a scheme outlining the integration between bioreactor technology and tissue engineering and will illustrate the role of bioreactors at every stage of artificial tissue fabrication. We next provide examples of bioreactors for mammalian cell culture, scaffold fabrication, and cellularization and bioreactors for perfusion, stretch, and electrical stimulation. In each case, we provide design considerations, effects of stimuli on tissue function, and examples of specific bioreactor systems that are currently in use at research laboratories.

6.1 INTRODUCTION TO BIOREACTORS

The development of bioreactor technology is a fundamental component of tissue engineering. The field of bioreactor technology has progressed in parallel with tissue engineering, as a scientific discipline. During the early phases of tissue engineering, fabrication of artificial tissue involved culture of cells within a 3D support matrix with cell-cell interactions and cell-matrix interactions leading to development of functional 3D artificial tissue. This remains the first step for 3D artificial tissue fabrication, even in current research. However, it was not long before researchers recognized the need for bioreactor technology in the fabrication, culture, and maintenance of artificial tissue. As is often the case in tissue engineering, inspiration is drawn from nature, and the development of bioreactor technology is no exception.

During normal human function, tissues are exposed to biomechanical forces, which are important in maintaining form and function of the particular tissue. For example, the heart is a dynamic organ that continuously generates force in response to a depolarization wave, and every time the heart beats, there is a change in the local stress environment. *What does this mean in terms of bioreactor technology?* This has clear implications in the development of artificial heart muscle, which needs to be cultured in the presence of continuous electrical stimulation and mechanical stretch, and subjected to continuous media perfusion. During the fabrication of artificial heart muscle, bioreactor technology will need to be implemented to simulate *in vivo* physiological conditions during controlled *in vitro* culture. This is the case with any tissue system under development—*in vivo* conditions need to be recapitulated *in vitro* by the development of bioreactor technology. Simply stated, bioreactor technology is important to support tissue development and maturation by simulating *in vivo* conditions during *in vitro* culture.

We can obtain a better understanding of the importance of bioreactor technology by looking at one specific example: changes in human physiology of astronauts during space missions. When astronauts are deployed to space missions, there is limited compression on their musculoskeletal systems due to the low gravity environment in space. Compression is required to maintain bone and skeletal muscle mass during normal physiological function. As a result of the low gravity environment in space, when astronauts return from space missions, they have a deficit in bone and skeletal muscle mass due to the lack of compressive forces on the musculoskeletal system. This loss of bone mass is reversible and can be recovered by physiotherapy, which involves subjecting skeletal muscle to gradually increasing compressive forces. This example shows the adaptive nature of tissue and the ability to modulate tissue function based on external stimuli: physiological signals in the human body and bioreactors in tissue engineering.

Let us look at some examples of biomechanical conditioning in nature. We have already seen that heart muscle is constantly exposed to electrical stimulation and mechanical stretch. We have also seen that bone is under constant compression and that compressive forces are important to maintain critical bone mass. Skeletal muscle is also very dynamic and an increase in muscle activity in the form of exercise results in an increase in muscle mass; the opposite is also true, as skeletal muscle undergoes atrophy in times of inactivity. Blood vessels are constantly exposed to pulsatile fluid flow on the luminal surface, and this stress environment is important for normal function.

We can view bioreactors in the context of physiological conditioning, designed to replicate and deliver specific stimuli to 3D artificial tissue at various stages of development to support formation, development, and maturation of functional tissue that is similar in form and function to mammalian tissue.

6.2 BIOREACTORS DEFINED

In the previous section, we introduced the concept of bioreactors by looking at the role of biomechanical forces in nature and the role of these forces in maintaining homeostasis during normal function. We also stated the essential role of these biomechanical forces during the fabrication and culture of artificial muscle, and stated that bioreactors are required to achieve this objective. *But what exactly is a bioreactor and how do we define it?* Unfortunately, there is no universally accepted definition for bioreactors in the field of tissue engineering. This is due to the diversity of functions for which bioreactors are used, and the degree of customization required from one application to another. This often results in different views on the definition and applications of bioreactors for tissue engineering. Although these differences exist, we believe that it is important to have a unifying definition for bioreactors.

Bioreactors have been used for a wide range of functions, including cell culture and proliferation, scaffold fabrication, artificial tissue formation, and for providing controlled physiological conditioning (mechanical stretch, electrical stimulation, continuous media perfusion) for 3D artificial tissue development and maturation. We will provide additional details on these specific applications for bioreactor technology in the next section. The use of bioreactor technology is intertwined with the process of 3D artificial tissue fabrication. Based on the specific application, bioreactors have either been categorized as enabling technology or supporting technology in tissue engineering. Enabling technology refers to any process that "enables" the formation of 3D artificial tissue; such technology is used at any stage prior to the fabrication of 3D artificial tissue. Enabling technology for bioreactors refers to systems that support cell culture and proliferation, scaffold fabrication, and tissue fabrication. In all of these cases, bioreactors are used to enable the process; they are used prior to the formation of 3D artificial tissue. Supporting technology refers to bioreactors used after formation of 3D artificial tissue; they are used to "support" the development, growth, and maturation of artificial tissue.

Since a universally accepted definition of bioreactors does not exist, we will build our own definition. We can develop a definition of bioreactors by assessing the components of the word bioreactor—"bio" and "reactor". "Bio" refers to the application of technology for biological purposes, and in tissue engineering, everything is biologically based. "Reactor" is often used in chemical engineering and refers to specific devices that are used for a chemical reaction. Bioreactors are generally not used to support any chemical reactions or reactions of any other sort during the fabrication and/or culture of artificial tissue. The use of the term "reactor" is not directly linked to bioreactor applications in tissue engineering, but due to its extensive use in the literature, the term will be retained.

What have we learned thus far about bioreactors that can assist in developing a working definition? Bioreactors are used extensively in tissue engineering for all steps in the tissue fabrication process, they are considered either enabling or supporting technologies, and in most cases, they do not involve chemical reactions. Based on this framework, we propose the following definition for bioreactors:

"Bioreactors are devices used extensively in tissue engineering to *enable* the fabrication of artificial of 3D artificial tissue and *support* the growth, maturation, and development of artificial tissue during controlled *in vitro* culture." This definition encompasses the breadth of applications for bioreactors in tissue engineering and embodies the use of bioreactors as enabling and supporting technologies. This is illustrated in Figure 6.1.

CLASSIFICATION OF BIOREACTORS

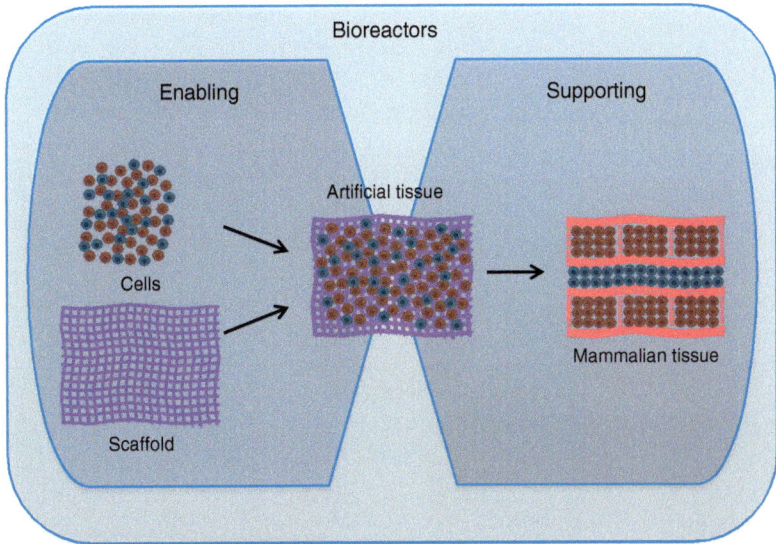

Figure 6.1 Definition of Bioreactors–**Enabling**—Technology required for the fabrication of artificial tissue. **Supporting**—Technology required for development and maturation of 3D artificial tissue. **Bioreactors = Enabling + Supporting**.

6.3 CLASSIFICATION OF BIOREACTORS

We introduced the concept of bioreactors in the first section by looking at biomechanical forces in nature and the need to simulate these forces during fabrication and culture of artificial tissue. In the previous section, we provided a unifying definition of bioreactors and described many applications of bioreactor technology in tissue engineering, including cell culture, scaffold fabrication, scaffold cellularization, and bioreactors for stretch, perfusion, and electrical stimulation. In this section, we describe each of these and provide a classification scheme for bioreactors.

Bioreactors for Cell Culture and Expansion—A large number of cells are required for most tissue engineering experiments. Cells are cultured in tissue culture plates and/or flasks, and once the cellsbecome confluent, they are subpassaged and cultured on additional tissue culture plates to increase cell yield. This process is done manually and has several limitations. Culture and expansion of cells is expensive and time-consuming, and working with a large number of culture plates can increase the risk of contamination. Many of the limitations of traditional cell culture can be overcome by using automated systems to perform all the tasks associated with the culture and expansion of mammalian cells (1–6). Cell culture bioreactors have the capacity to perform all tasks associated with the maintenance of cells, including media changes and supplementation of the media with growth factors and cytokines (Figure 6.2a). These bioreactors can also undertake all tasks

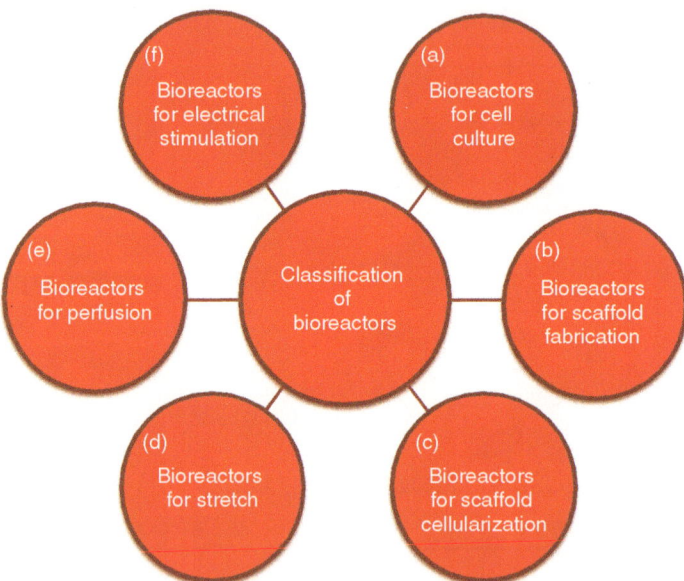

Figure 6.2 Classification of Bioreactors—**(a) Bioreactors for Cell Culture**—Robotic nozzles are used for delivery and aspiration of cell culture media. **(b) Bioreactors for Scaffold Fabrication**—Electrospinning is used for scaffold fabrication and leads to the formation of nanofibers, which can be assembled to form porous 3D scaffolds. **(c) Bioreactors for Scaffold Cellularization**—The cell suspension is perfused through a porous scaffold to promote scaffold cellularization. **(d) Bioreactors for Stretch**—Artificial tissue is subjected to continuous cyclic strain. **(e) Bioreactors for Perfusion**—Input and output ports are engineered within the culture vessel to support continuous media perfusion. **(f) Bioreactors for Electrical Stimulation**—Electrodes are engineered within the culture vessel to deliver controlled electrical stimulation.

associated with subpassaging, including trypsinization, dilution, and replating of these cells. All processes take place under sterile conditions using robotic arms and guidance control via software algorithms. The use of these bioreactors minimizes human effort and reduces the probability of contamination.

Bioreactors for Scaffold Fabrication—Electrospinning has been used extensively for the fabrication of scaffolds with very tightly regulated fiber architecture and diameter, ranging from nanometer to micrometer size (7–12). During the electrospinning process, the polymer is first solubilized in a suitable solvent and then, in response to a high voltage electric charge, discharged as a fluid jet. This fluid jet travels in air. During this process, the solvent in which the polymer is solubilized evaporates, leaving behind a thin microfibrous structure. The thin fibers are collected in a mandrel or some other collection device and are used to form complex scaffolds with individual fibers as the starting material (Figure 6.2b). This process has resulted in the fabrication of very complex scaffolds, which have been used to support a wide variety of tissue engineering applications.

Bioreactors for Scaffold Cellularization—This category of bioreactors is designed to support the fabrication of 3D artificial tissue in a controlled sterile environment. One of the steps in the process of tissue fabrication involves cellularization of 3D scaffolds. Scaffold cellularization can be achieved manually by suspending isolated cells in culture media and injecting the cells into the 3D scaffold using a syringe. However, direct cell injection does not result in a uniform cell distribution throughout the scaffold; it is an imprecise process and very user-dependent. In order to circumvent this problem, bioreactors have been developed to promote scaffold cellularization using many different techniques, including perfusion based technologies (13) (Figure 6.2c). In the case of perfusion bioreactors, cells are suspended in culture media and then perfused through a porous scaffold; perfusion seeding of cells results in uniform cell distribution throughout the 3D scaffold.

During normal human function and development, all cells and tissues are constantly exposed to a myriad of signals that guide tissue maturation and development, and alter tissue function. These signals are in the form of electrical currents, mechanical stretch, compression, fluid shear stress, and changes in the chemical environment (14–17). Bioreactors for stretch, perfusion, and electrical stimulation are designed to recapitulate *in vivo* stimulation protocols during *in vitro* culture and maturation of 3D artificial tissue. Collectively, this group of bioreactors, designed to deliver controlled signals for physiological conditioning, are the largest group of bioreactors in tissue engineering.

Bioreactors for Stretch—Stretch is an important modulator of tissue function, particularly for muscle tissue. During normal function of mammalian cardiac muscle, cells are exposed to repetitive cycles of lengthening and shortening; this stimulation is important for normal cardiac function. Bioreactors for stretch are designed to replicate these continuous cycles of lengthening and shortening of muscle tissue, and to support the development and maturation of artificial tissue (Figure 6.2d).

Bioreactors for Perfusion—In the human body, the circulatory system provides a distribution network for blood flow and serves to deliver nutrients to all cells and tissue and remove waste products. All cells in the human body are located within a couple hundred microns to a capillary; this property is critical to support cell and tissue viability. Bioreactors for perfusion are designed to do the same thing for 3D artificial tissue by providing continuous flow of cell culture media to deliver nutrients to artificial tissue and remove metabolic waste products (Figure 6.2e).

Bioreactors for Electrical Stimulation—During normal function of excitable tissue like cardiac muscle, skeletal muscle, and neural tissue, changes in voltage play an important role in maintaining and supporting cell and tissue level function. As an example, depolarization waves in the heart are due to voltage changes, which in turn provide the trigger for cardiac muscle contraction. Bioreactors for electrical stimulation are designed to replicate these changes in voltage and other electrical properties observed during normal mammalian function to support the growth and development of artificial tissue (Figure 6.2f).

6.4 DESIGN CONSIDERATIONS

The process flow chart for bioreactor design is presented in Figure 6.3 and consists of four steps: 1) definition of stimuli, 2) control of processing variables, 3) sensor technology, and 4) stimulation protocol. Let us look at each of these in more detail.

Definition of Stimuli—The most important design consideration that needs to be addressed during bioreactor design and fabrication is the stimuli that needs to be delivered to the cells and/or tissue. All other steps in the design process revolve around the nature of the stimuli. In most cases, the objective is to recapitulate *in vivo* conditions *in vitro*, and this is particularly the case when designing bioreactors for cell culture/proliferation and physiological stimulation of 3D artificial tissue. Some examples of stimuli that are used in bioreactors include mechanical stretch, electrical stimulation, continuous media perfusion, and compressive forces. The specific stimulation will depend on the type of tissue being fabricated, with compressive forces being important for bone tissue and electrical stimulation and stretch being important parameters for artificial heart muscle.

Control of Culture Variables—During culture of artificial tissue using bioreactor stimulation, processing variables need to be regulated. These culture variables include temperature, pH, and composition of the gaseous environment. The can be done in one of two ways. In the first case, the bioreactor and 3D artificial tissue are all maintained in a cell culture incubator, which has all the necessary mechanisms in place for control of culture variables. This makes use of a commercially available system, which has been tested and validated, to control the culture environment. This process also simplifies bioreactor design and fabrication considerably. The second strategy is to engineer control systems within the bioreactor system. This is a complex process, requires advanced engineering and sensor technology, and increases the time and cost of bioreactor development. However, in-house control of the culture environment provides a high degree of process control, something which is not available in a cell culture incubator.

In the case of in-house regulation of processing variables, temperature regulation has been achieved by heat exchange from water-jacketed chambers or by use of heated metal plates with feedback control. The regulation of pH often involves injection of CO_2 into the culture environment with feedback control to regulate pH at 7.4 and CO_2 at 5%, similar to what is seen in a cell culture environment. A variety of buffers are also available for pH regulation.

A balanced strategy could involve building a first-generation bioreactor that is used in a cell culture incubator; this can be used to test and validate bioreactor functionality. Once successful, a second-generation system can be developed to incorporate control of processing variables in the environment; this way, bioreactor development has been completed in an earlier phase.

Sensor Technology—Another important design consideration is sensor technology, also one of the least developed technologies in the bioreactor field. There are two categories of sensor technology. First, sensors are needed to monitor environmental conditions like temperature, pH, and gaseous environment and

DESIGN CONSIDERATIONS

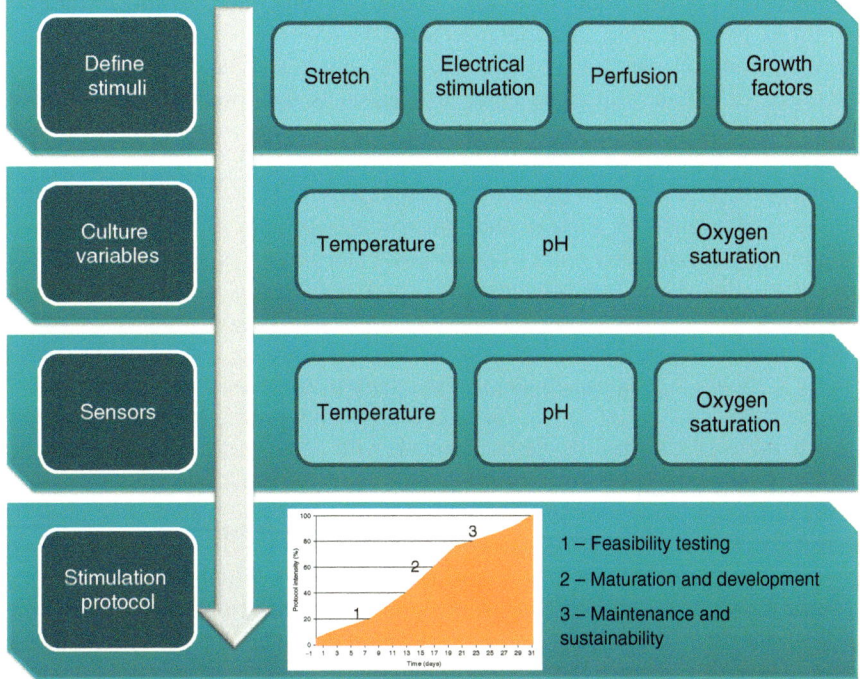

Figure 6.3 Design Consideration for Bioreactors—The flow chart outlines a decision-making process for bioreactor fabrication. **(a) Define Stimuli**—The first step in the process requires identification of specific stimuli to be delivered to 3D artificial tissue. Specific stimuli could be stretch, electrical stimulation, perfusion, and/or compression. **(b) Culture Variables**—During culture of artificial tissue, processing variables need to be regulated. These include temperature, pH, and gaseous composition. Control of processing variables can be accomplished in one of two ways. The bioreactor can be placed inside of a cell culture incubator, which has the capability to control these processing variables. Alternatively, bioreactors can be engineered with these capabilities; in this case, the bioreactor can be operated independent of a cell culture incubator. **(c) Sensors**—Sensor technology is required to monitor culture variables and to measure changes in the functional performance of 3D artificial tissue over time. There are two categories of sensors. First, there are sensors to monitor culture variables like temperature, pH, and gaseous composition, and second, there are sensors to monitor tissue function. **(d) Stimulation Protocol**—The final decision that needs to be made is the specific stimulation protocol to be used to condition 3D artificial tissue. The stimulation protocol is divided into three phases, which include feasibility testing, maturation and development, and finally, maintenance and sustainability.

use this information in a feedback control loop. This area of sensor technology is fairly well-established. The second area involves functional assessment of artificial tissue and again, use of this information for feedback control of tissue development. This area of sensor technology is poorly developed. The most common approach involves removing artificial tissue from the bioreactor at regular intervals for functional assessment. At this time, artificial tissue is sacrificed and the functional data acquired is not used in a feedback control loop. In addition, sampling points are few, infrequent, and are invasive and interface with tissue function. What is required is noninvasive real time monitoring of 3D artificial tissue function and use of this information in a feedback control loop.

Stimulation Protocol—Another important variable is the stimulation protocol—for example, *if we develop bioreactors for uniaxial stretch for cardiac patches, what stretch protocol should we use?* The parameter space can be defined by stretch frequency, percentage stretch, and duration of stretch. One may be tempted to initiate a stretch protocol that replicates stretch protocols observed under normal physiological function. While this may be a reasonable starting point, it should be cautioned that artificial tissue may be at an early stage of development. This means that artificial tissue may not be able to sustain a rigorous stretch protocol; it may be better to start with a milder stretch protocol and gradually progress toward replicating harsher *in vivo* conditions. The stimulation protocol should be divided into three phases. The first phase should be designed to test the feasibility of using any specific stimuli with 3D artificial tissue and involves very low stimulation protocols, designed as initial feasibility testing. Once artificial tissue sustains a low level of stimulation, stimulation intensity can be slowly increased to support development and maturation of 3D artificial tissue; this refers to the growth and maturation phase of 3D artificial tissue. Finally, as the stimulation protocol approaches values of *in vivo* conditions, there is concern of excessive stimulation, which can lead to tissue fatigue or damage. In this final phase, any increase in stimulation intensity needs to be incremental and carefully graduated, and the tissue response must be tightly monitored.

A Final Note on Design Considerations—As we develop bioreactor technology, a critical area needs to be addressed. *In vivo*, mammalian tissue is not exposed to a single variable at any given time, but rather, it responds to a diverse array of signals; in addition, there are temporal and spatial variations in these signals. However, most bioreactors are designed to deliver a single stimuli, for example: stretch, electrical stimulation, or perfusion. Further, in most cases, there are no temporal or spatial variations in the stimulation signals. This is an area that needs to grow as the field of tissue engineering evolves, and advanced bioreactors need to be developed that have the capability of controlling multiple stimulation protocols at the same time and supporting temporal and spatial variations in signals.

6.5 IDEALIZED BIOREACTOR SYSTEM

In this section, we discuss an idealized bioreactor system for the growth and maturation of artificial heart muscle (Figure 6.4). While components of such a system have been developed, a complete bioreactor that has all of the capabilities listed here has not been fabricated. As we have seen from our discussion thus far and will see in the reminder of this chapter, technological innovation in the bioreactor

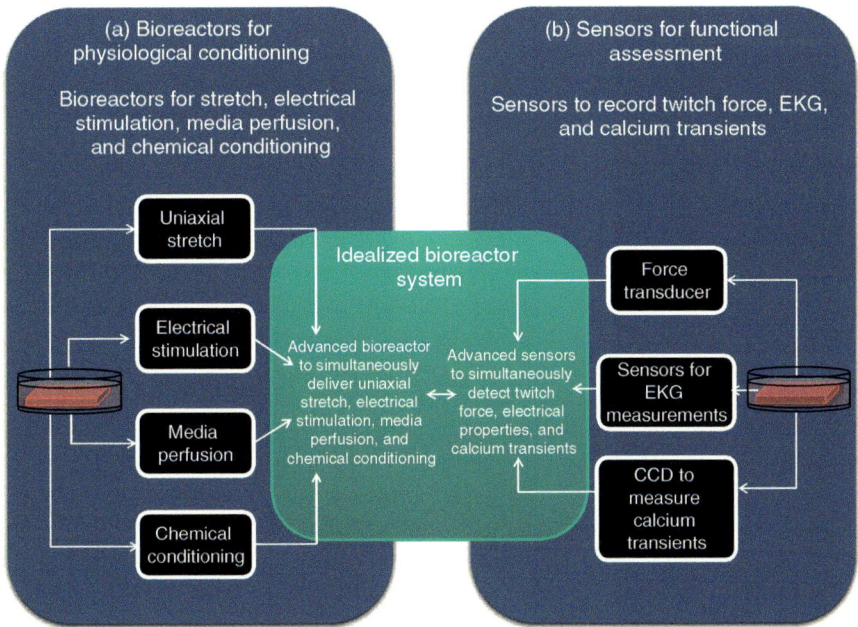

Figure 6.4 Idealized Bioreactor System for 3D Artificial Heart Muscle—(a) Bioreactors for Physiological Conditioning—Artificial heart muscle is fabricated in a tissue culture plate and stimulated using bioreactors for stretch, electrical stimulation, media perfusion, and chemical factors. In most cases, this is done using independent bioreactors for each stimuli; in the idealized bioreactor system, a single bioreactor will have the capabilities to deliver all four stimuli to condition 3D artificial heart muscle. (b) Sensors for Functional Assessment—Functional properties of 3D artificial heart muscle are recorded using sensors for twitch force, electrical properties, and calcium transients. The functional measurements are done independently using different sensors arrays for each measurement. In the idealized bioreactor system, functional properties of 3D artificial heart muscle can be measured using a single set of sensor arrays. In addition, the functional assessment can take place in real time and can make use of noninvasive sensor technology.

space is rapidly progressing to the point where advanced bioreactors are becoming a reality and fabrication of our idealized bioreactor is not far away.

The objective of the idealized bioreactor system is to provide physiological conditioning for 3D artificial heart muscle in order to support growth and maturation. The functional properties of artificial heart muscle from any research laboratory are lower than that of mammalian heart muscle. It is hypothesized that bioreactor conditioning will bridge the gap between the functional performance of artificial and mammalian heart muscle. Based on this hypothesis, our idealized bioreactor system has capabilities to deliver controlled electrical stimulation, uniaxial mechanical stretch, continuous media perfusion, and regulated growth factor stimulation (Figure 6.4a). These conditions accurately represent the culture environment of the heart during normal mammalian function. The last of the four, growth factor stimulation, does not require device fabrication and therefore may not be considered a bioreactor; however, it is important in driving heart muscle phenotype.

Systems for individual stimuli are first shown independently for illustrative purposes and then combined into a single bioreactor. One can easily appreciate the degree of complexity associated with the idealized bioreactor system. This system is designed to condition artificial heart muscle using uniaxial stretch, electrical stimulation, continuous media perfusion, and chemical conditioning. As can be seen by the schematic, this system is extremely complicated, delivery of multiple signals is difficult, and device prototyping is challenging, to say the least.

Temporal and spatial variations in stimuli are needed. As an example, during the early stages of artificial heart muscle development, chemical conditioning needs to stimulate extracellular matrix production in order to support tissue growth and development. Therefore, the composition of growth factors would be optimized to support ECM production and include compounds like ascorbic acid and TGF-β. This needs to be followed by stretch protocols to support alignment of the newly formed ECM, and the intensity of the stretch protocol needs to be increased with the increase in ECM production. The rate of ECM production needs to be monitored and used as an input to adjust the stretch protocol. This is true for electrical stimulation and perfusion, and continuous exchange of information needs to take place between all four stimulation protocols.

The second part of our idealized bioreactor system consists of real time functional assessment of 3D artificial heart muscle (Figure 6.4b). The functional performance of artificial heart muscle can be evaluated by measuring the contractile properties, electrical properties, and calcium transients. In order to measure contractile properties, force transducers need to be attached to 3D artificial heart muscle, and in order to measure electrical properties, EKG catheters need to be engineered into the system. Calcium transients are measured by voltage-sensitive dyes, with changes in dye intensity correlating to changes in intracellular calcium concentrations. Measurement of any one of these functional properties is a challenging task; needless to say, the task of measuring all three functional performance metrics is a formidable challenge.

The final element of our idealized bioreactor system is communication between the two major systems—bioreactors and sensors. The sensors need to monitor functional performance of 3D artificial heart muscle using real time noninvasive monitoring technology; this data then serves as feedback to control stimulation protocols. As one example, embedded sensors will measure changes in calcium conduction velocity and can correlate this to intercellular connectivity. If a decrease in calcium conduction velocity is measured, it can be due to a decrease in intracellular connectivity. This data can then be used as an input for the bioreactors, which can modulate the stimulation protocol to support increased intracellular connectivity. One way of doing this process is by increasing the intensity of the stretch protocol to support intercellular alignment, connectivity, and as a result, support an increase in calcium conduction velocity.

6.6 BIOREACTORS AND TISSUE ENGINEERING

Bioreactor technology is critical for tissue engineering (Figure 6.5) and has been used to support every stage of the tissue fabrication process. The two are interrelated, and tissue fabrication is dependent on bioreactor technology to the extent that it is difficult to fabricate 3D artificial tissue in the absence of bioreactors. The significance of bioreactor technology in the tissue fabrication process cannot be underestimated, and the interdependent relationship has to be at the forefront of technological innovation in tissue engineering. Bioreactors have allowed the field of tissue engineering to progress by leaps and bounds, and continued progress in the field will depend on advancements in bioreactor technology.

Tissue fabrication can, and often does, take place in the absence of any bioreactors. This has been the case during the early years in the development of the field, when bioreactor technology was not as prevalent. However, significant improvements in process efficiency can occur through the implementation of bioreactor technology, and this improvement can in turn, lead to the fabrication of artificial tissue that is closer in form and function to mammalian tissue.

In order to better understand the relationship between bioreactor technology and tissue engineering, let us revisit the process of tissue fabrication. The first step required in tissue fabrication requires isolation, purification, and expansion of cells. This is followed by scaffold fabrication and cellularization of the scaffold

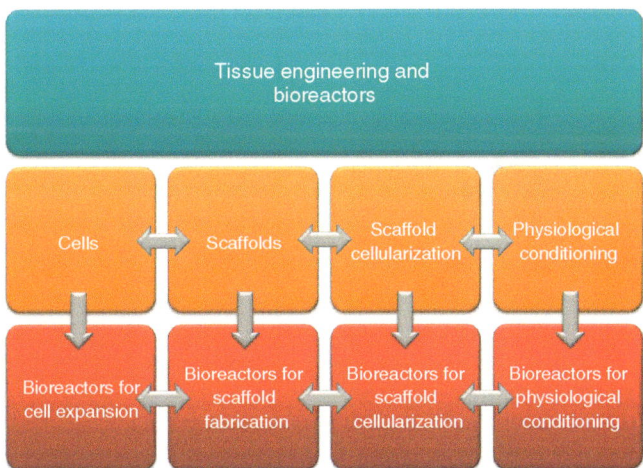

Figure 6.5 Integration of Bioreactor Technology with Tissue Engineering—This figure shows the close integration of bioreactor technology with tissue engineering and tissue fabrication. The figure illustrates core technologies required for tissue fabrication, including cells, scaffolds, scaffold cellularization, and physiological conditioning. Bioreactors can then be used to support core technologies necessary for tissue fabrication; this includes bioreactors for cell culture and expansion, bioreactors for scaffold fabrication, bioreactors for scaffold cellularization, and bioreactors for physiological conditioning (stretch, electrical stimulation, and continuous media perfusion).

with isolated cells. At this stage of the process, combination of cells with scaffolds results in the formation of functional 3D tissue. Artificial tissue at this early stage of development can be viewed as generation-one tissue, as it has some properties of mammalian tissue, but not all. The next step in the process is to support development and maturation of artificial tissue so that it is similar in form and function to mammalian tissue. One strategy to achieve this objective is by conditioning 3D artificial tissue using bioreactors to provide controlled stimulation for mechanical stretch, electrical stimulation, and continuous media flow.

Now that we have revisited the process of tissue fabrication, let us look at how bioreactors can be used at each step of the tissue fabrication process. During cell isolation and expansion, automated cell culture systems have been developed that have the capability of performing all related tasks using advanced robotics and control systems, thereby reducing the risk of contamination. Similarly, bioreactors are very often used for scaffold fabrication, and the process of electrospinning has gained popularity. Electrospinning is a tightly regulated process that supports fabrication of microfibers ranging in diameter from nanometers to micrometers (Figure 6.6).

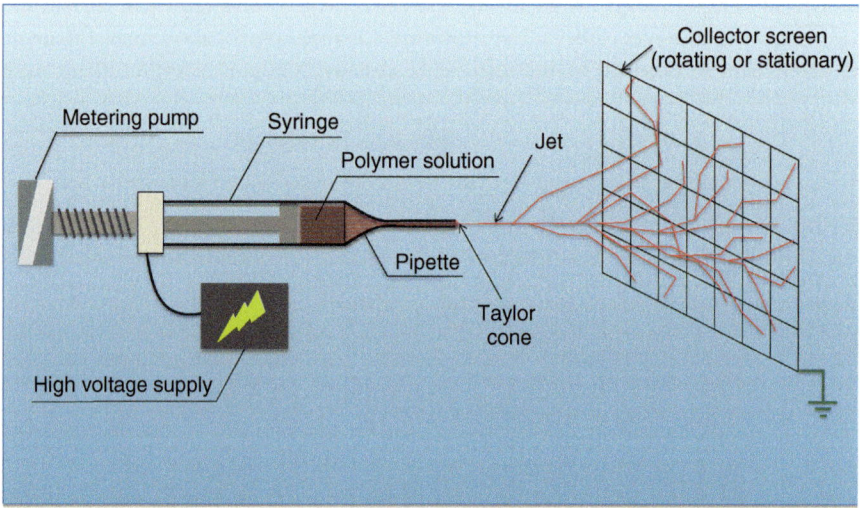

Figure 6.6 Electrospinning for Scaffold Fabrication—The process begins by solubilizing a polymer in a solvent and placing the polymer solution in a syringe. The polymer solution is held at the tip of the syringe by surface tension. One electrode is placed in the polymer solution and a second electrode is placed in the collection device. A high voltage power supply is used to propel the polymer solution out of the syringe by overcoming the surface tension that holds it in place. Gradually increasing the voltage results in the formation of a Taylor cone, which refers to the attachment of the polymer solution to the tip of the syringe in the shape of a cone. As the voltage is increased further, the fluid is ejected from the Taylor cone toward the collection device. As the polymer travels toward the collection device, the solvent evaporates, resulting in the formation of fibers. The fibers are collected in a collection device that can be configured to support the formation of 3D scaffolds.

Scaffold cellularization has also been supported by bioreactors using perfusion systems and other mechanisms for cell delivery within 3D scaffolds. This in turn results in increased cell retention and uniformity of cells distributed throughout the scaffold. Once 3D artificial tissue has been fabricated, bioreactors are used to deliver controlled stimulation to support the growth, development, and maturation of 3D artificial tissue. Bioreactors are used for mechanical stretch, electrical stimulation, and for continuous media perfusion.

6.7 BIOREACTORS FOR MAMMALIAN CELL CULTURE

Mammalian cell culture is at the heart of tissue engineering. Isolation, culture, and expansion of mammalian cells are critical prerequisites for tissue fabrication. Cell culture is one of the first techniques learned by new entrants to the field, including graduate and undergraduate students, and there are hundreds of cell culture facilities at large public universities. Mammalian cell culture is conducted using manual techniques, and for the most part, these manual techniques have been extremely successful. Traditional cell culture involves attachment and proliferation of adherent mammalian cells on tissue culture plates or flasks, with cell culture media being replaced every 1–2 days. Once cells reach confluency, trypsin is used to detach the cells from the culture surface; these cells are then replated at a lower density on multiple tissue culture plates. All of the techniques associated with mammalian cell culture have been conducted manually with a very high degree of success. Cell culture techniques are considered standard and routine in most tissue engineering research laboratories.

Based on the relative significance of mammalian cell culture in tissue engineering, it comes as no surprise that automated bioreactors for cell culture have been developed and are now commercially available. Automated cell culture bioreactors are designed to undertake all functions of mammalian cell culture using robotic technology. This involves culture and maintenance of cells on 2D monolayer surfaces, frequent media changes, enzymatic treatment to detach cells from the adherent surface, and subplating onto different monolayer surfaces. Automated mammalian cell culture systems are very sophisticated and require advanced engineering, control technology, and robotic systems for functionality. They offer clear advantages over manual cell culture by reducing overhead costs, personnel requirements, and the risk of contamination. However, these systems are expensive and therefore have not found wide acceptance in academic research laboratories.

An example of a bioreactor system for automated mammalian cell culture is the commercially available system, Cellerity™ from Tecan Group Ltd. The information presented here is based on publicly available information from the company website and from posters at conferences and other publications. The author does not endorse this product in any way, does not have any financial interest in the company, and does not have inside information about detailed engineering design for the system.

Let us begin by looking at some of the major components of the Cellerity system. Cellerity can be viewed as a cell culture laboratory condensed into a single unit with advanced robotic and control systems. Atmospheric air enters the workspace and is filtered through a HEPA filter for sterilization, similar to the process of using a laminar flow hood. Cell culture plates are maintained in a CO_2 regulated incubator, which has robotic arms to transfer tissue culture plates from one location to another, depending on functional requirements. The robotic arms require advanced control systems and are designed to replicate physical movement of tissue culture plates that would otherwise be conducted by a human operator. Fluid handling requires several components, including refrigerators for bulk storage of media, robotic nozzles for fluid delivery, and vessels for waste collection. Cellerity also has automated cell counters and shakers. All processes are controlled using an advanced software program that has the capacity to control the frequency and volume of media changes, trypsinization, and subpassaging protocols, including regulation of plating density.

Tissue culture plates used for manual mammalian cell culture are not compatible with Cellerity and other bioreactors. A new generation of cell and tissue culture plates have been fabricated to support automated cell culture and are now commercially available. An example of such plates is roboflasks. These culture vessels have a large culture surface and have a lid, which vents to the atmosphere to promote gaseous exchanges and serves as an entry point for fluid exchange nozzles (required for media changes). During normal mammalian cell culture, roboflasks are positioned on their sides; however, they are used in an upright position during fluid exchange. There is also a venting port, which provides a sterile barrier to support gaseous exchange, built into the roboflasks. Fluid handling is done by robotic nozzles that pierce the lids on the surface of roboflasks and serve as an entry point for media aspiration and/or delivery.

Let us compare Cellerity with manual cell culture for maintaining cells on a 2D monolayer culture surface. We can assume that cells have been plated on the surface of tissue culture plates and are being maintained in a cell culture incubator, either as a part of Cellerity, or a stand-alone system in a traditional cell culture laboratory. In our example, we will compare the protocol for a single media change using Cellerity with the protocol for a single media change using manual methods.

Let us begin this discussion by providing a protocol for maintaining cells in culture the traditional way: using an operator based system. In this case, the operator will remove tissue culture plates from the incubator and transfer them to a laminar flow hood that has been sterilized with ethanol. Spent media will then be aspirated from the culture plates using a vacuum system, and fresh media, which has been pre-heated to 37°C, will be added to the culture plates. Once all culture plates have been processed, they will be transferred from the laminar flow hood to the cell culture incubator. Depending on the cell type, this process will be repeated every 1–2 days for an average of 7–10 days.

Now let us look at the same protocol using Cellerity. Roboflasks will be maintained in an incubator chamber and transferred to the culture surface using robotic arms. The culture plates will be secured in the housing, and spent media will be aspirated using robotic nozzles while fresh media, which has been perfused through a heating coil, will be delivered to the plates, again using robotic nozzles.

The roboflasks will then be returned to the incubator chamber using robotic arms. The frequency of media changes, along with the volume, will be programmed using software.

Let us compare the relative advantages and disadvantages of the manual versus automated methods for media changes. The most significant advantage is obvious—the automated process does not require direct human control and therefore is more cost effective (excluding capital expenditure) and less prone to contamination. The biggest disadvantage of the automated system is indeed the initial capital investment, which is lower for a traditional cell culture experimental setup than automated bioreactors. If large-scale studies are being conducted, an automated system will be more cost-effective, and the extra work of having to train multiple personnel to perform traditional cell culture techniques can be significantly reduced or eliminated altogether.

We will end our discussion on bioreactors for mammalian cell culture by answering one question—*can these systems provide a valuable tool for tissue engineering?* From a tissue engineering standpoint, the objective is often, if not always, to expand the number of cells to obtain large cell numbers for tissue fabrication. The expansion of mammalian cells requires numerous tissue culture plates, often reaching in excess of one hundred. Handling such a large number of tissue culture plates manually can prove to be challenging and labor intensive, and due to the repetitive nature of media changes, this manual handling can increase the likelihood of contamination. Automated cell culture systems provide a valuable tool for such large-scale studies by eliminating or reducing operator-dependent errors and therefore lead to process efficiencies and cost reduction.

6.8 BIOREACTORS FOR SCAFFOLD FABRICATION

In this section, we look at electrospinning as one example of bioreactors that have been used for scaffold fabrication. Electrospinning is a method fabricating individual fibers of a polymer that can be combined in different configurations to promote 3-dimensional scaffold fabrication (7–12). Scaffolds have been fabricated using electrospinning with a wide array of polymers, including polylactic acid, poly(glycolide), polyurethane, polystyrene, collagen, fibrinogen, gelatin, hyaluronic acid, and chitosan. In the electrospinning process, the polymer is first solubilized, and polymer solution is introduced into a syringe. The polymer solution is held in position at the end of the syringe tip by surface tension. A high voltage is used to deliver an electric field to the polymer solution; this step results in the polymer being charged from the syringe tip. The electric field acts in the opposite direction of the surface tension, and when it overcomes surface tension, a charged jet of polymer solution is extruded from the syringe. Ejection of the polymer solution from the syringe results in formation of a Taylor cone—attachment of the polymer solution to the tip of the syringe in the shape of a cone. As the voltage is increased further, fluid is ejected from the Taylor cone and travels toward the collection device. As the fluid travels, the solvent evaporates, resulting in the formation of thin fibers. These fibers are collected in a collection device that can be configured to support the formation of 3D scaffolds (Figure 6.6).

What are the important processing parameters affecting the electrospinning process? We now look at ways to control the electrospinning process which determine properties of the scaffold. The most important variables that regulate the electrospinning process are: 1) viscosity and concentration of polymer solution, 2) molecular weight of polymer solution, 3) conductivity of polymer solution, 4) surface tension at tip of syringe, 5) voltage of the electric field, 6) distance between syringe tip and collector, and 7) design and configuration of the collector. Let us look at these seven variables further.

Viscosity and Concentration of the Polymer Solution—Viscosity and concentration of the polymer solution are related, as increasing polymer concentration leads to an increase in solution viscosity. Increasing viscosity of the polymer solution translates to having more polymer per unit volume of solution and less solvent. As the polymer jet travels from the syringe tip toward the collector, evaporation of the solvent takes place; since there is less solvent at the start of the electrospinning process, evaporation results in excessive drying of the polymer fiber, which adversely alters 3D scaffold fabrication. Now let us look at the reverse argument. If we begin the electrospinning process with a low viscosity polymer solution, this translates to having less polymer per unit volume of solution and more solvent. As the jet travels from the syringe tip to the collector, the solvent is not able to completely evaporate due to the presence of a greater solvent per volume of polymer solution. When the jet of polymer solution reaches the collector, it is wet, and as before, this wetness adversely affects 3D scaffold fabrication.

Molecular Weight of Polymer Solution—This relationship can be explained based on the relationship between the molecular weight of the polymer and the viscosity of the polymer solution; the higher the molecular weight of the polymer, the higher the viscosity of the solution with all other variables remaining constant. There is a correlation between the molecular weight and fiber diameter; increasing molecular weight of the polymer leads to an increase in fiber diameter, while decreasing molecular weight of the polymer leads to a decrease in fiber diameter or bead formation. Bead formation refers to the inability of the electrospinning process to produce polymer fibers; rather than formation of fibers, beads are formed. This bead formation is a negative outcome that adversely affects 3D scaffold formation.

Conductivity of Polymer Solution—Conductivity of the polymer solution affects the ability of any given applied electric field to eject the polymer solution from the syringe tip. Polymers with low conductivity are ejected from the syringe tip with a lower force compared to polymers with high conductivity. While there are some discrepancies in the relationship between polymer conductivity and fiber diameter, within a critical range of polymer conductivities, it has been found that there is a decrease in fiber diameter with increasing conductivity of the polymer solution. Within the critical range of polymer conductivity, increasing polymer conductivity leads to an increase in ejection velocity, which in turn increases solvent evaporation leading to a decrease in fiber diameter. If a polymer solution has low conductivity outside of the critical range when an electric field is applied, the force of ejection is low, and this results in inadequate elongation of the polymer fiber; this elongation in turn leads to a reduction in fiber stability, uniformity, and can prevent fiber formation altogether. On the other hand, if the polymer solution has conductivity higher than the critical range, there is a high force of ejection when the electric field is applied; this high force of ejection leads to unstable fibers and can prevent fiber formation altogether.

Surface Tension at Syringe Tip—Surface tension serves to hold the polymer in place prior to application of an electric field. Once the electric field is applied, the surface tension has to be overcome and the polymer is ejected from the collector. If the surface tension of the polymer solution increases with all other variables remaining the same, the polymer solution will "hold on" to the syringe tip with a greater force; this means that application of a given electric field will not be sufficient to eject the polymer solution from the syringe tip. If the polymer solution is retained at the polymer tip or if less polymer is ejected toward the collection vessel, it will lead to instability in fiber formation and can also lead to the formation of beads. The higher the surface tension of the polymer, the greater the magnitude of the problem.

Effect of Applied Electric Field—The properties of the electric field play an important role in the electrospinning process and serve to overcome surface tension at the syringe tip and eject the polymer solution toward the collection device. An increase in applied voltage will result in an increase in the amount of polymer being ejected from the syringe tip; this ejection will in turn result in an increase in fiber diameter. While an increase in the electric field results in an increase in fiber diameter, this relationship is dependent on the conductivity of the polymer. In the case of highly conductive polymers, the reverse relationship applies: an increase in applied electric field results in a decrease in fiber diameter. In the case of highly conductive polymers, the electric field serves to destabilize the polymer due to an increase in electrostatic repulsive forces within the polymer solution; this in turn can result in a decrease in fiber diameter or disrupt fiber formation altogether.

Distance between Tip and Collector—The distance the polymer jet travels is determined by the placement of the collector relative to the syringe. If this distance is too small or too large, it will hamper scaffold formation and can result in wet fiber or bead formation. If the collector is placed too close to the syringe, there is a reduction in the time available for the polymer jet to travel to the collector. This in turn results in reduced solvent evaporation resulting in the formation of wet fibers. If the collector is placed too far from the syringe, excessive solvent evaporation takes place and can lead to excessive drying, which will result in bead formation.

Design and Configuration of Collector—The orientation of the fibers is dictated by the design configurations of the collector. Planar scaffolds can be fabricated with planar collection vessels, while cylindrical scaffolds can be fabricated with rotating collection vessels.

6.9 BIOREACTORS FOR SCAFFOLD CELLULARIZATION

Fabrication of 3D artificial tissue requires cellularization of scaffolds. In the simplest embodiment, cells are injected into complex 3D scaffolds using a syringe as the delivery vehicle, with multiple injections being required for cellularization. Success of the cellularization strategy is measured based on cell viability, cell retention, cellular alignment, uniform distribution of cells, and functional coupling between adjacent cells and with extracellular matrix. As can be envisioned, direct cell injection strategies are not always the most efficient at achieving the stated objectives due to many problems, including low cell retention within the scaffold. Bioreactors have been developed to aid the cellularization process, and in this section we will discuss six cellularization methods: 1) direct cell injection, 2) cell entrapment using hydrogels, 3) perfusion seeding, 4) surface acoustic waves, 5) centrifugal force, and 6) magnetic nanoparticles. These cellularization strategies are presented in Figure 6.7.

Direct Cell Injection—Direct cell injection is the simplest strategy for scaffold cellularization and is the oldest and most widely used method (18) (Figure 6.7a). In this method, primary cells are suspended in cell culture media and transferred to a syringe or other delivery vehicle. These cells are then directly injected on the surface of the scaffold or injected within the fibers of the 3D scaffold. In most cases, multiple injections on several sites on the scaffold are required for complete cellularization. The scaffold is left undisturbed for 1–2 hours to allow cells to populate the scaffold and then is transferred to an incubator for culture. The advantages of direct cell transplantation are that it is simple, does not require expensive instrumentation or training for personnel, and can be completed within a short time period. There are, however, some limitations with these methods. Delivery of the cells through a needle or syringe can lead to cell damage and loss of viability, particularly if the orifice size is small. Direct injection of cells within the scaffold does not always result in uniform cell distribution, as the process does not involve any degree of precision. Cell retention is low, and the cells have a tendency to be washed out either immediately upon cellularization or after the first media change. Although there are known limitations with direct cell injection methods, these methods continue to be preferred for scaffold cellularization to support tissue fabrication due to the simplicity of the method and ease of implementation.

Cell Entrapment using Hydrogels—One of the major limitations with direct cell injection methods is low cell retention, as the injected cells have a tendency to leak out of the scaffold. A second category of cellularization strategies has been developed to address this limitation and is known as cell entrapment using hydrogels. The idea is to trap the cells within a biodegradable hydrogel during cellularization, thereby reducing the likelihood that cells will leak out of the scaffold (19). The hydrogel is designed to secure the cells in place within the 3D scaffold and is designed with tunable degradable kinetics. During the initial stages of scaffold cellularization, the biodegradable hydrogel serves to hold the cells within the scaffold. As cells populate and cellularize the 3D scaffold to form functional tissue, the hydrogel degrades over time using hydrolysis, enzymatic treatment, or some other degradation method. By this time, cells generate their own extracellular matrix,

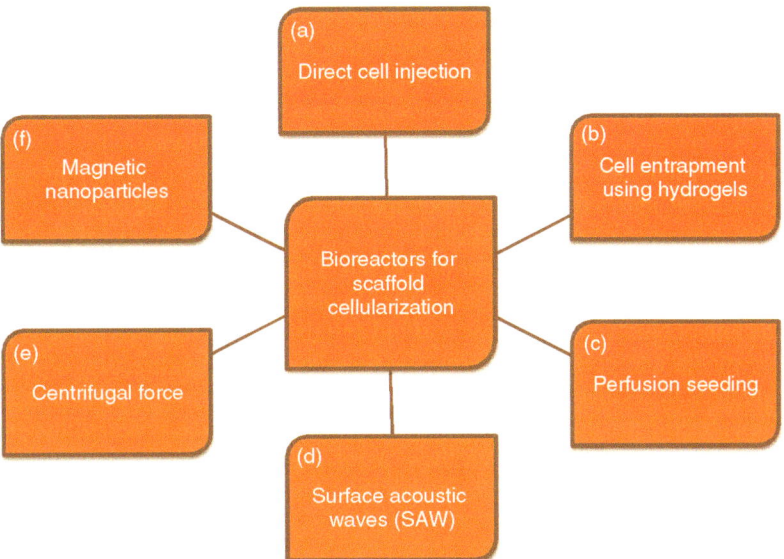

Figure 6.7 Bioreactors for Scaffold Cellularization—(a) Direct Cell Injection—Cells are suspended in culture media, placed in a syringe, and directly injected into the 3D scaffold. **(b) Cell Entrapment using Hydrogels**—Cells are suspended in a thrombin solution, placed in a syringe, and directly injected into the 3D scaffold, which has been presoaked in a fibrinogen solution. The addition of thrombin promotes polymerization of the fibrinogen and results in the formation of fibrin gel. **(c) Perfusion Seeding**—Porous channels are fabricated within 3D scaffolds. Cells are suspended in cell culture media and perfused through the scaffold using a peristaltic pump. Perfusion of the cell suspension results in cells entering into the porous channels of the scaffold, which supports scaffold cellularization. **(d) Surface Acoustic Waves (SAW)**—Cells are suspended in culture media and are placed in a chamber. IDT electrodes are used to generate SAW, which guides movement of cells towards the 3D scaffold. **(e) Centrifugal Force**—3D scaffold is positioned within a conical tube. Cells are suspended in culture media and added to the conical tube. The conical tube is placed in a centrifuge and subjected to a centrifugal force, which supports cellularization of the 3D scaffold. **(f) Magnetic Nanoparticles**—Magnetic nanoparticles are combined with a cell suspension and directly added to a scaffold. Magnetic force is used to guide and position cells within the 3D scaffold.

which is used to retain cells and support formation of 3D tissue. The biodegradable hydrogel, therefore, acts as a temporary support matrix.

One example of cell entrapment technology is fibrin being used as the biodegradable hydrogel for cell entrapment (19) (Figure 6.7b). Fibrin is a commonly used hydrogel in tissue engineering. It is formed by the polymerization of fibrinogen by thrombin. In this example, cells are suspended in a solution that contains thrombin while the 3D scaffold is soaked in fibrinogen. The cell suspension is added to the scaffold using a syringe. As the thrombin comes in contact with fibrinogen, fibrin gel forms and traps cells in place.

The two methods of scaffold cellularization that have been described so far, direct cell transplantation and cell entrapment using hydrogels, are not based on bioreactor technology. These two methods are included in our discussion for the sake of completion and to compare these methods with methods that rely on bioreactor technology, as we will see for the remainder of this discussion.

Perfusion Seeding—Perfusion bioreactors have been used to support scaffold cellularization, and here we look at one specific example from the literature (20). In this case, porous scaffolds were designed with pore sizes in the range of 75–100 μm and flow channels with a diameter of 250 μm. The flow channels were designed to serve as the entry point for cells within the 3D scaffold, and the pores were designed as attachment points for cells. For scaffold cellularization, cells were suspended in culture media and perfused through the flow channels; the direction of the flow was changed at regular intervals. As the cell suspension was perfused through the flow channels, cells traversed from the flow channels to the pores within the 3D scaffold and were trapped within these pores. This method of perfusion seeding has been shown to result in uniform cell distribution with seeding efficiencies of up to 80%. Perfusion seeding is advantageous, as it uses fluid flow to support entry of cells within 3D scaffolds rather than passive seeding, which in effect, forces cells into the scaffold. This way, perfusion seeding increases seeding efficiency and regulates spatial distribution of cells based on flow channel geometry. The major limitations of perfusion seeding are the need for specialized perfusion apparatus, custom designed scaffolds, and specialized training for personnel, all of which lead to an increase in cost and time.

Surface Acoustic Wave Technology—Surface acoustic waves (SAWs) are acoustic waves that are generated on the surface of a piezoelectric substrate and travel along the surface of the material. Their amplitude typically decays exponentially with depth into the material (21). SAWs are frequently used in electronic circuits as filters, oscillators, and transformers. SAWs are generated by interdigital transducer (IDT) electrodes and travel from an input transducer toward an output transducer (21) (Figure 6.7d). In this case, IDT electrodes were engineered onto a piezoelectric substrate and used to generate SAWs by application of a radio frequency. In this example, the SAW travels from the input transducer to the output transducer. A 3D scaffold is positioned in the path of the SAW, and a droplet of fluorescent beads is positioned in the path of the SAW in close proximity to the scaffold. As the SAW travels from the input to the output transducer, the fluorescent beads travel toward the scaffold and populate the 3D scaffold. This process has shown to be very rapid, and cellularization can be accomplished within seconds, compared to hours for other cellularization strategies like direct cell transplantation. SAW bioreactors for scaffold cellularization is a fairly new technology and is limited to a few specialized laboratories with the required engineering capabilities. However, SAW technology is promising and has the potential to be used in beneficial ways for the tissue fabrication process.

Centrifugal Force—Centrifugal force has been used as a strategy for cellularization of scaffolds. In one example, scaffolds were positioned within a conical tube and then submerged in cell suspension (22). The conical tube with the scaffold and cell suspension was placed in a centrifuge and subjected to centrifugal force (Figure 6.7e). The centrifugal force resulted in transfer of the cells to the interior of the scaffold. After cellularization, the scaffold was removed from the conical tube and cultured within a cell culture incubator. This strategy was shown to significantly improve cellularization efficiency when compared to direct cell transplantation.

Magnetic Nanoparticles—Another technique that has been used for scaffold cellularization is known as magnetic force-based tissue engineering, or Mag-TE (23). Mag-TE methods make use of magnetite cationic liposomes (MCLs), which are nanoparticles with a diameter of approximately 150 nm that contain magnetite nanoparticles (Fe_3O_4). MCLs interact with cells, and the magnetite nanoparticles result in cells being charged; this property of cells is then used as a tool for scaffold cellularization (23) (Figure 6.7f). In one example, MCLs were first mixed with a cell population. Magnetically labeled cells were then added to the surface of a 3D scaffold that was placed on the surface of a magnet. The cells travelled through the scaffold starting on the surface and gradually progressing inward; this movement was controlled by the magnetic field created by the underlying magnet. This process resulted in uniform cell distribution throughout the 3D scaffold. Mag-TE is a novel method to support scaffold cellularization and can have a significant impact on the field of tissue engineering. However, high levels of expertise are required for the fabrication of MCLs, which are not commonly used in many labs. Investment is also required for personnel training, adding to the time and cost of implementation of Mag-TE.

6.10 PERFUSION SYSTEMS

Need for Perfusion Systems—In the human body, the circulatory system acts as a distribution network for the delivery of nutrients to cells and tissues while at the same time removing waste products. All cells in the human body are within a couple hundred microns from a capillary; this characteristic is essential to support cell viability. During 2D monolayer culture, cells are cultured on the surface of a tissue culture plate and cell culture media is replaced every second or third day, depending on cell type. This strategy is also used to support the culture of 3D artificial tissue. While adequate to support the culture of cells and tissue, these culture conditions do not replicate *in vivo* physiological flow conditions. For example, media changes every second or third day does not mimic blood flow conditions and is not adequate to support cell viability during culture of 3D artificial tissue. This alludes to the fact that bioreactor systems, which are capable of delivering continuous fluid flow to support the metabolic activity of cells and artificial tissue during controlled *in vitro* culture, are needed. The need for perfusion systems is greater for 3D artificial tissue than for cells during 2D monolayer culture due to an increase in metabolic activity resulting from an increase in tissue mass of 3D artificial tissue (24–36).

Control of Processing Variables in Perfusion Systems—Perfusion systems can be designed to operate inside of a cell culture incubator or completely independent of a cell culture incubator, with the former strategy being more common.

The advantage of designing a perfusion system to function within cell culture incubators is the ability to regulate processing variables like temperature, pH, and CO_2/O_2 levels using feedback control loops within the cell culture incubator. There is, however, a significant limitation with the use of cell culture incubators, as these devices have fixed configurations and do not allow significant user control over processing variables. Design and fabrication of such perfusion systems that rely on the use of cell culture incubators to regulate processing variables is simple. However, if user control over processing variables is required, the perfusion system has to be fabricated with onboard sensors and feedback loops for regulation of temperature, pH, and CO_2/O_2 levels. In this case, perfusion systems are more complex, but they have the advantage of user control over processing variables, which can be changed based on the specific tissue fabrication application. While this process is lengthy and expensive, in the long run, it delivers the level of control that is necessary for tissue engineering applications. The main advantage of embedding sensors within bioreactors is the degree of flexibility in controlling processing variables and the ability to fine-tune the specific processing parameters based on tissue-specific requirements.

There are many processing variables that need to be regulated during perfusion culture of 3D artificial tissue, the most important of which include temperature, pH and gaseous composition. If perfusion bioreactors are being designed to function within a cell culture incubator, then these processing variables are accurately regulated within the incubator. If the perfusion bioreactors are being designed as stand-alone systems, then control of temperature, pH, and gaseous composition is performed by the bioreactor. Temperature can be controlled in one of two ways: either by the use of water-jacketed reservoirs or temperature-regulated metallic plates (Figure 6.8a).

In the first case, water-jacketed reservoirs are fabricated around the chamber or reservoir to be regulated; hot water is perfused through these reservoirs using a thermocirculator, which is a piece of equipment with a heating element and pump. Heat exchange from the hot water to the chamber or reservoir results in a temperature increase within the culture chamber/reservoir. The second approach involves the use of a metallic plate, attached to a heating element with feedback control, to regulate temperature of the culture environment. The feedback control loops serve to regulate the temperature at a preset value. Both methods are fairly accurate for temperature regulation and have been used extensively.

Regulation of pH is achieved in one of two ways (Figure 6.8b). The first method involves the use of CO_2 to regulate pH of the culture media, just as is the case in a cell culture incubator. CO_2 is injected within the culture chamber or media reservoir and acts to lower the pH to physiological values of 7.4. Once the pH reaches a user-defined preset value, say 7.3, CO_2 flow to the media is stopped. As a result, the pH of the cell culture media or culture environment begins to rise, and once the pH reaches a second user defined preset value, say 7.5, the flow of CO_2 is re-initiated; this serves to lower the pH of the culture media. Using this strategy, the pH of the cell culture media or culture chamber can be maintained with a narrow user-defined range, typically 7.3–7.5. The use of CO_2 for pH regulation

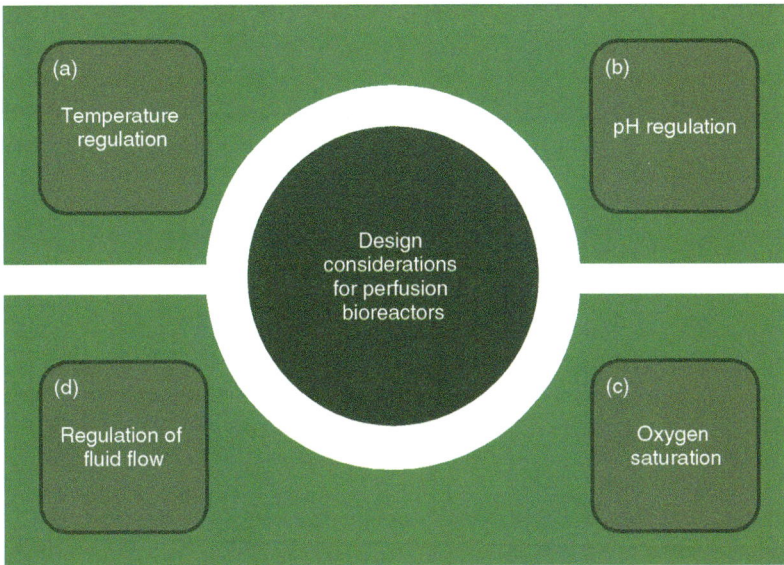

Figure 6.8 Design Considerations for Perfusion Bioreactors—(a) Temperature Regulation—Water-jacketed reservoirs are used for temperature regulation in cell culture reservoirs and incubation chambers. Alternatively, a heating element with a thermostat for temperature regulation can be used. **(b) pH Regulation**—CO_2 injection with feedback control can be used to regulate the pH of cell culture media at physiological levels for long-term studies. For short-term studies, pH regulation can be achieved using buffers. **(c) Oxygen Saturation**—Oxygen saturation can be regulated by direct bubbling of oxygen to the cell culture media. Alternatively, a cell culture incubator with oxygen regulation capabilities can be used with perfusion bioreactors. **(d) Regulation of Fluid Flow**—A single peristaltic pump can be used for media delivery and aspiration, or a dual pump configuration can be used, with a single peristaltic pump being used for media delivery and a second one being used for media aspiration.

is complex and requires CO_2 sensors, control valves, and feedback control algorithms. CO_2 regulation as a strategy to maintain pH is used in perfusion systems only when the specific application requires long-term culture, usually ranging from weeks to months. For short-term culture applications, ranging from hours to days, there are many commercially available buffers that are excellent for pH regulation, including 4-(2-hydroxyethyl)-1-piperazineethanesulfonic acid (HEPES) and sodium bicarbonate.

Regulation of a gaseous environment requires control of CO_2 level for pH control (as we have seen before) and oxygen saturation (Figure 6.8c). Oxygen saturation of the cell culture media is an experimental variable and varies from application to application. In some cases, hypoxic conditions are required and oxygen saturation has to be adjusted to less than 1%, and in other cases, higher oxygen saturation is needed, sometimes in excess of 50%. The atmospheric oxygen concentration is 21% (and 78% nitrogen). If any tissue engineering application requires oxygen concentration in excess of 21%, oxygen is injected in the culture chamber or bubbled in the cell culture media; conversely, if any given application requires lower oxygen saturation, nitrogen is injected into the culture chamber and acts to displace oxygen from the culture environment. The exact oxygen concentration needs to be measured with sensors and embedded feedback control systems. This is not a trivial task and requires advanced sensor and control technology.

Components of Perfusion Systems—Many components are required to assemble a perfusion system, one of which is the housing for 3D artificial tissue; several design variables need to be taken into consideration when fabricating housing for 3D tissue. The most important variable is the transfer of 3D artificial tissue from the culture vessel to the perfusion chamber; *will the entire culture plate will be transferred or just the 3D artificial tissue?* The former strategy is preferred, as it does not require physical handling of artificial tissue, which can lead to damage and loss of function. Another important decision is the time at which perfusion is initiated. When artificial tissue is cultured in the presence of continuous media flow, perfusion is initiated a few days after tissue fabrication. This is to allow time for the cells to attach to the scaffold and support 3D tissue formation; if perfusion is started too early, cells can be washed away, leading to tissue damage.

Pulsatile pumps are frequently used to replicate the pulsatile behavior of blood flow; programmable pumps are used to deliver advanced flow regimes like sinusoidal waves (Figure 6.8d). At times, a single pump is sufficient to support media delivery and aspiration to the culture plates, while for some applications, two pumps are required: one for media aspiration and a second pump for media delivery. If a single pump is used, the rate of fluid delivery and aspiration needs to be balanced to prevent media overflow; this is sometimes a difficult task to achieve, and therefore, dual pump models are frequently used.

An Example of a Perfusion System—The system consists of a custom biochamber designed to accommodate 10 tissue culture plates with each plate stacked vertically on an independent stage (16). Cell culture media was maintained in a temperature-regulated reservoir and was delivered to the tissue culture plates via a peristaltic pump. Two specialized manifolds were precision-fabricated to allow simultaneous placement of the fluid flow ports to each chamber. A single peristaltic pump drives the inflow, drawing from a water-jacketed, temperature-controlled reservoir. Media oxygenation was accomplished by a membrane oxygenator. Media was aspirated from each plate utilizing custom-designed manifolds attached to a vacuum line. The aspirated media was recycled to the media reservoir. High relative humidity was maintained in the biochamber by delivering moist air from a humidification chamber. The inflow and outflow fluid manifolds were configured with inline measurement of O_2, pH, and temperature. Threaded Luer-Lock ports at the top of the chamber, in conjunction with single direction check valves, allowed for gas inflow and outflow without the possibility of reverse flow contamination.

6.11 BIOREACTORS FOR STRETCH

Introduction—Stretch is an important modulator of physiological function. All mammalian cells have biological machinery to sense and respond to changes in the stretch environment. This is particularly important in the cardiovascular system where cells are constantly exposed to a variety of hemodynamic forces and must respond to these changes to maintain normal physiological function. Cells have biological force sensors, which respond to changes in the force environment, embedded within the cell membrane; these biological force sensors are known as stretch-activated channels (SACs) (37–48). In response to changes in the extracellular stretch environment, SACs undergo a conformational change leading to a cascade of intracellular signaling events. The intracellular signaling cascade leads to changes in the biological functions of the cell by inducing changes in the gene and protein expression pattern. In this manner, SACs are able to respond to the extracellular stress environment by eliciting an intracellular response.

In the next section, we will look at the effect of stretch on cells of the cardiovascular system, namely vascular smooth muscle cells, endothelial cells, and cardiac myocytes. We will then discuss bioreactors that have been developed to deliver controlled stretch of cells/tissue for the cardiovascular system. Let us begin our discussion by looking at the effect of stretch on vascular smooth muscle cells.

Effect of Stretch on Vascular Smooth Muscle Cells—All blood vessels are constantly subjected to fluid shear stresses in response to pulsatile blood flow, and vascular smooth muscle cells (SMCs) and endothelial cells (ECs) constantly remodel in response to these pulsatile fluid stresses. In response to stretch, SMCs are known to exhibit a host of phenotypic changes, including an increase in the rate of proliferation and expression of contractile proteins (49–58). In SMCs, zyxin is an important mechanosensitive protein that binds to the cytoskeleton proteins actin and vinculin to form a focal adhesion complex. In response to stretch, zyxin rapidly accumulates at the point at which stress is applied and promotes recruitment of α-actinin and vasodilator-stimulated phosphoprotein (VASP), which serve to stabilize intracellular actin fibers. Many of the downstream effects of stretch are mediated by RhoA, which is a GTPase protein that acts on rho-associated coiled-coil-containing *protein (*ROCK), and this signaling pathway leads to an increase in the expression of SMC contractile proteins.

Effect of Stretch on Endothelial Cells—Endothelial cells (ECs) line the luminal surface of blood vessels within the vascular system and are constantly exposed to the hemodynamic force resulting from pulsatile blood flow. ECs play an important role in regulating vascular tone and act to regulate the properties of SMCs. In effect, ECs are the first responders to changes in the physiological hemodynamic environment and therefore have an extensive biological system in place for sensing and transmitting mechanical forces. ECs have several ion channels located on the membrane surface that are known to respond to changes in fluid hemodynamics, some of which include transient receptor potential channels (TRPs), P2X4 purinoreceptors, potassium channels, and chlorine channels (59–68). In addition to ion channels, ECs also have other types of mechanosensors, including integrins, platelet endothelial cell adhesion molecule-1 (PECAM-1), VE-cadherin, caveolae, G proteins, glycocalyx, and the endothelial cell cytoskeleton.

In response to stretch, these mechanosensors act to increase intracellular calcium contractions, which in turn results in an increase in the vasodilator, nitric oxide (NO); this increase in NO leads to vasodilation. Shear stress is also known to result in an increase in the expression of many proteins in ECs, including cytoskeletal and matrix proteins.

Effect of Stretch on Cardiac Myocytes—Just as blood vessels are constantly exposed to pulsatile fluid flow, the heart is also under constant hemodynamic stress, and cardiac cells, including fibroblasts and myocytes, respond to these changes by regulating intracellular molecular and cellular events (69–75). Under normal physiological conditions, the heart responds to changes in hemodynamic loads by increasing extracellular matrix components and by hypertrophic and hyperplasic growth of cardiac cells. In addition, pressure overload in the heart leads to an increase in the rate of proliferation of cardiac fibroblasts associated with an increase in expression of collagen type I. Changes in hemodynamic loading resulting from volume overload are also associated with cardiac myocyte hypertrophy in addition to increases in collagen deposition.

Stretch and Tissue Engineering—Our discussion until this point has served to demonstrate the role of mechanical stretch and/or pulsatile fluid flow on modulating the cellular and molecular properties of cardiovascular cells, including SMCs, ECs, and cardiac myocytes. This serves to illustrate the critical role of stretch in maintaining and modulating tissue function by regulating cellular and molecular properties. Moving on from this discussion, we need to ask one question - *how can this information be applied to tissue engineering?* Our objective in tissue engineering is to fabricate artificial tissue and develop *in vitro* culture conditions that mimic *in vivo* physiological conditions in order to support growth and maturation of 3D artificial tissue. As can be seen from our previous discussion, stretch can be used as a modulator of cell and tissue function. Therefore, applied to tissue engineering, the objective is to design and fabricate of custom bioreactors to deliver controlled stretch protocols to support growth and maturation of 3D artificial tissue. Let us look at one example of a bioreactor that has been developed to deliver controlled stretch to modulate the function of 3D artificial heart muscle.

Design Variables for Stretch—We begin by looking at some of the important design variables that need to be taken into consideration. Any given stretch protocol can be defined by percentage stretch, frequency, and duration of stretch. The percentage stretch refers to the displacement of the tissue construct in response to the stretch protocol; for example, if artificial tissue is planar and measures 20 mm × 20 mm, a 10% stretch would require a displacement of 2 mm for every stretch cycle. Frequency of stretch refers to the number of cycles per second and is measured in Hz—a frequency of 1 Hz refers to one cycle per second. Stretch duration refers to the time for which the stretch protocol is implemented, ranging from minutes to weeks. Uniaxial stretch refers to stretch protocols in one direction, while biaxial stretch refers to stretch in two directions (x-y plane). Stretch velocity refers to the time required to complete a single stretch cycle. For example, if a displacement of

2 mm is required and this process required 1 second to complete, the stretch velocity is 2 mm/s. Rest periods are also integrated within the stretch protocols, particularly when working with systems involving skeletal muscle, which has a high degree of fatigability.

Now that we have some indication of the variables that define stretch protocols, the next question to address is: *what stretch protocol should we begin with for any given 3D artificial tissue?* It is instinctive to use stretch protocols that are based on *in vivo* parameters for *in vitro* conditioning of artificial tissue. For example, if we have to develop a stretch protocol for artificial heart muscle, we will instinctively draw inspiration from stretch parameters in the heart and replicate these conditions *in vitro*. Often, this is not the best strategy, as 3D artificial tissue does not have the same architecture as mammalian tissue and cannot withstand harsh stretch protocols. When developing stretch protocols for 3D artificial tissue, the objective should always be to start with mild stretch protocols, validate tissue compatibility, and then gradually increase the intensity of the stimulation protocols.

An Example of a Stretch Bioreactor—We describe a stretch bioreactor designed to readily accommodate a range of self-organizing tissues, including skeletal, cardiac, and smooth muscle, tendon and ligament, and bone (14). Adjacent to each 35 mm plate is a movable post. One end of each tissue specimen is affixed to this post via a stainless steel wire that is shaped to allow it to pass under the cover of the 35 mm dish and attach to the stainless steel insect pins that affix the ends of each tissue specimen. The 35 mm plates do not require modification; the lids are simply tipped forward to accommodate the stainless steel wire. The culture plates are held stationary above a moving platform to which the movable posts are affixed below. Thus, during mechanical movement of the posts, the tissue specimen length is changed, but the base of the culture dish remains motionless. This arrangement minimizes sloshing of the culture media that would occur if the culture dish were moved cyclically or dimensionally deformed. The mechanical strain mechanism is comprised of a linear actuator that drives the moving platform beneath the culture plates. The linear actuator is a direct-drive ACME miniature power screw affixed to the output shaft of the gear-head step motor. Mechanical strain of the same amplitude is applied to all specimens simultaneously. In general, different mechanical stimulus protocols would be carried out on separate but identical bioreactors. Mechanical strain is controlled by a gear-head stepper motor driven by an open-loop stepper motor driver.

6.12 ELECTRICAL STIMULATION

Introduction—In this section, we will look at bioreactors for controlled electrical stimulation of 3D artificial tissue. The use of electrical stimulation as a treatment modality has been around for decades and has been used in numerous applications, ranging from cardiac pacemakers to vagus nerve stimulation for treatment of seizures during epilepsy. During normal mammalian function, changes in voltage are used as a trigger to modulate cell and tissue level function. These changes in voltage are brought about by changes in ion concentration within the cellular and/or extracellular environment, which leads to conformational changes in voltage-sensitive channels. We begin this section with a discussion of specific applications of electrical stimulation as a treatment modality outside of the tissue engineering space.

Electrical Stimulation as a Treatment Modality—Cardiac pacemakers are used to synchronize cardiac contraction in cases of arrhythmias or tachycardia (76–79) (Figure 6.9a). Pacemakers consist of a control unit, which is designed to monitor activity of the native heart and deliver specific electrical impulses to the electrodes, and the actual electrodes, which are implanted in contact with the right atrium, ventricle, and heart muscle. The pacemaker is implanted under the collarbone, and

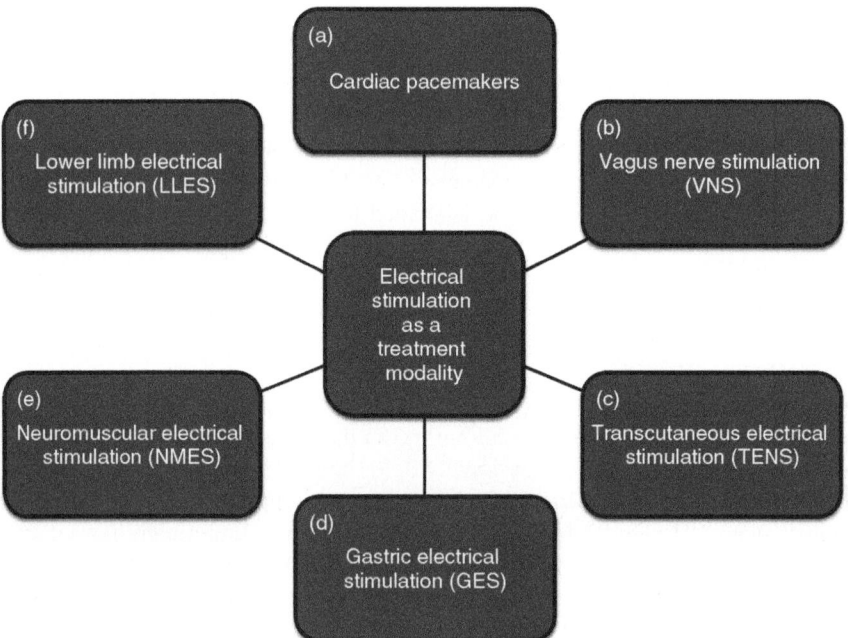

Figure 6.9 Electrical Stimulation as a Treatment Modality—(a) Cardiac Pacemakers—Cardiac pacemakers have been used for decades to synchronize electrical activity of the heart. **(b) Vagus Nerve Stimulation (VNS)**—Electrodes are implanted in direct contact with the vagus nerve and have proven to be effective in the management of epilepsy. **(c) Transcutaneous Electrical Stimulation(TENS)**– Low voltage electrical stimulation on the surface of the skin is used for the treatment of pain. **(d) Gastric Electrical Stimulation (GES)**—GES is used in cases of gastroparesis by delivering a high frequency electrical stimulation to the muscles in the stomach. **(e) Neuromuscular Electrical Stimulation (NMES)**—NMES is used in rehabilitation engineering to condition skeletal muscles for cases of muscle atrophy related to aging or trauma. **(f) Lower Limb Electrical Stimulation (LLES)**—LLES is a specialized case of NMES used for patients with chronic heart failure as a mechanism to increase muscle activity.

leads are inserted through a vein that runs under the collarbone. Since the electrodes are in direct contact with heart muscle, they monitor the electrical activity of the heart. If the electrical activity of the heart is normal, the control unit does not deliver any signal. However, if the electrical activity of the heart deviates from normal behavior, the electrodes are able to sense this change and deliver an electrical impulse that resynchronizes the cardiac conduction system. This process continues for every heartbeat. Therefore, cardiac pacemakers serve as sensors for cardiac electrical conduction and as on-demand delivery vehicles for electrical stimulation protocols.

Vagus nerve stimulation (VNS) is another example in which electrical stimulation has been used as a treatment modality, in this case for epilepsy (80–83) (Figure 6.9b). Epilepsy is a condition in which patients suffer from spontaneous seizures that vary in frequency, duration, and intensity. The effect of these seizures varies among patients, but in most cases, there is some loss of awareness, physical control, and/or alteration in communication. Seizures are caused by random electrical activity from neurons in the brain that lose their ability to act in a synchronized manner. The exact cause of epilepsy is unknown, though genetic and environmental factors are known to have a role. Pharmacological agents are commonly used to control seizures, and in severe cases of epilepsy, VNS can also be used as a treatment modality. The vagus nerve runs from the brain stem to the chest and abdomen area and is the target for VNS. In VNS, a control unit, similar to that for cardiac pacemakers, is implanted in the collarbone area. Electrodes are placed in direct contact with the vagus nerve and provide long-term electrical stimulation of the vagus nerve; this process has proven to be effective in controlling seizures in cases of epilepsy.

Transcutaneous electrical nerve stimulation (TENS) is a process by which low-voltage electrical stimulation is delivered to the surface of the skin as a therapy for pain relief (84–89) (6–9C). As in the previous two examples, a control unit is used to regulate the electrical stimulation protocol, and the output is via two leads attached to the site of delivery, which is a specific point on the surface of the skin. TENS is used extensively as a therapy for relief of muscle, joint, or bone pain, especially for lower back and neck pain.

Gastroparesis is a medical condition in which partial paralysis of the stomach results in delayed emptying of the stomach (Figure 6.9d). Muscle contraction in the stomach causes food to move through the digestive tract and is under the regulation of the vagus nerve. In gastroparesis, muscle contraction in the stomach is compromised, causing food to remain in the stomach for extended periods of time. This can affect a patient's nutrient absorption and can also lead to nausea and vomiting. In such cases, high-frequency gastric electrical stimulation (GES) is used to support muscle contractions in the stomach (90–93). As in the previous cases, a control unit is implanted in a subcutaneous pocket with electrical leads touching the surface of the stomach. High-frequency electrical stimulation of the muscle has been shown to reduce symptoms associated with gastroparesis.

Neuromuscular electrical stimulation (NMES) is a technique used extensively in rehabilitation engineering, which involves the use of electrical stimulation to condition superficial skeletal muscle (94,95) (Figure 6.9e). As we have seen in all the cases before, NMES involves the use of a control unit and electrodes—in this case, the electrodes are positioned in contact with the skin superficial to the muscle of interest. NMES is used extensively in physical therapy and rehabilitation clinics in cases associated with muscle injury, trauma, or aging; the objective of this therapy is to preserve, restore, and/or improve skeletal muscle function. Lower limb electrical stimulation (LLES) is used for patients with chronic heart failure as a rehabilitation tool to increase skeletal muscle activity, which can complement pharmacological treatment modalities for these patients (Figure 6.9f).

Electrical Stimulation and Tissue Engineering—Our discussion so far has focused on looking at examples of electrical stimulation as a treatment modality for different conditions. We have looked at cardiac pacemakers, VNS, TENS, GES, and NMES. It should become abundantly clear that electrical stimulation has been extensively used as a treatment or therapeutic modality. There is an entire field known as functional electrical stimulation (FES), which is based on electrical stimulation for functional recovery and encompasses all of the examples that have been discussed. Our next objective is to move from this broad overview of electrical stimulation to a specific application involving 3D artificial tissue. We will study the role of electrical stimulation in guiding 3D artificial tissue formation and function, and we will examine the design and development of bioreactors to achieve this objective.

There have been several applications that study the effect of electrical stimulation in tissue engineering. Some examples include neural tissue engineering applications, the culture and differentiation of stem cells, and the development and maturation of 3D artificial heart muscle. We will look at these applications in the following sections.

In one study, PC-12 cells, obtained from rat pheochromocytoma in the adrenal gland, were plated on surface of a thin film of oxidized polypyrrole (PPy), which is a conducting polymer (96) (Figure 6.10a).

In this study, cells were expanded in culture and plated on the surface of the PPy film with a 24-hour attachment period. After this initial attachment period, cells were exposed to one of two electrical protocols, using a constant voltage of 100 mV for 2 hours or a constant current of 10 μAmp, also for 2 hours. After the 2-hour stimulation protocol, the cells were retained in culture for an additional 24 hours and processed for histological assessment. Electrical stimulation resulted in an increase in differentiation phenotype of PC-12 cells toward a neuronal lineage, as was evident by an increase in neurite length and an increase in cell spreading.

The previous study showed the positive effects of electrical stimulation on PC-12 on a 2D monolayer surface, while the current study was conducted in a 3D polyaniline (PANI) scaffold (97) (Figure 6.10b). Electrospinning was used to fabricate 3D PANI scaffolds with a fiber diameter of 112–189 nm and pore size in the range of 0.27–1.5 μm. The 3D scaffolds were placed in a 24-well tissue culture plate and seeded with nerve stem cells (NSCs) at a density of 1.5×10^4 cells/plate. After a 24-hour culture period, the PANI scaffolds with the NSCs were subjected to an electrical stimulation protocol consisting of 1.5V for 15, 30, and 60 minutes and then

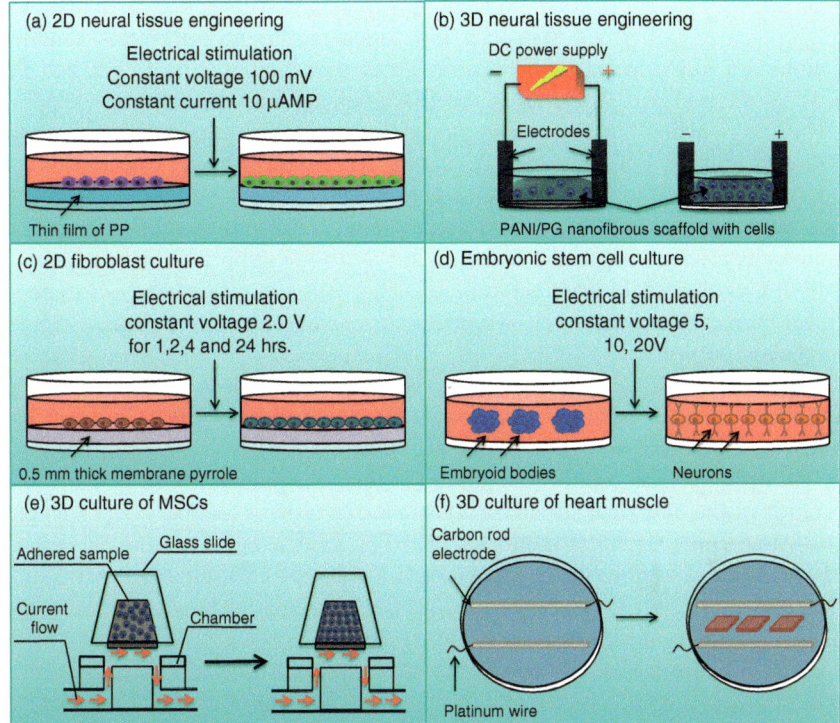

Figure 6.10 Bioreactors for Electrical Stimulation in Tissue Engineering—(a) **2D Neural Tissue Engineering**—Differentiation of PC-12 cells toward a neural phenotype in response to electrical stimulation during 2D culture. (b) **3D Neural Tissue Engineering**—Differentiation of NSCs toward a neural phenotype during 3D culture and in response to electrical stimulation. (c) **2D Fibroblast Culture**—Increase in rate of fibroblast proliferation during 2D culture and electrical stimulation. (d) **Embryonic Stem Cell Culture**—Differentiation of ESCs toward a neuronal phenotype in response to electrical stimulation. (e) **3D Culture of MSCs**—Increase in MSCs adhesion to substrate in response to electrical stimulation during 3D culture. (f) **3D Culture of Heart Muscle**—Improvement in electrophysiological properties of artificial heart muscle in response to electrical stimulation.

maintained in culture for an additional 24 hours prior to histological assessment. As in the previous example, electrical stimulation resulted in an increase in neurite outgrowth, which was also accompanied by an increase in the rate of NSC proliferation.

In another study, a 0.5 mm thick membrane pyrrole was fabricated, and human skin fibroblasts were cultured on the surface of the membrane at a concentration of 6×10^4 cells/cm^2 (98) (Figure 6.10c). Immediately after cell seeding, an electrical stimulation protocol, which consisted of a constant voltage of 2V for 1, 2, 4, and 24 hours, was initiated. Electrical stimulation resulted in a significant increase in the rate of fibroblast proliferation, as measured by the MTT assay.

One interesting application of electrical stimulation has been to drive the differentiation fate of embryonic stem cells (99) (Figure 6.10d). ES cells were maintained in culture for 3 days to support formation of embryoid bodies (EB). Controlled electrical stimulation was delivered to the EBs after this initial 3 day culture period, and histological and functional assessment of the cells was conducted after another 10 days in culture. For electrical stimulation, a constant voltage protocol was used at 5, 10, and 20 volts while the cells were cultured in a 4 mm gap cuvette. This study demonstrated that electrical stimulation of ES cells resulted in differentiation of the cells toward a neuronal lineage. This is a very important finding since the ability to regulate the differentiation fate of ES cells toward any given lineage is a critical barrier that needs to be overcome prior to the utilization of these cells for therapeutic purposes.

Another interesting application has been the use of electrical stimulation for 3D culture of bone marrow-derived mesenchymal stem cells (100) (Figure 6.10e). MSCs were cultured with fibroblasts in a 3D collagen gel using a cell density of 2×10^5 cells/ml. The cells were cultured in the collagen gel for 12 hours prior to initiation of the electrical stimulation protocol. In order to deliver controlled electrical stimulation, a custom-fabricated chamber was used to house the 3D collagen gel with MSCs and fibroblasts, and a constant voltage of 7 V/cm was used. Electrical stimulation resulted in an increase in cellular orientation and an increase in the adhesion strength of MSCs to the collagen scaffold.

Electrical stimulation has been used extensively in cardiac tissue engineering applications to support the development and maturation of artificial heart muscle (101) (Figure 6.10f). Artificial heart muscle was fabricated by culturing primary neonatal cardiac myocytes within a 3D collagen sponge using 6×10^6 cells per sponge. The collagen sponges were maintained in culture for 3 days prior to electrical stimulation and then moved to custom bioreactors to initiate the stimulation protocol for an additional 5 days. The stimulation protocol consisted of square monophasic pulses with a duration of 2 ms, frequency of 1 Hz, and a variable voltage in the range of 0–12.5V. Electrical stimulation of artificial heart muscle resulted in a significant increase in cellular organization and alignment, as shown by histological data.

SUMMARY

Current State of the Art—Bioreactor technology is very well-developed and has broad applications in tissue engineering. Bioreactor technology has been incorporated in all facets of the tissue engineering process, from cell culture and expansion to physiological conditioning of artificial tissue. The current generation of bioreactors is advanced and has significant functional capabilities engineered within the system. During the course of this chapter, we have seen examples of bioreactor technology used to support fabrication of 3D artificial tissue and to deliver controlled stimuli to guide tissue formation and maturation. In addition, there is extensive literature that demonstrates a positive correlation between bioreactor conditioning and improvement in functional performance of 3D artificial tissue.

Thoughts for Future Research—There are three areas in the development of bioreactor technology: 1) bioreactors to simultaneously deliver multiple signals (electrical stimulation, stretch and perfusion), 2) bioreactors with embedded sensors for real time monitoring of artificial tissue function and using these signals for feedback control of bioreactor signals, and 3) bioreactors capable of regulating culture variables like temperature and pH without the need for cell culture incubators.

PRACTICE QUESTIONS

1. Explain the role of biomechanics during normal physiological function of mammalian tissue. Provide three specific examples, not described in the chapter, explaining the relationship between biomechanical forces and normal tissue function in mammalian tissue.

2. Discuss the role of bioreactors during the tissue fabrication process. What exactly are bioreactors and why are they important in tissue engineering?

3. Explain the difference between bioreactors for enabling technology and bioreactors for supporting technology. Identify any tissue or organ system that you would like to fabricate. Describe how bioreactors will be used to enable and support the fabrication of the selected tissue/organ.

4. How would the tissue fabrication process change in the absence of bioreactors? What impact would this have on the functional performance of 3D artificial tissue?

5. During our discussion of bioreactor classification, we provided six categories of bioreactors: bioreactors for cell culture, bioreactors for scaffold fabrication, bioreactors for scaffold cellularization, bioreactors for stretch, bioreactors for perfusion, and bioreactors for electrical stimulation. Explain each of these categories.

6. During our discussion of bioreactor classification, we provided six categories of bioreactors: bioreactors for cell culture, bioreactors for scaffold fabrication, bioreactors for scaffold cellularization, bioreactors for stretch, bioreactors for perfusion, and bioreactors for electrical stimulation. Identify any tissue or organ system that you would like to fabricate. For the selected tissue/organ, explain how these six categories of bioreactors will be used to support the tissue fabrication process.

7. We provided a process flow chart describing the design of bioreactors. The following variables were discussed: definition of the specific stimuli, identification of culture variables, development of sensor technology, and development of a specific stimulation protocol. Explain each of these steps.

8. We provided a process flow chart describing the design of bioreactors. The following variables were discussed: definition of the specific stimuli,

identification of culture variables, development of sensor technology, and development of a specific stimulation protocol. For this question, we will focus on the fabrication of artificial heart muscle; apply the process flow chart to 3D artificial heart muscle development. Explain how each of these four steps will be applied during different stages of the tissue development; end your discussion by providing specific variables for each of the four steps in the bioreactor process flow chart as they apply to the development of artificial heart muscle.

9. During our discussion of idealized bioreactors for artificial heart muscle, we described many critical components of the system. Explain the idealized bioreactor system for the culture of 3D artificial heart muscle.

10. During our discussion of idealized bioreactors for artificial heart muscle, we described many critical components of the system. Pick any tissue engineering application and develop an idealized bioreactor system for your selected application.

11. During our discussion of the integration of bioreactor technology with tissue engineering, which is illustrated in Figure 6.5, we described points along the tissue fabrication pathway where bioreactor technology has been incorporated: cell culture, scaffold fabrication, scaffold cellularization, and physiological conditioning. While this list encompasses many aspects of the tissue fabrication process, it is not exhaustive. There are several additional areas during the tissue fabrication process where bioreactor technology will be beneficial. Identify and discuss one such area where you believe that bioreactor technology will benefit the tissue fabrication process.

12. We discussed the potential use of bioreactors for the culture and expansion of mammalian cells. Compare the relative advantages and disadvantages of manual versus automated cell culture techniques.

13. Explain the process of electrospinning for scaffold fabrication. What variables affect the scaffold fabrication process?

14. During our discussion of bioreactors for scaffold cellularization, we looked at the use of direct cell transplantation and five additional strategies: cell entrapment using hydrogels, perfusion seeding, surface acoustic waves, centrifugal force, and magnetic nanoparticles. Each one of these technologies has relative advantages and disadvantages. Direct cell transplantation is extensively used to support the scaffold cellularization during the tissue fabrication process. Do you believe that direct cell transplantation is the most effect strategy for scaffold cellularization? Explain your answer.

15. During our discussion of bioreactors for scaffold cellularization, we studied six methods used for the cellularization of 3D scaffolds. Describe these six methods and discuss the relative advantages and disadvantages of each.

16. During our discussion of bioreactors for scaffold cellularization, we studied six methods used for the cellularization of 3D scaffolds. Select any tissue engineering application and explain which scaffold cellularization method is best suited for your application.

17. We discussed the role of perfusion systems to support the culture of 3D artificial tissue systems. Perfusion systems can be designed to operate within or independent of a cell culture incubator. What are the relative advantages and disadvantages of the two approaches?

18. What are some important processing variables that need to be controlled during perfusion of 3D artificial tissue? How would you control these processing variables?

19. We discussed the role of perfusion systems to support the culture of artificial tissue systems. Pick any tissue/organ system and develop a perfusion system to support the culture of the artificial tissue/organ. Explain why perfusion is important for your selected application. Describe what components will be parts of your perfusion system. Provide a schematic of your perfusion system. What will be your test variables? How will you identify optimal fluid flow conditions? End your discussion by providing a list of specific perfusion variables that you will use to support the culture of your selected tissue system.

20. Explain why stretch is important for the development and maturation of 3D artificial tissue.

21. How do cells within the cardiovascular system sense and response to stretch in the external culture environment?

22. What design variables are important for stretch bioreactors?

23. We discussed the role of stretch to support the culture of artificial tissue systems. Pick any tissue/organ system and develop a system to deliver controlled stretch to support the culture of the artificial tissue/organ. Explain why stretch is important for your selected application. Describe what components will be a part of your stretch system. Provide a schematic of your system. What will be your test variables? How will you identify optimal stretch conditions? End your discussion by providing a list of specific stretch variables that you will use to support the culture of your selected tissue system.

24. We have discussed the role of electrical stimulation to support the culture of artificial tissue systems. Pick any tissue/organ system and develop a system to deliver controlled electrical stimulation to support the culture of the artificial tissue/organ. Explain why electrical stimulation is important for your selected application. Describe what components will be a part of your system. Provide a schematic of your system. What will be your test variables? How will you identify the optimal electrical stimulation conditions? End your discussion

by providing a list of the specific electrical stimulation variables that you will use to support the culture of your selected tissue system.

25. During the course of this chapter, we have extensively discussed the role of bioreactors during the tissue fabrication process. All of our applications have been focused on the use of bioreactors *in vitro*. However, bioreactors are also used extensively *in vivo* during the tissue engineering process. Discuss how bioreactors can be used *in vivo*. How will bioreactors be used *in vivo* and what role will these *in vivo* bioreactors serve during the tissue fabrication process?

REFERENCES

1. Nielsen LK. Bioreactors for hematopoietic cell culture. Annu. Rev. Biomed. Eng. 1999;1:129–52.
2. Xing Z, Kenty BM, Li ZJ, Lee SS. Scale-up analysis for a CHO cell culture process in large-scale bioreactors. Biotechnol. Bioeng. 2009 Jul 1;103(4):733–46.
3. Ducos JP, Terrier B, Courtois D. Disposable bioreactors for plant micropropagation and mass plant cell culture. Adv. Biochem. Eng Biotechnol. 2010;115:89–115.
4. Wendt D, Riboldi SA, Cioffi M, Martin I. Potential and bottlenecks of bioreactors in 3D cell culture and tissue manufacturing. Adv. Mater. 2009 Sep 4;21(32–33):3352–67.
5. Mandenius CF, Bjorkman M. Scale-up of cell culture bioreactors using biomechatronic design. Biotechnol. J. 2012 Aug;7(8):1026–39.
6. Tharakan JP, Gallagher SL, Chau PC. Hollow-fiber bioreactors for mammalian cell culture. Adv. Biotechnol. Processes 1988;7:153–84.
7. Nair LS, Bhattacharyya S, Laurencin CT. Development of novel tissue engineering scaffolds via electrospinning. Expert. Opin. Biol. Ther. 2004 May;4(5):659–68.
8. Sill TJ, von Recum HA. Electrospinning: applications in drug delivery and tissue engineering. Biomaterials 2008 May;29(13):1989–2006.
9. Ramachandran K, Gouma PI. Electrospinning for bone tissue engineering. Recent Pat Nanotechnol. 2008;2(1):1–7.
10. Wang HS, Fu GD, Li XS. Functional polymeric nanofibers from electrospinning. Recent Pat Nanotechnol. 2009;3(1):21–31.
11. Ashammakhi N, Wimpenny I, Nikkola L, Yang Y. Electrospinning: methods and development of biodegradable nanofibres for drug release. J. Biomed. Nanotechnol. 2009 Feb;5(1):1–19.
12. Bhardwaj N, Kundu SC. Electrospinning: a fascinating fiber fabrication technique. Biotechnol. Adv. 2010 May;28(3):325–47.
13. Rauh J, Milan F, Gunther KP, Stiehler M. Bioreactor systems for bone tissue engineering. Tissue Eng Part B Rev. 2011 Aug;17(4):263–80.
14. Birla RK, Huang YC, Dennis RG. Development of a novel bioreactor for the mechanical loading of tissue-engineered heart muscle. Tissue Eng. 2007 Sep;13(9):2239–48.
15. Hecker L, Khait L, Radnoti D, Birla R. Development of a microperfusion system for the culture of bioengineered heart muscle. ASAIO J. 2008 May;54(3):284–94.

16. Hecker L, Khait L, Radnoti D, Birla R. Novel bench-top perfusion system improves functional performance of bioengineered heart muscle. J. Biosci. Bioeng. 2009 Feb;107(2):183–90.

17. Khait L, Hecker L, Radnoti D, Birla RK. Micro-perfusion for cardiac tissue engineering: development of a bench-top system for the culture of primary cardiac cells. Ann. Biomed. Eng. 2008 May;36(5):713–25.

18. Blan NR, Birla RK. Design and fabrication of heart muscle using scaffold-based tissue engineering. J. Biomed. Mater. Res. A 2008 Jul;86(1):195–208.

19. Huang YC, Khait L, Birla RK. Contractile three-dimensional bioengineered heart muscle for myocardial regeneration. J. Biomed. Mater. Res. A 2007 Mar 1;80(3):719–31.

20. Maidhof R, Marsano A, Lee EJ, Vunjak-Novakovic G. Perfusion seeding of channeled elastomeric scaffolds with myocytes and endothelial cells for cardiac tissue engineering. Biotechnol. Prog. 2010 Mar;26(2):565–72. PMCID:PMC2854846.

21. Li H, Friend JR, Yeo LY. A scaffold cell seeding method driven by surface acoustic waves. Biomaterials 2007 Oct;28(28):4098–104.

22. Godbey WT, Hindy SB, Sherman ME, Atala A. A novel use of centrifugal force for cell seeding into porous scaffolds. Biomaterials 2004 Jun;25(14):2799–805.

23. Shimizu K, Ito A, Honda H. Enhanced cell-seeding into 3D porous scaffolds by use of magnetite nanoparticles. J. Biomed. Mater. Res. B Appl. Biomater. 2006 May;77(2):265–72.

24. Mironov V, Kasyanov VA, Yost MJ, Visconti R, Twal W, Trusk T, Wen X, Ozolanta I, Kadishs A, Prestwich GD, et al. Cardiovascular tissue engineering I. Perfusion bioreactors: a review. J. Long Term Eff. Med. Implants 2006;16(2):111–30.

25. Kim SS, Penkala R, Abrahimi P. A perfusion bioreactor for intestinal tissue engineering. J. Surg. Res. 2007 Oct;142(2):327–31.

26. Radisic M, Marsano A, Maidhof R, Wang Y, Vunjak-Novakovic G. Cardiac tissue engineering using perfusion bioreactor systems. Nat. Protoc. 2008;3(4):719–38. PMCID:PMC2763607.

27. Brown MA, Iyer RK, Radisic M. Pulsatile perfusion bioreactor for cardiac tissue engineering. Biotechnol. Prog. 2008 Jul;24(4):907–20.

28. Kasper FK, Liao J, Kretlow JD, Sikavitsas VI, Mikos AG. Flow perfusion culture of mesenchymal stem cells for bone tissue engineering. 2008.

29. Grayson WL, Marolt D, Bhumiratana S, Frohlich M, Guo XE, Vunjak-Novakovic G. Optimizing the medium perfusion rate in bone tissue engineering bioreactors. Biotechnol. Bioeng. 2011 May;108(5):1159–70. PMCID:PMC3077473.

30. Hidalgo-Bastida LA, Thirunavukkarasu S, Griffiths S, Cartmell SH, Naire S. Modeling and design of optimal flow perfusion bioreactors for tissue engineering applications. Biotechnol. Bioeng. 2012 Apr;109(4):1095–9.

31. Dahlin RL, Meretoja VV, Ni M, Kasper FK, Mikos AG. Design of a high-throughput flow perfusion bioreactor system for tissue engineering. Tissue Eng Part C Methods 2012 Oct;18(10):817–20. PMCID:PMC3460612.

32. Song L, Zhou Q, Duan P, Guo P, Li D, Xu Y, Li S, Luo F, Zhang Z. Successful development of small diameter tissue-engineering vascular vessels by our novel integrally designed pulsatile perfusion-based bioreactor. PLoS One 2012;7(8):e42569. PMCID:PMC3411804.

33. Shachar M, Benishti N, Cohen S. Effects of mechanical stimulation induced by compression and medium perfusion on cardiac tissue engineering. Biotechnol. Prog. 2012 Nov;28(6):1551–9.
34. Knapp Y, Deplano V, Bertrand E. Flow dynamics characterisation of a novel perfusion-type bioreactor for bone tissue engineering. Comput. Methods Biomech. Biomed. Engin. 2012;15 Suppl 1:116–9.
35. Gaspar DA, Gomide V, Monteiro FJ. The role of perfusion bioreactors in bone tissue engineering. Biomatter. 2012 Oct;2(4):167–75. PMCID:PMC3568103.
36. Yu HS, Won JE, Jin GZ, Kim HW. Construction of mesenchymal stem cell-containing collagen gel with a macrochanneled polycaprolactone scaffold and the flow perfusion culturing for bone tissue engineering. Biores. Open. Access. 2012 Jun;1(3):124–36. PMCID:PMC3559226.
37. Sackin H. Stretch-activated ion channels. Kidney Int. 1995 Oct;48(4):1134–47.
38. Yeung EW, Allen DG. Stretch-activated channels in stretch-induced muscle damage: role in muscular dystrophy. Clin. Exp. Pharmacol. Physiol. 2004 Aug;31(8):551–6.
39. Ducret T, Vandebrouck C, Cao ML, Lebacq J, Gailly P. Functional role of store-operated and stretch-activated channels in murine adult skeletal muscle fibres. J. Physiol. 2006 Sep 15;575(Pt 3):913–24. PMCID:PMC1995676.
40. Han JH, Bai GY, Park JH, Yuan K, Park WH, Kim SZ, Kim SH. Regulation of stretch-activated ANP secretion by chloride channels. Peptides 2008 Apr;29(4):613–21.
41. Ninio DM, Saint DA. The role of stretch-activated channels in atrial fibrillation and the impact of intracellular acidosis. Prog. Biophys. Mol. Biol. 2008 Jun;97(2–3):401–16.
42. Ward ML, Williams IA, Chu Y, Cooper PJ, Ju YK, Allen DG. Stretch-activated channels in the heart: contributions to length-dependence and to cardiomyopathy. Prog. Biophys. Mol. Biol. 2008 Jun;97(2–3):232–49.
43. Irnaten M, Barry RC, Quill B, Clark AF, Harvey BJ, O'Brien CJ. Activation of stretch-activated channels and maxi-K+ channels by membrane stress of human lamina cribrosa cells. Invest Ophthalmol. Vis. Sci. 2009 Jan;50(1):194–202.
44. Liu X, Huang H, Wang W, Wang J, Sachs F, Niu W. Stretch-activated potassium channels in hypotonically induced blebs of atrial myocytes. J. Membr. Biol. 2008 Nov;226(1–3):17–25.
45. Sachs F. Stretch-activated ion channels: what are they? Physiology (Bethesda.) 2010 Feb;25(1):50–6. PMCID:PMC2924431.
46. Youm JB, Han J, Kim N, Zhang YH, Kim E, Leem CH, Kim SJ, Earm YE. Role of Stretch-activated Channels in the Heart: Action Potential and Ca2+ Transients. 2005;.
47. Baumgarten CM, Browe DM, Ren Z. Swelling- and Stretch-activated Chloride Channels in the Heart: Regulation and Function. 2005.
48. Schubert R, Brayden JE. Stretch-activated Cation Channels and the Myogenic Response of Small Arteries. 2005.
49. Haga JH, Li YS, Chien S. Molecular basis of the effects of mechanical stretch on vascular smooth muscle cells. J. Biomech. 2007;40(5):947–60.
50. Li F, Guo WY, Li WJ, Zhang DX, Lv AL, Luan RH, Liu B, Wang HC. Cyclic stretch upregulates SDF-1alpha/CXCR4 axis in human saphenous vein smooth muscle cells. Biochem. Biophys. Res. Commun. 2009 Aug 14;386(1):247–51.
51. Wang BW, Chang H, Shyu KG. Regulation of resistin by cyclic mechanical stretch in cultured rat vascular smooth muscle cells. Clin. Sci. (Lond) 2010 Feb;118(3):221–30.

52. Cheng WP, Wang BW, Chen SC, Chang H, Shyu KG. Mechanical stretch induces the apoptosis regulator PUMA in vascular smooth muscle cells. Cardiovasc. Res. 2012 Jan 1;93(1):181–9.

53. Song L, Duan P, Guo P, Li D, Li S, Xu Y, Zhou Q. Downregulation of miR-223 and miR-153 mediates mechanical stretch-stimulated proliferation of venous smooth muscle cells via activation of the insulin-like growth factor-1 receptor. Arch. Biochem. Biophys. 2012 Dec 15;528(2):204–11.

54. Song J, Hu B, Qu H, Bi C, Huang X, Zhang M. Mechanical stretch modulates microRNA 21 expression, participating in proliferation and apoptosis in cultured human aortic smooth muscle cells. PLoS One 2012;7(10):e47657. PMCID:PMC3474731.

55. Zhao Y, Koga K, Osuga Y, Izumi G, Takamura M, Harada M, Hirata T, Hirota Y, Yoshino O, Fujii T, et al. Cyclic stretch augments production of neutrophil chemokines and matrix metalloproteinase-1 in human uterine smooth muscle cells. Am. J. Reprod. Immunol. 2013 Mar;69(3):240–7.

56. Shah MR, Wedgwood S, Czech L, Kim GA, Lakshminrusimha S, Schumacker PT, Steinhorn RH, Farrow KN. Cyclic stretch induces inducible nitric oxide synthase and soluble guanylate cyclase in pulmonary artery smooth muscle cells. Int. J. Mol. Sci. 2013;14(2):4334–48. PMCID:PMC3588102.

57. Luo DY, Wazir R, Tian Y, Yue X, Wei TQ, Wang KJ. Integrin alphav Mediates Contractility Whereas Integrin alpha4 Regulates Proliferation of Human Bladder Smooth Muscle Cells via FAK Pathway Under Physiological Stretch. J. Urol. 2013 Apr 12.

58. Tian Y, Yue X, Luo D, Wazir R, Wang J, Wu T, Chen L, Liao B, Wang K. Increased proliferation of human bladder smooth muscle cells is mediated by physiological cyclic stretch via the PI3KSGK1Kv1.3 pathway. Mol. Med. Rep. 2013 Jul;8(1):294–8.

59. Hishikawa K, Luscher TF. Pulsatile stretch stimulates superoxide production in human aortic endothelial cells. Circulation 1997 Nov 18;96(10):3610–6.

60. Lacolley P. Mechanical influence of cyclic stretch on vascular endothelial cells. Cardiovasc. Res. 2004 Sep 1;63(4):577–9.

61. Ali MH, Mungai PT, Schumacker PT. Stretch-induced phosphorylation of focal adhesion kinase in endothelial cells: role of mitochondrial oxidants. Am. J. Physiol Lung Cell Mol. Physiol. 2006 Jul;291(1):L38–L45.

62. Barron V, Brougham C, Coghlan K, McLucas E, O'Mahoney D, Stenson-Cox C, McHugh PE. The effect of physiological cyclic stretch on the cell morphology, cell orientation and protein expression of endothelial cells. J. Mater. Sci. Mater. Med. 2007 Oct;18(10):1973–81.

63. Korff T, Aufgebauer K, Hecker M. Cyclic stretch controls the expression of CD40 in endothelial cells by changing their transforming growth factor-beta1 response. Circulation 2007 Nov 13;116(20):2288–97.

64. Lee YU, Drury-Stewart D, Vito RP, Han HC. Morphologic adaptation of arterial endothelial cells to longitudinal stretch in organ culture. J. Biomech. 2008 Nov 14; 41(15):3274–7. PMCID:PMC2823635.

65. Raaz U, Kuhn H, Wirtz H, Hammerschmidt S. Rapamycin reduces high-amplitude, mechanical stretch-induced apoptosis in pulmonary microvascular endothelial cells. Microvasc. Res. 2009 May;77(3):297–303.

66. Iwaki M, Ito S, Morioka M, Iwata S, Numaguchi Y, Ishii M, Kondo M, Kume H, Naruse K, Sokabe M, et al. Mechanical stretch enhances IL-8 production in pulmonary microvascular endothelial cells. Biochem. Biophys. Res. Commun. 2009 Nov 20;389(3):531–6.
67. Zhao H, Hiroi T, Hansen BS, Rade JJ. Cyclic stretch induces cyclooxygenase-2 gene expression in vascular endothelial cells via activation of nuclear factor kappa-beta. Biochem. Biophys. Res. Commun. 2009 Nov 27;389(4):599–601. PMCID:PMC2763434.
68. Wojtowicz A, Babu SS, Li L, Gretz N, Hecker M, Cattaruzza M. Zyxin mediation of stretch-induced gene expression in human endothelial cells. Circ. Res. 2010 Oct 1;107(7):898–902.
69. Zeng T, Bett GC, Sachs F. Stretch-activated whole cell currents in adult rat cardiac myocytes. Am. J. Physiol Heart Circ. Physiol. 2000 Feb;278(2):H548–H557.
70. van Wamel JE, Ruwhof C, van der Valk-Kokshoorn EJ, Schrier PI, van der Laarse A. Rapid gene transcription induced by stretch in cardiac myocytes and fibroblasts and their paracrine influence on stationary myocytes and fibroblasts. Pflugers Arch. 2000 Apr;439(6):781–8.
71. Shyu KG, Chen CC, Wang BW, Kuan P. Angiotensin II receptor antagonist blocks the expression of connexin43 induced by cyclical mechanical stretch in cultured neonatal rat cardiac myocytes. J. Mol. Cell Cardiol. 2001 Apr;33(4):691–8.
72. Torsoni AS, Marin TM, Velloso LA, Franchini KG. RhoA/ROCK signaling is critical to FAK activation by cyclic stretch in cardiac myocytes. Am. J. Physiol Heart Circ. Physiol. 2005 Oct;289(4):H1488–H1496.
73. Lal H, Verma SK, Smith M, Guleria RS, Lu G, Foster DM, Dostal DE. Stretch-induced MAP kinase activation in cardiac myocytes: differential regulation through beta1-integrin and focal adhesion kinase. J. Mol. Cell Cardiol. 2007 Aug;43(2):137–47. PMCID:PMC2039913.
74. Shyu KG. Cellular and molecular effects of mechanical stretch on vascular cells and cardiac myocytes. Clin. Sci. (Lond) 2009 Mar;116(5):377–89.
75. Matsui H, Yokoyama T, Tanaka C, Sunaga H, Koitabashi N, Takizawa T, Arai M, Kurabayashi M. Pressure mediated hypertrophy and mechanical stretch up-regulate expression of the long form of leptin receptor (ob-Rb) in rat cardiac myocytes. BMC. Cell Biol. 2012; 13:37. PMCID:PMC3543168.
76. Mitrani RD, Simmons JD, Interian A, Jr, Castellanos A, Myerburg RJ. Cardiac pacemakers: current and future status. Curr. Probl. Cardiol. 1999 Jun;24(6):341–420.
77. Sarko JA, Tiffany BR. Cardiac pacemakers: evaluation and management of malfunctions. Am. J. Emerg. Med. 2000 Jul;18(4):435–40.
78. Kusumoto FM, Goldschlager N. Implantable cardiac arrhythmia devices--part I: pacemakers. Clin. Cardiol. 2006 May;29(5):189–94.
79. Jung W, Rillig A, Birkemeyer R, Miljak T, Meyerfeldt U. Advances in remote monitoring of implantable pacemakers, cardioverter defibrillators and cardiac resynchronization therapy systems. J. Interv. Card Electrophysiol. 2008 Oct;23(1):73–85.
80. Colicchio G, Montano N, Fuggetta F, Papacci F, Signorelli F, Meglio M. Vagus nerve stimulation in drug-resistant epilepsies. Analysis of potential prognostic factors in a cohort of patients with long-term follow-up. Acta Neurochir.(Wien.) 2012 Dec; 154(12):2237–40.

81. Hoppe C. Vagus nerve stimulation: urgent need for the critical reappraisal of clinical effectiveness. Seizure. 2013 Jan;22(1):83–4.
82. Krahl SE, Clark KB. Vagus nerve stimulation for epilepsy: A review of central mechanisms. Surg. Neurol. Int. 2012;3(Suppl 4):S255–S259. PMCID:PMC3514919.
83. Lehtimaki J, Hyvarinen P, Ylikoski M, Bergholm M, Makela JP, Aarnisalo A, Pirvola U, Makitie A, Ylikoski J. Transcutaneous vagus nerve stimulation in tinnitus: a pilot study. Acta Otolaryngol. 2013 Apr;133(4):378–82.
84. McNearney TA, Sallam HS, Hunnicutt SE, Doshi D, Chen JD. Prolonged treatment with transcutaneous electrical nerve stimulation (TENS) modulates neurogastric motility and plasma levels of vasoactive intestinal peptide (VIP), motilin and interleukin-6 (IL-6) in systemic sclerosis. Clin. Exp. Rheumatol. 2013 Feb 7.
85. Cho HY, In TS, Cho KH, Song CH. A single trial of transcutaneous electrical nerve stimulation (TENS) improves spasticity and balance in patients with chronic stroke. Tohoku J. Exp. Med. 2013;229(3):187–93.
86. Loh J, Gulati A. The Use of Transcutaneous Electrical Nerve Stimulation (TENS) in a Major Cancer Center for the Treatment of Severe Cancer-Related Pain and Associated Disability. Pain Med. 2013 Feb 25.
87. Ding L, Song T, Yi C, Huang Y, Yu W, Ling L, Dai Y, Wei Z. Transcutaneous electrical nerve stimulation (TENS) improves the diabetic cytopathy (DCP) via up-regulation of CGRP and cAMP. PLoS One 2013;8(2):e57477. PMCID:PMC3585412.
88. Andrade SC, Freitas RP, de Brito Vieira WH. Transcutaneous electrical nerve stimulation (TENS) and exercise: strategy in fibromyalgia treatment. Rheumatol. Int. 2013 Mar 31.
89. Simpson PM, Fouche PF, Thomas RE, Bendall JC. Transcutaneous electrical nerve stimulation for relieving acute pain in the prehospital setting: a systematic review and meta-analysis of randomized-controlled trials. Eur. J. Emerg. Med. 2013 Jul 7.
90. Chu H, Lin Z, Zhong L, McCallum RW, Hou X. Treatment of high-frequency gastric electrical stimulation for gastroparesis. J. Gastroenterol. Hepatol. 2012 Jun;27(6):1017–26.
91. Soffer EE. Gastric electrical stimulation for gastroparesis. J. Neurogastroenterol. Motil. 2012 Apr; 18(2):131–7. PMCID:PMC3325298.
92. Guerci B, Bourgeois C, Bresler L, Scherrer ML, Bohme P. Gastric electrical stimulation for the treatment of diabetic gastroparesis. Diabetes Metab. 2012 Nov;38(5):393–402.
93. Lahr CJ, Griffith J, Subramony C, Halley L, Adams K, Paine ER, Schmieg R, Islam S, Salameh J, Spree D, et al. Gastric electrical stimulation for abdominal pain in patients with symptoms of gastroparesis. Am. Surg. 2013 May;79(5):457–64.
94. Stowe AM, Hughes-Zahner L, Barnes VK, Herbelin LL, Schindler-Ivens SM, Quaney BM. A pilot study to measure upper extremity H-reflexes following neuromuscular electrical stimulation therapy after stroke. Neurosci. Lett. 2013 Feb 22;535:1–6. PMCID:PMC3592334.
95. Maddocks M, Gao W, Higginson IJ, Wilcock A. Neuromuscular electrical stimulation for muscle weakness in adults with advanced disease. Cochrane. Database. Syst. Rev. 2013;1:CD009419.
96. Schmidt CE, Shastri VR, Vacanti JP, Langer R. Stimulation of neurite outgrowth using an electrically conducting polymer. Proc. Natl. Acad. Sci. U.S.A 1997 Aug 19;94(17):8948–53. PMCID:PMC22977.

97. Ghasemi-Mobarakeh L, Prabhakaran MP, Morshed M, Nasr-Esfahani MH, Ramakrishna S. Electrical stimulation of nerve cells using conductive nanofibrous scaffolds for nerve tissue engineering. Tissue Eng. Part A. 2009 Nov;15(11):3605–19.
98. Shi G, Rouabhia M, Meng S, Zhang Z. Electrical stimulation enhances viability of human cutaneous fibroblasts on conductive biodegradable substrates. J. Biomed. Mater. Res. A 2008 Mar 15;84(4):1026–37.
99. Yamada M, Tanemura K, Okada S, Iwanami A, Nakamura M, Mizuno H, Ozawa M, Ohyama-Goto R, Kitamura N, Kawano M, et al. Electrical stimulation modulates fate determination of differentiating embryonic stem cells. Stem Cells 2007 Mar;25(3):562–70.
100. Sun S, Titushkin I, Cho M. Regulation of mesenchymal stem cell adhesion and orientation in 3D collagen scaffold by electrical stimulus. Bioelectrochemistry. 2006 Oct;69(2):133–41.
101. Tandon N, Marsano A, Cannizzaro C, Voldman J, Vunjak-Novakovic G. Design of electrical stimulation bioreactors for cardiac tissue engineering. Conf. Proc. IEEE Eng Med. Biol. Soc. 2008;2008:3594–7. PMCID:PMC2771167.

7

TRACHEAL TISSUE ENGINEERING

Learning Objectives

After completing this chapter, students should be able to:

1. Describe the structure of the trachea, including vascularization and innervation.
2. Discuss congenital tracheal stenosis, including symptoms and classification schemes.
3. Describe the genetic regulation of tracheal development, including the role of various genes and changes in branching patterns.
4. Describe tracheal stenosis in cases of tracheal intubation and tracheostomy.
5. Discuss the use of balloon dilation, stents, and trachea resection and reanastomosis for the treatment of tracheal stenosis.
6. Describe design considerations for tracheal tissue engineering.
7. Describe the process of bioengineering 3D artificial tracheas.
8. Discuss tissue engineering models to support the fabrication of artificial tracheas.
9. Describe examples of artificial tracheas and clinical applications in adult and pediatric patients.

Introduction to Tissue Engineering: Applications and Challenges, First Edition. Ravi Birla.
© 2014 The Institute of Electrical and Electronics Engineers, Inc. Published 2014 by John Wiley & Sons, Inc.

CHAPTER OVERVIEW

In this chapter, we begin with a discussion of the structure and function of the trachea, including organization and distribution of various cell types. We also discuss the formation of tracheas during embryogenesis and the genetic signals that regulate tracheal development. We then discuss common tracheal disorders, including congenital tracheal stenosis and tracheal stenosis during tracheal intubation and tracheostomy. We then discuss some of the treatment modalities that are used in cases of tracheal disorders. We describe the use of balloon dilation and stents and the use of surgical reconstruction for the treatment of tracheal disorders. After providing this framework, we next move on to discuss the role of tissue engineering in the development of artificial tracheal tissue. We start with a discussion of design criteria for tracheal tissue engineering and then present a process flow chart identifying the steps required to fabricate 3D artificial tracheas. We then provide specific examples of artificial tracheas that have been developed in research laboratories, and we relate these examples to our design criteria and process flow chart. The field of tracheal tissue engineering has been one of the success stories in tissue engineering, with two examples of clinical applications for 3D artificial tracheas, one in adult patients and one in pediatric patients. We conclude our chapter by presenting this seminal work in the field of tracheal tissue engineering.

7.1 STRUCTURE AND FUNCTION OF THE TRACHEA

Introduction—The trachea, commonly referred to as the windpipe, is part of the respiratory system and serves as a conduit for air from the larynx to the bronchioles within the lungs (1–4). The trachea extends from the sixth cervical vertebra to the fifth thoracic vertebra at which point it bifurcates into the left and right bronchi, which feed into each of the two lungs. The trachea is a rigid cylindrical tube with a complex organization of epithelial cells, extracellular matrix, and cartilaginous support matrix. The length of the trachea in humans is 10–12 cm with a diameter in the range of 20–25 mm (5).

Ciliary Escalator—The organization of cells and the extracellular matrix in the trachea is very complex. The trachea consists of several cell types, including pseudostratified epithelial cells and goblet cells. The part of the trachea containing these two cell types is known as the ciliary escalator or the mucociliary escalator. The epithelial cells contain ciliary structures, which are hair-like protrusions from the surface of the cell. The goblet cells produce mucus, which consists of a protein known as mucin. The function of the mucus is to trap foreign bodies, including bacteria and other microbes, from the inhaled air and prevent the foreign bodies from entering into the respiratory system. Once these microbes are trapped, coordinated movement of the epithelial cells, including the ciliary escalator, functions to transport the foreign bodies to the pharynx where they can either be swallowed or expectorated. Thus, the ciliary escalator serves as a defense mechanism against entry of foreign bodies to the respiratory system, with epithelial cells, ciliary structures, and mucus-producing goblet cells acting in tandem to achieve this function.

Lamina Propria—The next layer in the trachea is known as the lamina propria, which consists of many cell types, like macrophages and mast cells, and extracellular matrix components like elastin. The elastin provides flexibility to the trachea, thereby supporting the changes in lumen diameter during inhalation and exhalation of air. The macrophages remove any foreign matter that has entered the respiratory system by escaping the first level of defense provided by the mucus and the ciliary structures.

Submucosa—The next layer of tracheal tissue is known as the submucosa; it contains numerous mucous and serous glands. The mucous glands in the submucosal layer of the trachea produce mucus, which travels to the luminal surface of the tissue and combines with the mucus produced by the goblet cells. The serous glands contain serous cells, which are the primary defensive cells of the submucosal layer and secrete antimicrobials to target airborne pathogens.

Cartilaginous Rings—The next layer in the trachea is a thick layer of hyaline cartilage. The trachea consists of 16–20 rings of hyaline cartilage. The rings of the hyaline cartilage are designed to provide structural support and are known as tracheal rings. The hyaline cartilage in the tracheal rings allows the trachea to remain open in the absence of positive air pressure; in the absence of tracheal rings, the trachea would collapse during respiration, due to changes in air pressure and the force exerted by the surrounding tissue. These tracheal rings are macroscopic structures, are visible with the naked eye, and can also be felt by touching the neck. They measure about 1 mm in thickness and are separated by narrow distances, although they merge to form a continuous cartilaginous layer. The first and last tracheal rings have distinct characteristics, with the former being broader then the central tracheal rings and the latter being broad in the middle.

Adventitia—The outermost layer of the trachea is known as the adventitia; it consists of loose connective tissue which provides structural support and connects the trachea to the esophagus and other neighboring organs.

Smooth Muscle Tissue—The trachea also has smooth muscle tissue, which functions to regulate the diameter of the trachea based on oxygen demand. Smooth muscle relaxation results in an increase in tracheal diameter, thereby increasing airflow to the lungs to meet an increase in oxygen demand.

Vascularization—The vascularization structure of the trachea is very complex, as there is not a single major artery feeding the entire trachea structure (6,7). Rather, the vascular organization of the trachea has multiple major arteries feeding different parts of the tissue. The cervical portion of the trachea, which is the part closer to the larynx, receives its blood supply either from the inferior thyroid artery or from the subclavian artery. These major arteries branch into several smaller vessels known as the tracheoesophageal branches, which feed both the trachea and the esophagus. The blood supply to the thoracic trachea is fed by the bronchial arteries or the supreme intercostal artery. As in the case for the cervical trachea, these major arteries form tracheoesophageal branches that feed the trachea and the esophagus. The tracheoesophageal branches feeding the trachea are further divided into smaller vessels known as the primary tracheal arteries. The primary tracheal arteries divide further to form intercartilaginous arteries, which feed directly to the tissue between the cartilage rings. The trachea also has an extensive capillary network feeding the mucosal and submucosal tissue. The cartilage rings do not have blood vessels directly feeding into the tissue, but instead depend on diffusion from the capillaries in the submucosal tissue.

Innervation—The trachea, and the bronchus which enters the lungs, are innervated by branches of the vagus nerve (8–11), which serves to regulate smooth muscle contraction and relaxation. Smooth muscle contraction and relaxation in turn regulate the airflow to the lungs. Smooth muscle relaxation occurs in response to stimulation by sympathetic division of the autonomic nervous system, resulting in an increase in airflow to the lungs. Smooth muscle contraction occurs in response to stimulation by the parasympathetic division of the autonomic nervous system, which in turn causes a decrease in diameter and reduced airflow to the lungs.

7.2 CONGENITAL TRACHEAL STENOSIS

Congenital tracheal stenosis (CTS) is a condition in which variable segments of the trachea are narrow, a condition which can prove to be fatal in neonates and infants (12–18). CTS can be caused by complete tracheal cartilage rings and thickening of the submucosal tissue, leading to narrowing in the internal diameter of the trachea. Under normal physiological conditions, the cartilage rings in the trachea are disk-shaped and have a specific configuration that allows airflow during respiratory function. However, in cases of CTS, the cartilage rings are circular in configuration and have a reduced diameter, thereby constricting the trachea and limiting airflow during respiratory function. Long segment CTS (LCTS) refers to cases of CTS in which greater than half the length of the trachea undergoes stenosis.

CTS is associated with other congenital disorders including those of the cardiovascular, pulmonary, and gastrointestinal system, and if left untreated, CTS can be fatal. Until about two decades ago, most cases of CTS were fatal due to obstruction of the respiratory tract. However, with recent advances in medical technology, the survival rate of patients with CTS has significantly improved. While variable between hospitals, the survival rate has been reported to be in the range of 78–92% (12).

CTS is considered to be a rare disease by the Office of Rare Diseases Research at the National Institute of Health, which defines a rare disease as any disease having a prevalence of less than 200,000 in the entire US population. While the prevalence of CTS is not very high in the US population, the complications that arise due to obstruction of the airflow track can be fatal; therefore, treatment and management of CTS has received significant attention in the recent literature.

Classification—There are three classifications of CTS based on anatomical characteristics of the cartilage rings. This system has given rise to type I, type II, and type III CTS (15). In type I CTS, the entire trachea is affected, and there is fairly uniform stenosis throughout the length of the trachea. The stenosis in the airflow tract is uniform and is caused by deformations in all or almost all of the cartilage rings. In type II CTS, there is a tapering pattern associated with the trachea with an increase in stenosis along the length of the trachea. Terms like funneling and tapering have been used to describe this condition. In this condition, the upper parts of the trachea are not affected and the cartilage rings at the upper part of the trachea are normal. However, the cartilage rings closer to the ends of the trachea are smaller in size, leading to a tapering in the diameter of the airflow track. In type III CTS, there is no regular pattern for the tracheal cartilage rings, and only a few regions of the trachea are affected. In this case, only a small portion of the respiratory tract is affected.

Symptoms and Diagnosis—The symptoms of CTS can manifest within the first few weeks after birth or can take several months (15). Some symptoms include problems with breathing or wheezing, chest congestion, and pneumonia (15). Another symptom of CTS is stridor, which refers to noisy breathing resulting from turbulent airflow in the narrowed trachea. CTS can be diagnosed by imaging techniques like x-rays, CT scans, or MRIs and may also require microlaryngoscopy and bronchoscopy (MLB). MLB is a direct visualization technique in which an endoscope is placed within the larynx (microlaryngoscopy) and/or bronchi (bronchoscopy), through the mouth to visualize the airway. Photographs and videos allow direct visualization of the larynx, trachea, and the bronchi; image processing and analysis tools can be used to extract valuable information about airway stenosis from the endoscope.

CTS and Development of the Trachea—During the third to fourth week of development, the respiratory primordium dilates and undergoes bifurcation, leading to the initial stages of tracheal development (19). By week 8 of development, early rudiments of tracheal cartilage can be found; these develop to form mature cartilage rings by week 10, including the addition of smooth muscle tissue to the trachea. It is believed that there are two stages of abnormalities during tracheal development that can give rise to CTS (19). Any deficiencies in tracheal development during the 4-week time period will affect formation of the entire respiratory system and can lead to a severe form of CTS with associated disorders of the respiratory system (19). However, if abnormalities in tracheal development occur between the 8- and 10-week time period, it is believed that the process of cartilage formation will be affected, leading to deficiencies in the formation and organization of cartilage rings in the trachea.

7.3 GENETIC REGULATION OF TRACHEAL DEVELOPMENT

Tracheal development is a very complex process that is regulated by changes in the expression of specific genes at specific time points during the developmental path (20,21). The drosophila has been used extensively to dissect the genetic pathways that regulate tracheal formation. In the drosophila, the trachea is used to deliver oxygen to all tissues; it consists of a single monolayer of epithelial cells without the complex organizational structure of mammalian tracheas. The process of tracheal development begins with the commitment of ectodermal stem cells to a tracheal lineage, followed by reorganization of these cells to form clusters or sacs around the peripheral end of the embryos. This is followed by migration of these cells to the central region of the embryo, followed by invagination to form tracheal placodes, which are the earliest semblance of a cylindrical structure with a luminal surface; these placodes can be viewed as a very early precursor to a trachea. The committed tracheal cells in the placodes then initiate a complex pattern of branching, starting with the formation of primary and secondary branches and ending with the formation of terminal branches. Tracheal development is regulated by changes in the expression of several genes, some of which include *trachealess, branchless, breathless, pantip-2, and blistered* while also involving the participation of the EGF and FGF signaling pathways (20,21).

Formation of Tracheal Placodes—Tracheal cells are formed by differentiation of ectodermal cells to generate clusters or sacs of cells, known as tracheal placodes, which are lined up on either side of the embryo. On average, there are 10 tracheal placodes that are formed on either side of the embryo, and each one contains about 80 cells. The differentiation of ectodermal cells to tracheal cells is regulated by the expression of the trachealess gene, which converts the planar ectodermal cells to sacs of tracheal cells (22–27). Mutations in the trachealess gene have inhibited the formation of tubes in drosophila and therefore have prevented the formation of the trachea (28). One very interesting fact about tracheal development in drosophila is that the total number of cells remains constant throughout the process. Although there are significant changes in cell distribution and organization, there is no accompanying cell proliferation. The formation of branching networks to support tracheal development is a result of cell migration and reorganization.

Invagination of Tracheal Placodes—Once sac formation is complete, the next step in the process is migration of tracheal placodes to the center of the embryo and invagination of the tracheal placodes, which refers to reorganization of the tracheal cells within the sacs to form a hollow structure. The cells within the tracheal placodes are still retained within this structure, although they organize to form a hollow chamber at the center. This process is due to migration and reorganization of the existing cells that have populated the tracheal placodes. It has been shown that epidermal growth factor signaling plays an important role in the invagination of tracheal cells in the placodes (29,30). It has been proposed that upregulation in the expression of rhomboid (rho) in the tracheal cells leads an increase in EGF signaling, which in turn affects invagination of tracheal placodes. It has also been shown that mutation in rho leads to defects in cell migration and hampers the invagination of tracheal placodes.

Formation of Primary Branches—After the formation of the tracheal placodes, the next step in the process is the formation of primary branches, which is followed by the formation of second and terminal branches. The primary branching pattern is initiated by the expression of branchless, which has been shown to serve as a critical determinant of primary branching patterns (31). The expression of branchless is turned on within clusters of cells that are about to initiate primary branches, and is turned off once the branching pattern has been completed. Branchless encodes a protein that functions as a ligand for the breathless receptor, and subsequent signaling pathways provide instructive cues for the development of primary branching patterns. The role of branchless has further been demonstrated by mutations in the gene, which restrict the formation of primary branches from tracheal placodes. The protein encoded by branchless is homologous to fibroblast growth factors (FGFs), thereby suggesting that FGF signaling may be involved in regulating the formation of primary branches during tracheal development.

Formation of Secondary Branches—The next step in the process is extension of the primary branches to form secondary branches, which appear a few hours after the primary branches. The formation of the secondary branches is also regulated by the relative expression of several genes, with Pantip-2 being one such example (20). Pantip-2 has been shown to be expressed in many cells in the primary branches, which eventually lead to the formation of secondary branches. As the secondary branches form and continue to expand, the expression of Pantip-2 is restricted to the lead cells, which are located at the ends of the primary branches. The expression of Pantip-2 was shown to be progressively limited to fewer cells at the leading tip of the primary branches as the branching pattern continued.

Formation of Terminal Branches—The final step of tracheal development is the formation of terminal branches from the secondary branches, a process that is regulated by the expression of several genes, including a gene known as blistered (32). Blistered encodes a protein known as the Drosophila Serum Response Factor, the expression of which has been shown to be upregulated in all tracheal cells that undergo terminal branching. It has also been shown that the expression of drosophila serum response factor is regulated by the expression of branchless. As we have seen in the formation of primary tracheal branches, branchless encodes a protein that is homologous to FGF, leading to the activation of the FGF signaling pathway. Activation of the FGF signaling pathway is an earlier event in the formation of primary branches, which has been shown to lead to the activation of blistered and the formation of terminal branches.

7.4 POST INTUBATION AND POST TRACHEOSTOMY TRACHEAL STENOSIS

Introduction—In a previous section, we studied CTS and the complications that can lead to tracheal resection/reanastomosis or the need for complete tracheal replacement. Examples of some other conditions that may require similar intervention include post-intubation and post-tracheostomy tracheal stenosis (33–38). Post-intubation and post-tracheostomy tracheal stenosis do not have a very high incidence, estimated to be 4.9 cases per year for every one million of the general population (39). The incidence of these disorders has been reported to be in the range of 10–22% of all patients that need to be intubated. Further, only 1–2% of these patients suffer severe complications. Just as we saw in the previous case when we discussed CTS, the prevalence of cases in the United States is not very high; nonetheless, patients who have severe complications require care in tertiary centers, and many times, these complications can lead to fatalities. Therefore, there is an imperative need to understand the progression of such disorders and develop effective tools for the management and treatment of these conditions, some of which we will study in subsequent sections of this chapter.

Endotracheal Intubation and Tracheostomy—Endotracheal intubation requires insertion of a flexible tube in the trachea (40–45). This tube is connected to a mechanical ventilator to support respiration. The endotracheal tube is inserted through the mouth, and once inserted, it is secured in place by inflation of a balloon cuff at the end of the tube. Mechanical ventilators are used routinely during surgical procedures as a way to regulate respiratory parameters. While endotracheal intubation is a safe procedure, the pressure inserted by the balloon cuffs on the trachea can lead to cell necrosis and cuff stenosis. In cases when there is trauma to the respiratory tract or large blockages that disrupt the airflow to the lungs, a tracheostomy can be performed by creating a direct interface between atmospheric air and the trachea. In this case, an incision is made in the neck region of the patient, which is followed by cutting through a small portion of the thyroid gland. This is followed by a small incision in the cartilaginous rings of the trachea, and a tracheostomy tube is then placed and secured in the trachea. This creates a direct pathway for atmospheric air to flow through the trachea and to the bronchi and the lungs. While the procedure is considered to be safe, complications like tracheal stenosis can occur with long-term tracheostomies.

Post-Intubation and Post-Tracheostomy Tracheal Stenosis—One of the potential complications of tracheal intubation and tracheostomy is stenosis, narrowing of the airflow track that leads to compromised respiratory function (39). In the case of tracheal intubation, stenosis can arise from the balloon cuff, leading to a condition known as cuff stenosis. The purpose of inserting an inflated balloon during tracheal intubation is to secure the tracheal tube within the trachea; increasing the pressure within the balloon cuff provides direct contact with the luminal surface of the trachea. If the pressure in the balloon cuff is increased beyond the pressure in the local blood vessels feeding into the submucosal tissue, the blood flow to the tissue will be significantly reduced. Cartilaginous rings of the trachea do not have an independent blood supply, but rely on diffusion of nutrients from the submucosa tissue. Any reduction in the blood supply to the submucosa tissue will adversely affect the cartilaginous rings and can lead to cell necrosis and loss of structural integrity of the tracheal tissue; this necrosis and loss of integrity in turn can lead to tracheal stenosis, which in case is referred to as cuff stenosis.

Post-tracheostomy tracheal stenosis is a result of tissue damage resulting from insertion of the tracheostomy tube. Granulation tissue, which is formed as a result of the wound healing process, can directly lead to stenosis by blocking the flow of air in the trachea. In addition, structural damage to the cartilaginous rings by direct physical contact with the tracheostomy tube can lead to loss of structural integrity of the trachea and lead to stenosis. Also, formation of granulation tissue from the wound healing process around the cartilaginous rings can further accelerate stenosis.

7.5 TREATMENT MODALITIES FOR TRACHEAL STENOSIS

As we discussed in the previous sections, tracheal stenosis can occur due to congenital disorders like CTS or can be a result of endotracheal intubation or tracheostomy. These conditions affect infants and adults, although the prevalence of congenital disorders is lower compared to complications arising from tracheal intubation or tracheostomy. In order to treat tracheal stenosis, several strategies have been developed, some of which include the use of stents, balloon dilation, tracheal resection, and reanastomosis.

Endoscopic Balloon Dilation for Tracheal Stenosis—This procedure involves insertion through the mouth of a balloon catheter within the trachea (46–50). Once positioned within the trachea, the balloon is gradually inflated to apply radial pressure to the luminal surface of the trachea. This increase in radial pressure acts to increase the internal diameter of the trachea and reduce the extent of stenosis. The pressure in the balloon is then reduced, causing it to deflate, and the catheter is then removed.

This procedure has been conducted extensively in the United States and other parts of the world; a recent report summarizing the findings from 209 patients has been prepared (51). The most significant findings was that endoscopic balloon dilation resulted in significant improvement of tracheal stenosis and was not associated with any major complications.

One example of a balloon dilation system that is currently available on the market is the CRE^{TM} Pulmonary Balloon Dilator by Boston Scientific. The catheter is available in different dimensions, with an internal diameter ranging from 8 mm to 18 mm at a pressure of 3 atmospheres (information from company website). The length of the catheter is fixed at 75 cm while the length of the balloon itself varies from 3.0 mm to 5.5 mm.

Airway Stents—A stent is a hollow, cylindrical prosthesis that maintains luminal patency and provides support (52–57). Stents can be inserted within the trachea to alleviate stenosis and reduce the extent of airway narrowing (58–65). The devices are implanted in the trachea and retained for an extended time period. This is unlike balloon dilation, in which a catheter is inserted into the trachea a single time and then removed from the patient (although the procedure may be repeated several times). According to a recent publication by Mehta, an ideal airway stent should have the following characteristics (54): 1) easy to insert and remove, 2) customized to fit the dimensions and shape of the stricture, 3) able to re-establish the airway and maintain luminal patency with minimum rate of migration, 4) made of an inert material that does not irritate the airway, precipitate infection, or promote granulation tissue formation, 5) able to exhibit similar clearance characteristics like the normal airway so that mobilization of secretions is not impaired, and 6) economically affordable.

Airway stents are considered to be a long-term solution for the treatment of tracheal stenosis because they can remain in place for months, with reports of stents remaining implanted for periods in excess of 50 months. While there are risks associated with all surgical interventions, implantation of stents in the trachea is safe. In one study in which 42 patients were treated for tracheal stenosis, the patients first underwent balloon dilation, followed by implantation of silicone stents (66). In this study, 5% of the patients suffered from complications resulting from granuloma formation, which were successfully treated, and an additional 5% of the patients required repeat surgery for stent replacement, which was successfully conducted.

There are two categories of airway stents: metal stents and tube stents. One example of an airway tube stent is the Montgomery Safe-T-Tube™ series, which is offered by Boston Medical Products. The stent is available in 5 different styles: pediatric, standard, thoracic, extra-long, and tapered. The standard stents are made from implant grade silicone and come in different dimensions, ranging from 10 mm to 16 mm in external diameter (information from company website).

The second category of airway stents are metallic mesh stents, which can be coated with a plastic covering and may also be balloon-expandable or self-expanding. Balloon-expandable stents are implanted in a deflated form and, once inserted at the desired position, are secured by inflation of the balloon. Self-expanding metallic stents are made with shape memory alloys like nitinol, and are maintained in a specific configuration at a lower temperature. Once inserted inside the body, the increase in temperature causes the material to change configuration, thereby allowing it to fit into its functional position. Metal stents are easier to work with and can be easily placed within the trachea via flexible bronchoscopy under local anesthesia in an outpatient setting. One example of a metal stent for tracheal stenosis is the Ultraflex™ tracheobronchial stent, made by Boston Scientific. Ultraflex stents are made with nitinol, an alloy of titanium and nickel that comes in different dimensions, ranging in external diameter from 146 mm to 20 mm and ranging in length from 40 mm to 80 mm (information from company website).

Tracheal Resection and Reconstruction—Tracheal resection and reconstuction is a surgical procedure which involves removal of a small portion of the trachea that has been affected by stenosis and reconstruction of the remaining tissue to form a complete, though shorter, trachea (67–74). Removal of up to 4–5 cm of the trachea can be conducted. Prior to conducting the surgery, it is important to recreate the geometry of the trachea and identify the region of stenosis; this identification can be achieved by CT scans of the neck and chest region. Direct laryngoscopy and bronchoscopy also need to be conducted in order to obtain an accurate assessment of the tracheal stenosis. Tracheal resection and reconstruction is an invasive surgical procedure that is only conducted after unsuccessful attempts using other strategies, like balloon dilation. For the duration of the surgical procedure, the patient needs to be placed on a mechanical ventilator due to the resection of the trachea.

The surgical procedure requires a low traverse cervical incision for tracheal stenosis in the upper portions, while stenosis in the lower parts of the tracheal may require an additional upper midline sternostomy. The strap muscles are retracted to expose the trachea, which is separated from the surrounding tissue and secured in place. The part of the trachea that is affected is excised, and the open ends of the tissue are sutured together.

Careful consideration and planning goes into the surgery due to potential complications that can occur. Mechanical ventilation is needed for 18 to 24 hours after the surgical procedure has been completed. Steroids are required to reduce edema for several weeks after the surgery. Other potential complications include the formation of granulation tissue, infection, and injury to the laryngeal nerve.

A recent study looked at the success rate of 110 patients after tracheal resection and reanastomosis (75). In this study, the length of the trachea resected ranged from 2.0 cm to 6.5 cm, with a median length of 3.5 cm. The most frequent post-surgical complication reported was recurrent nerve paralysis, observed in 5.5% of the cases. The long-term results of these surgeries were evaluated 12 to 226 months after the surgery, and 93.5% of the patients reported satisfactory results. The mortality rate related to the surgical procedure was reported to be 2.7%.

7.6 DESIGN CONSIDERATIONS FOR TRACHEAL TISSUE ENGINEERING

The design considerations for tracheal tissue engineering can be easily stated as *"bioengineered tracheas should be similar in form and functional to mammalian tracheas."* This is the overarching theme in tracheal tissue engineering, and specific requirements have been defined to meet this objective. This is not only the case for tracheal tissue engineering; it also applies across tissue systems, as we will come across in the later chapters.

The specific design considerations for tracheal tissue engineering have been eloquently defined by Macchiarini and Grillo as (76,77): 1) biocompatibility, 2) liquid- and air-tight, 3) nonimmunogenic and minimal inflammatory response, 4) nontoxic and noncarcinogenic, 5) avoidance of collapse by reasonable strength, 6) support cell engraftment, 7) support neovascularization, 8) possibility of growth, 9) resistance to fibroblastic and bacterial invasion, 10) standardized easy and short fabrication, 11) customizable and low-cost, 12) easy surgical handling, 13) provide physiological environment similar to ECM, 14) minimal necessity of donors and accessibility, 15) result in predictably successful engraftment, 16) provide or support epithelial resurfacing, 17) avoid stenosis or late buckling, 18) avoid accumulations of secretions, 19) must not dislocate or erode over time, and 20) permanent constructions.

7.7 PROCESS OF BIOENGINEERING ARTIFICIAL TRACHEAS

In Chapter 1, we presented a general scheme to bioengineer 3D artificial tissue. In this section, we will discuss a modified process scheme that has been adopted to bioengineer artificial tracheas. The process scheme presented here is based on

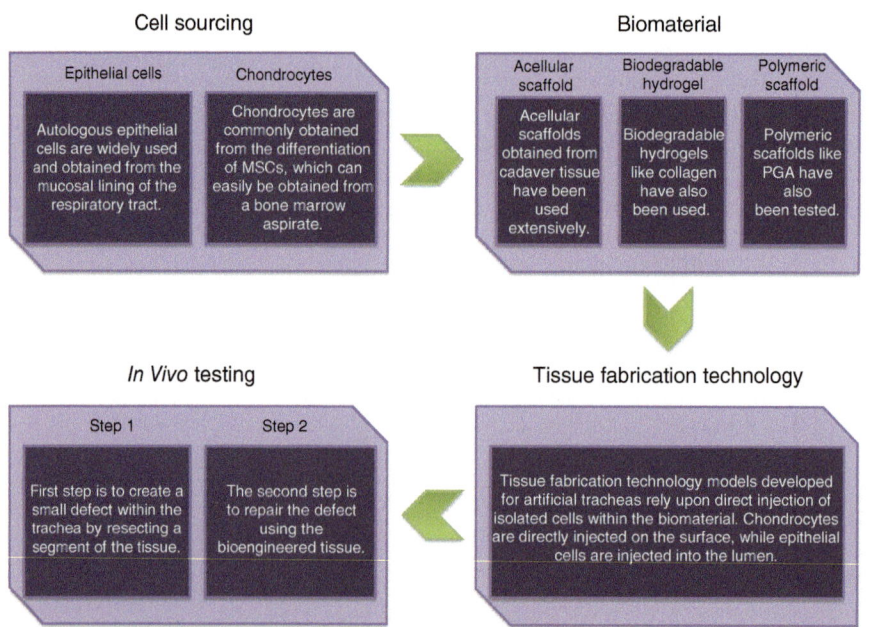

Figure 7.1 Process of Bioengineering Artificial Tracheas—Tubular scaffolds have been fabricated using many different biomaterials, including acellular grafts, biodegradable hydrogels, and polymeric scaffolds. Epithelial cells have been isolated from a tissue biopsy of the mucosal layer, while chondrocytes have routinely been obtained by differentiation of bone marrow MSCs. Tissue fabrication technologies for tracheal tissue engineering have commonly involved the use of direct injection strategies for incorporation of chondrocytes. Epithelial cells are injected into the lumen of the tissue graft. Once the tubular biomaterial has been cellularized, the artificial trachea is tested for *in vivo* efficacy using a small animal injury model, in which part of the trachea is resected and replaced with the artificial tissue.

recent publications that have presented varying strategies to bioengineer 3D trachea using different tissue engineering platforms (Figure 7.1).

As we discuss this process scheme, we will quickly see that attempts to bioengineer trachea have used different cell sources, many different biomaterials, and several tissue fabrication technologies. We will also see that bioreactor technology has not been extensively integrated in the process flow sheet for tracheal tissue engineering, as has been the case for several other tissue systems. The same is true for vascularization efforts, as there are not many examples of tissue-engineered tracheas with a blood vessel supply incorporated into the artificial tissue.

In this section, we provide a general scheme to bioengineer 3D artificial tracheas, and in the next section, we describe specific examples of tissue-engineered models for artificial tracheas. This is followed by two recent examples of tissue-engineered tracheas that have been used in clinical applications for the treatment of patients.

Cell Sourcing—It may be recalled that the trachea consists of several cell types, including epithelial cells on the luminal surface and chondrocytes within

the cartilaginous rings. Most efforts for bioengineering tracheas have focused on identifying suitable sources for these two cell types. Autologous epithelial cells have been widely used and are obtained from the mucosal lining of the respiratory tract. A tissue biopsy can be obtained using minimally invasive methods and the tissue biopsy can be digested to isolate epithelial cells. Chondrocytes have been commonly obtained from the differentiation of mesenchymal stem cells (MSCs), which can easily be obtained from a bone marrow aspirate. Once the MSCs are isolated and cultured *in vitro*, several protocols have been developed to effectively drive the differentiation of cultured MSCs toward a chondrocyte lineage. Autologous epithelial cells and MSC-derived chondrocytes have been a preferred source for tracheal cells, both for animal studies and for clinical studies.

Biomaterial Development—There has been considerable variation in the choice of biomaterials that have been used to bioengineer artificial tracheas. Acellular scaffolds obtained from cadaver tissue have been used extensively, including in clinical studies discussed later in this chapter. In addition, biodegradable hydrogels, like collagen, have also been used; and polymeric scaffolds like PGA have also been tested. Composite biomaterials with PCL and collagen have also been used for tracheal fabrication. Scaffold-free technologies have not been evaluated due to challenges in the formation of hollow structures that can support changes in fluid hemodynamics over long time periods.

Tissue Fabrication Technology—Tissue fabrication technology is perhaps one area that is lacking in the field of tracheal tissue engineering. Many of the models that have been developed for artificial tracheas have relied upon direct injection of isolated cells within the biomaterial. As we have discussed before, direct injection technology provides limited spatial resolution for the cells. There have not been any reports using cell/organ printing or solid freeform fabrication technologies to support the development of artificial tracheas.

Bioreactors—Bioreactors are another technology that have not been developed very effectively in the tracheal tissue engineering field. There have been few studies that report the use of bioreactors for the development of artificial tracheas. There was one study in which a bioreactor system was developed for scaffold cellularization, and the resulting tracheas were used clinically. In addition, there have been two studies that describe the use of bioreactors to support the fabrication of artificial trachea. In another study, a bioreactor was developed to provide physiological conditioning to an artificial trachea after fabrication of the tissue graft. In this case, an artificial trachea was fabricated by culturing chondrocytes on a composite biomaterial and cultured under static *in vitro* conditions for several days prior to bioreactor conditioning. The bioreactor was designed to provide continuous media flow to support the metabolic requirements of the artificial trachea. In this bioreactor system, the rotational speed of the bioreactor could be varied, thereby culturing the trachea under controlled shear stress regimes.

Vascularization—There have not been any reports of *in vitro* methods for vascularization of artificial tracheas. Instead, the strategy has been to utilize acellular grafts to bioengineer artificial tracheas, and once implanted, the acellular scaffolds supported neovascularization from host tissue. Indeed, this has been the case for the two clinical studies, both of which have demonstrated neovascularization after tracheal implantation. However, the need for *in vitro* vascularization still persists and continues to be an area that requires attention.

In Vivo Testing—There have been several groups that have demonstrated the efficacy of bioengineered tracheas using small animal models. While there have been variations in the specific models used, they all aim to create a small defect within the trachea by resecting a segment of the tissue and then replacing this defect using the bioengineered tissue.

7.8 TISSUE ENGINEERING MODELS FOR ARTIFICIAL TRACHEAS

In this section, we will look at several models that have been developed to fabricate artificial trachea using tissue engineering technologies (Figure 7.2).

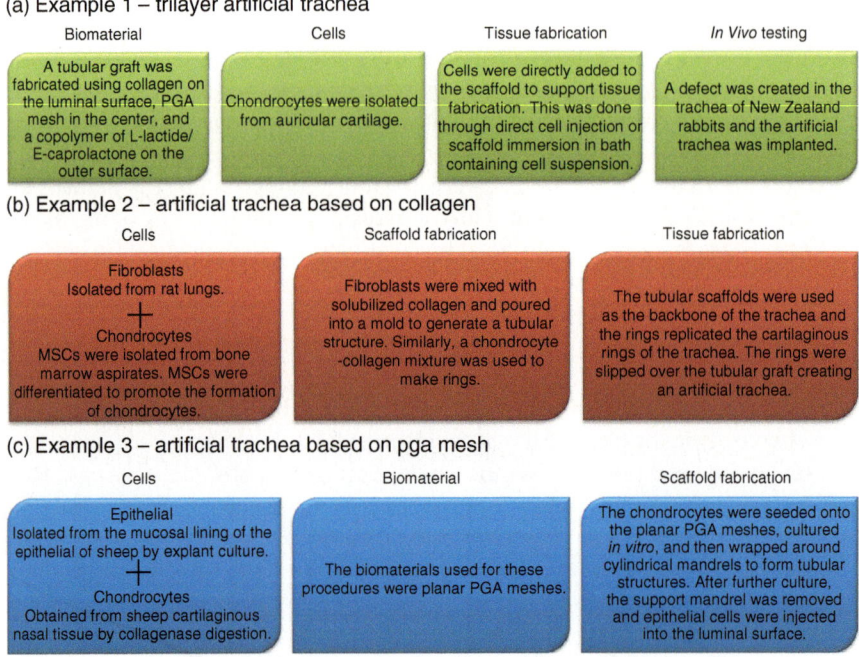

Figure 7.2 Models of Artificial Tracheas—(a) Tri-layer Artificial Trachea—A trilayer tubular scaffold was fabricated and populated with autologous chondrocytes, which were isolated from auricular cartilage. Once an artificial trachea was bioengineered, it was tested *in vivo* to assess efficacy. **(b) Artificial Trachea Based on Collagen**—Fibroblasts were mixed with collagen and used to form tubular grafts. The same strategy was used to fabricate ring structures, which were coupled with the tubular grafts to form an artificial trachea. **(c) Artificial Trachea based on PGA Mesh**—A planar PGA mesh was populated with chondrocytes isolated from a nasal septum tissue biopsy. The planar mesh was rolled up to form a hollow tubular structure, and epithelial cells were injected within the lumen, resulting in the formation of an artificial trachea.

As we review these models, we see that different researchers have tested different biomaterials, several cell sources have been used, and researchers have adopted various tissue fabrication technologies. In addition, we will look at the use of bioreactor technology to fabricate tracheal tissue or to provide physiological conditioning and tissue development post-fabrication. As we study these models, we will have an opportunity to relate many tissue engineering principles, namely those presented in earlier chapters, for the fabrication of artificial tracheas.

Example 1 for Tissue-Engineered Tracheas—A tri-layer tracheal graft was fabricated using collagen, nonwoven PGA mesh, and a copolymer of (L-lactide/ϵ-caprolactone) (78). A tubular graft was fabricated using these three materials, with collagen on the luminal surface, PGA mesh in the center, and the copolymer of L-lactide/ϵ-caprolactone was used on the outer surface. The tri-layer tracheal graft was designed with three separate materials to provide different functions. The collagen layer was designed to support the attachment and functionality of epithelial cells, the central PGA mesh was designed to support chondrocyte viability, and the outermost copolymer layer was designed to provide elasticity to the trachea. Chondrocytes were isolated from auricular cartilage, which was obtained from New Zealand rabbits. Biopsies were obtained and digested in a collagenase solution, and the isolated cells were maintained and expanded under controlled cell culture conditions. In order to support tissue fabrication, the cells were directly added to the scaffold. As the details of the cellularization protocol were not provided, it can be assumed that the cells were added either by direct injection into the scaffold using a syringe, or by immersing the scaffold within a bath containing the cell suspension. Once cellularization was complete, the scaffold was maintained in a cell culture incubator for 24 hours prior to *in vivo* testing using an animal model. It should be noted that the artificial trachea was not populated with epithelial cells during *in vitro* tissue fabrication; instead, this model is based on the hypothesis that *in vivo* implantation of the artificial trachea will lead to epithelization using autologous cells from the host.

The effectiveness of the artificial trachea was tested using a small animal injury model. Using New Zealand rabbits for *in vivo* testing, a defect was created and the artificial trachea was sutured in place. Controlled release of transforming growth factor was used to support chondrogenesis. Three months post-evaluation, epithelization of the luminal surface was demonstrated in the artificial trachea, along with accumulation of cartilage tissue. This clearly demonstrated the feasibility of tissue engineering strategies to bioengineer artificial tracheas.

Example 2 for Tissue-Engineered Trachea—A very interesting approach to bioengineer artificial tracheas was recently published. In the previous example and in others we have come across in the book, the scaffold fabrication and cellularization have been independent, though sequential, events in the tissue fabrication process. In this example, these two were coupled (79). Cells were obtained from rats, with fibroblasts being isolated from the lungs and MSCs isolated from bone marrow aspirates. MSCs were subjected to a differentiation strategy to promote the formation of chondrocytes; differentiation was induced using cell culture media supplemented with β-glycerophosphate, ascorbic acid phosphate, and dexamethasone.

Collagen was used as the scaffolding material to support the fabrication of artificial trachea.

Cells, either fibroblasts or MSC-differentiated chondrocytes, were mixed with solubilized collagen, and this cell-gel mixture was poured into a mold to generate tubular or ring-shaped structures. Tubular structures were constructed with fibroblasts and were designed to provide the backbone of the trachea or the major structural component. The rings that were fabricated using chondrocytes were designed to replicate the cartilaginous rings of the trachea. Once fabricated, the rings were slipped over the tubular grafts; the tubular grafts replicated the backbone of the trachea while the rings replicated the cartilage rings. This process supported the fabrication of an artificial trachea that was close in form to mammalian trachea.

Once fabricated, the artificial trachea was implanted in rats after a tracheal defect, and was shown to survive for very short time periods not exceeding two days. While the tissue fabrication technology adapted in this model was very interesting, there was no consideration for the formation of an epithelial lining that may have contributed to the limited *in vivo* success of the artificial trachea. The use of fibroblasts was also interesting; it appears the fibroblasts were used to provide structural support in the formation of a hollow tubular structure, which could not be easily accomplished with the collagen itself. Nonetheless, the model was novel and interesting, and upon further development and optimization, it could lead to a viable strategy to bioengineer artificial tracheas.

Example 3 for Tissue-Engineered Tracheas—In another very interesting and novel approach to bioengineer artificial tracheas, epithelial cells and chondrocytes were obtained from the nasal septum of sheep (80). Epithelial cells were isolated from the mucosal lining of the epithelial by explant culture, which means that tissue biopsies were plated on a tissue culture plate, resulting in the outgrowth of epithelial cells from the tissue sample. Chondrocytes were obtained from the cartilaginous tissue, which was subjected to a collagenase digestion procedure. The chondrocytes were seeded onto planar PGA meshes, cultured *in vitro*, and then wrapped around cylindrical mandrels to form tubular structures. These cellularized constructs were implanted in a subcutaneous pocket in nude mice to support the formation and development of the tissue graft. The tissue graft was implanted with the cylindrical mandrel in place. Upon explantation, the support mandrel was removed from the tissue graft, and epithelial cells were injected into the luminal surface. The tissue graft was cultured *in vitro* after injection of the epithelial cells, which led to the fabrication of an artificial trachea. While the artificial trachea was not tested for *in vivo* functionality, the anatomical and functional properties of the artificial tissue were shown to be comparable to that of mammalian tracheas.

Concluding Remarks—In this section, we have looked at several examples of tissue engineering strategies that have been used to fabricate artificial tracheas. We have seen the use of different materials, different sources of cells, and very different tissue fabrication technologies. The reader can easily relate many of these strategies to various topics that have been covered in previous chapters, including cell sourcing, stem cell differentiation, biomaterial development, and tissue fabrication technologies. In addition, based on the information that has been provided in earlier chapters, the reader can design custom strategies that can lead to the fabrication of artificial trachea. One limitation of the strategies that we have looked at is the inability to incorporate bioreactor technology in the fabrication of artificial tracheas, either for scaffold cellularization or for physiological conditioning. Indeed, this has been one of the limitations in the field of tracheal tissue engineering, as there are very few studies that describe the use of bioreactor technology to support the fabrication of artificial trachea.

7.9 TRACHEAL TISSUE ENGINEERING—AN EXAMPLE OF A CLINICAL STUDY

There have been two successful clinical studies relating to the transplantation of tissue-engineered tracheas, one of which is described in this section (Figure 7.3)

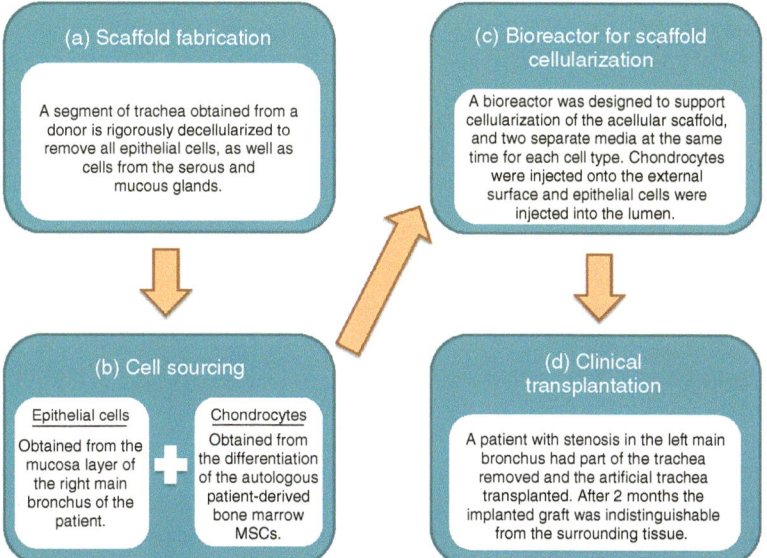

Figure 7.3 Fabrication of Artificial Trachea for Clinical Applications—(a) **Scaffold Fabrication**—Donor tracheal tissue was subjected to a decellularization protocol designed to remove the cells, leaving behind an intact extracellular matrix. (b) **Cell Sourcing**—Epithelial cells were isolated from a biopsy of the mucosal layer of the right main bronchus. Chondrocytes were obtained from differentiation of bone marrow MSCs. (c) **Bioreactor for Scaffold Cellularization**—Chondrocytes are added to the scaffold by direct injection of the cells onto the outer surface of the acellular scaffold using a microsyringe. Epithelial cells were added to the scaffold by delivery of the cells to the luminal surface of the acellular scaffold. (d) **Clinical Transplantation**—A segment of the left main bronchus was removed due to stenosis and replaced with the tissue-engineered trachea.

and a second one, which is described in the next section (81,82). Indeed, fabrication of artificial trachea has been one of the more successful areas of research in the field of tissue engineering. The clinical studies have been instrumental in validating the feasibility of tissue and organ engineering and have demonstrated many of the principles of tissue fabrication. The first of these clinical studies was reported in 2008 (82), while the second was reported in 2012 (81).

Scaffold Fabrication—The first successful human transplantation of tissue-engineered trachea was reported in a seminal publication in the year 2008 (81). An artificial trachea was fabricated using an acellular scaffold. A segment of trachea, which measured 7 cm in length, was obtained from a human donor and subjected to a very rigorous decellurization protocol. As we have seen before, decellularization strategies are designed to remove all cellular components from the tissue, leaving behind an intact extracellular matrix. The ECM has the right composition, distribution, and orientation of various proteins, making it suitable to support the tissue fabrication process. In this case, the decellularization process consisted of 25 cycles using sodium deoxycolate and deoxyribonuclease I. It was demonstrated that the decellularization protocol removed all epithelial cells, cells of the serous and mucous glands, and almost all of the chondrocytes associated with the cartilaginous rings. It was also demonstrated in this study that the decellularization process did not damage the properties of the ECM or change the mechanical properties in any significant manner.

Cell Sourcing—Patient-derived autologous cells were used to fabricate the artificial trachea, which completely removed the need for immunosuppression. In order to fabricate artificial trachea, epithelial cells were required to populate the luminal surface, while chondrocytes were required to populate regions of the trachea that give rise to the cartilaginous rings. In this study, the epithelial cells were obtained from the right primary bronchus, while the chondrocytes were obtained by differentiation of bone marrow mesenchymal stem cells. Both cell sources used in this study were autologous, which means they were obtained from the patient.

In order to source epithelial cells, a small tissue biopsy was obtained from the mucosal layer of the right main bronchus. The tissue biopsy was transported to the laboratory in phosphate-buffered saline containing penicillin and streptomycin. The tissue specimen was digested using trypsin, and the isolated epithelial cells were plated on a tissue culture surface and maintained in a cell culture incubator. The cells were cultured in a DMEM based media and supplemented with bovine pituitary extract and recombinant epidermal growth factors.

Chondrocytes were obtained by controlled differentiation of autologous patient-derived bone marrow mesenchymal stem cells (BMMSCs). BMMSCs were obtained from a bone marrow aspirate and were maintained in a DMEM based media supplemented with basic fibroblast growth factor. The cells were cultured in tissue culture flasks and subpassaged at 90% confluency. In order to differentiate BMMSCs to chondrocytes, the BMMSCs were cultured for 72 hours in complete media containing human transforming growth factor, recombinant parathyroid hormone-related peptide, dexamethasone, and insulin.

Bioreactors Design—A novel bioreactor was designed to support cellularization of the acellular scaffold with epithelial cells on the luminal surface and chondrocytes on the external surface. The bioreactor was housed within a polysulphone chamber, which provided anchoring points for the acellar scaffold. Chondrocytes were directly applied to the external surface of the acellular graft using a direct injection strategy with a microsyringe. Epithelial cells were injected into the lumen of the acellular scaffold through a separate access port engineered into the bioreactor. The concentration of chondrocytes and epithelial cells was adjusted to one million cells per milliliter for both cell types. The bioreactor was designed to accommodate two separate media formulations at the same time: one for the chondrocytes and one for the epithelial cells. The acellular graft was cultured in the bioreactor for 96 hours prior to implantation. Another novel feature of the bioreactor was the design of a component that rotated between a liquid phase and an air phase, which was a mechanism designed to enhance oxygenation, increase the supply of nutrients, and support waste removal.

Implantation of Tissue-Engineered Trachea—The artificial trachea was implanted in a patient; this is one of the few examples of successful clinical applications of tissue-engineered grafts. The artificial trachea was used to replace the left main bronchus in a patient. The patient was discharged from the hospital ten days after the surgery was performed and was monitored for up to 2 months post-surgery. Remarkably, the implanted graft was indistinguishable from the surrounding tissue and was vascularized by the host tissue and there were no signs of inflammatory cells 60 days after surgery. This is indeed a remarkable fate and one of the true accomplishments in the field of tissue engineering.

Concluding Remarks—The example presented here serves to demonstrate one of the most significant milestones in the field of tissue engineering: the clinical application of artificial tissue to improve a patient's quality of life. In addition, this example serves to demonstrate the role of biomaterials, cells, and bioreactors for the fabrication of artificial tissue; a clear validation of the building blocks of tissue engineering. Tissue engineering has been defined as tissue fabrication in Chapter 1 of this book; here, we see how this works in a clinical setting.

7.10 TRACHEAL TISSUE ENGINEERING—A SECOND EXAMPLE OF A CLINICAL STUDY

In this section, we discuss a second case of a tissue-engineered trachea being used for the treatment of a patient (82) (Figure 7.4). In this case, the patient was born with a birth defect known as congenital tracheal stenosis, which has been described earlier in this chapter. This patient was treated using several strategies, which included autologous patch tracheoplasty, implantation of balloon-expandable stainless steel stents, and transplantation of a tracheal homograft after complications resulting from the stent. At the age of 10, the patient suffered further complications and required additional surgical intervention. Therefore, in this case, a tissue-engineered trachea proved to be a viable treatment option for this patient.

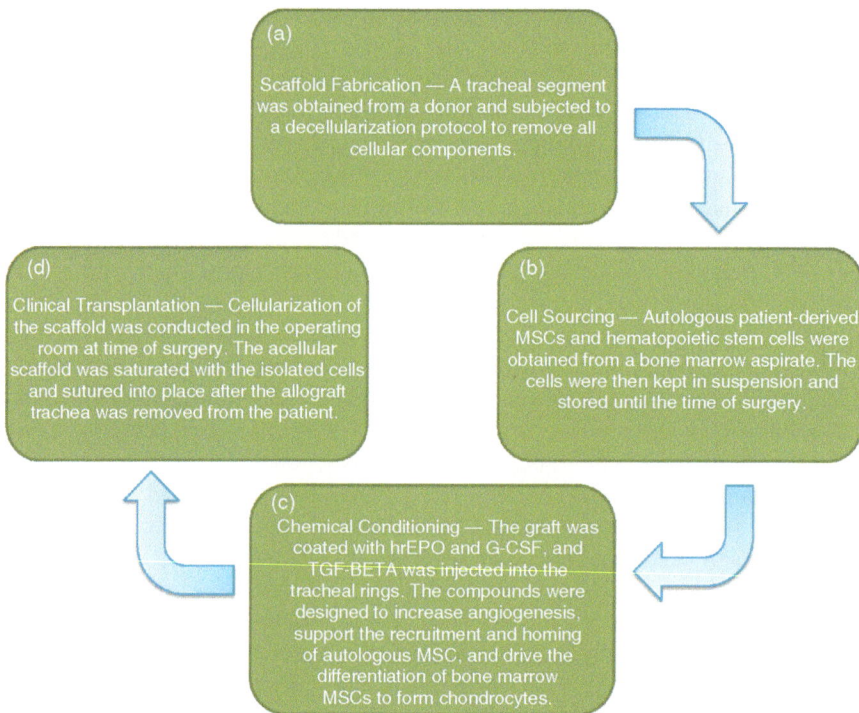

Figure 7.4 A Second Example: Fabrication of Artificial Tracheas for Clinical Applications—(a) **Scaffold Fabrication**—Donor tracheal tissue was subjected to a decellularization protocol designed to remove the cells, leaving behind an intact extracellular matrix. (b) **Cell Sourcing**—Bone marrow aspirate is used to source autologous MSCs and hematopoietic stem cells. (c) **Chemical Conditioning**—Several chemical compounds were used to drive the differentiation of MSCs to form chondrocytes, to support the recruitment of circulating MSCs, and to increase angiogenesis. (d) **Clinical Transplantation**—A segment of the trachea was removed due to stenosis and replaced with the tissue-engineered trachea.

Scaffold Fabrication—The strategy for scaffold fabrication was the same one used in the first example. A tracheal segment was obtained from a donor and subjected to a decellularization protocol to remove all cellular components, leaving behind an intact extracellular matrix.

Cell Sourcing—Autologous cells were used in this case; they consisted of a mixture of patient-derived mesenchymal stem cells and hematopoietic stem cells, both of which were obtained from a bone marrow aspirate. The isolated cells were not maintained and/or expanded in culture; on the contrary, the cells were retained in suspension and stored until the time of the surgery. This means that the number and proportion of cells at the time of isolation remained the same until the time of surgical implantation.

There are significant differences in the cell sourcing strategy used for the two clinical studies that we have presented. In the previous example, the isolated cells were maintained and manipulated in culture prior to surgical transplantation, whereas in the current example, the cells were stored in suspension prior to use. In the previous example, the epithelial cells were isolated from a biopsy of the right main bronchus, and the chondrocytes were obtained from differentiation of bone marrow MSCs. This is in contrast to the current example, which used unmodified MSCs and hematopoietic stem cells obtained from a bone marrow aspirate.

Surgical Engraftment—Cellularization of the acellular scaffold was conducted in the operating room at the time of surgery. The acellular scaffold was saturated with the isolated cells, and the cellularized scaffold was sutured into place along the length of the trachea after removal of the allograft trachea that was previously implanted into the patient. The graft was coated with erythropoietin and granulocyte-colony stimulating factor, and transforming growth factor was injected into the tracheal rings. These compounds were designed to increase angiogenesis, support the recruitment and homing of autologous MSC, and drive the differentiation of bone marrow MSCs to form chondrocytes.

Comment—There are a few significant differences between the two clinical examples presented in this chapter. In the current example, the cells and the scaffold were coupled together at the time of surgical engraftment, as will be described later in this section; there was no scaffold cellularization prior to surgical implantation. Along the same line of thought, this also means that the strategy described in this study does not require or make use of bioreactors, either for cellularization and/or for physiological conditioning. This is consistent with the tissue engineering model that has been adapted in the study: coupling of cells and scaffold at the time of surgical engraftment. This model is significantly different from the previous one presented in this chapter in which cellularization of the scaffold was conducted prior to surgical implantation; in this case, bioreactors were used to support scaffold cellularization and tissue fabrication. These are two completely different strategies, each of which has proven to be successful in the clinical setting.

There were some short-term surgical complications resulting from implantation of the artificial trachea. However, once the patient stabilized, steady recovery was reported. Fifteen months after surgery, the implanted trachea showed complete epithelialization on the lumen surface, and the patient had normal lung function. Two years after surgical intervention, the patient was reported to be *"well, growing, and had not needed medical intervention for 6 months."*

Discussion—As we saw in the previous example, this case study serves to demonstrate successful implantation of a tissue-engineered trachea and is a remarkable demonstration of accomplishment in the field. There are clear differences between the two cases, particularly with regard to the specific tissue engineering technology adapted. In the first case study, an artificial trachea was developed *in vitro* and then transplanted; scaffold cellularization and maturation took place under controlled *in vitro* culture conditions and made use of bioreactor technology. In the second case, scaffold cellularization was conducted at the time of surgery, and the scaffold was immersed in the cell suspension and immediately transplanted; several compounds were added to the trachea to promote development and maturation during *in vivo* culture of the implanted tissue graft. While different, both of these strategies proved to be successful as judged by the most important criteria, which is the ability to improve patient quality of life. Indeed, both of these studies are seminal in the field and clearly demonstrate the tremendous potential for the field of tissue engineering.

SUMMARY

Current State of the Art—In this chapter, we have looked at several examples for the development of 3D artificial tracheas. The models presented in this chapter served to illustrate many of the principles we have studied during the course of this book, including cells, biomaterials, bioreactors, and tissue fabrication technology. We have seen that different combinations of these fundamental principles in tissue engineering lead to very different results and very different models for 3D artificial tracheas. We have also seen that careful selection of design variables, like the source of cells, type of biomaterial, cellularization strategy, and culture conditions, including bioreactor conditioning, have a significant impact on the form and function of 3D artificial trachea. Perhaps most importantly, we have described the application of tissue engineering technology in a clinical setting. We studied examples of 3D artificial tracheas that have been transplanted in patients to significantly enhance the quality of life for these patients. These clinical studies are described as seminal publications, and serve to demonstrate and validate the potential impact of tissue engineering.

Thoughts for Future Research—There has been considerable progress made in the field of tracheal tissue engineering. However, the development of biomaterials to support fabrication of 3D artificial tracheas remains one area of research that has not fully matured. Mammalian tracheas are hollow structures that have a high demand on the mechanical properties of the extracellular matrix. In tissue engineering, acellular scaffolds have been used extensively to support fabrication of artificial tracheas; this is due to the high mechanical stability of acellular scaffolds. While acellular scaffolds have proven to be effective for fabrication of transplantable tissue, a generation of new biomaterials needs to be developed; materials which have mechanical properties that are comparable to acellular grafts are necessary. A generation of biomaterials needs to be synthesized to support the fabrication of 3D artificial tracheas with mechanical properties comparable to that of mammalian tracheas.

PRACTICE QUESTIONS

1. Describe the structure of the mammalian trachea. Include a description of the different cell types and the functions of these cells. Also describe the extracellular matrix components that constitute the mammalian trachea.
2. What is congenital tracheal stenosis (CTS)? Explain the classification scheme for CTS. What are some of the symptoms associated with CTS? How can CTS be diagnosed?

3. Describe the genetic regulation of tracheal development.

4. What are the causes of post-intubation and post-tracheostomy tracheal stenosis?

5. What are some treatment modalities for tracheal stenosis? Discuss the relative advantages and disadvantages of the treatment modalities that you describe.

6. In Chapter 4, we studied two strategies for tissue fabrication using scaffold-free technology: self-organization and cell sheet engineering. Start by providing a brief description of these two strategies to support tissue fabrication. Do you believe either one of these can be used to support the fabrication of artificial tracheas? If so, which one do you think is more suitable and why?

7. During our discussion of tracheal tissue engineering, there was an evident lack of interest in the development of bioreactor technology to support the formation and culture of artificial tracheas. If you were to develop bioreactors for tracheal tissue engineering, where would you use the bioreactors in the tissue fabrication process? Give one specific example, including a design for the bioreactor, that can be used to support the fabrication of artificial tracheas.

8. During our discussion on stem cells for tracheal tissue engineering, we studied a few examples using primary bone marrow MSCs to support the fabrication of artificial tracheas. Instead of using bone marrow MSCs, we can also use iPS cells to fabricate artificial tracheal tissue. Do you believe that iPS cells will be advantageous for development of artificial tracheal tissue? Explain your answer.

9. During our discussion in Chapter 2, we studied iPS cells and hES cells and their potential application during the tissue fabrication process. Which one of these two will you choose to fabricate artificial tracheas and why?

10. Acelullar scaffolds have been used extensively to support the fabrication of artificial tracheas. What are some advantages of acellular scaffolds? What are some of the disadvantages of acellular scaffolds for the fabrication of tracheal tissue?

11. In Chapter 4, we discussed cell and organ printing for the fabrication of artificial tracheas. Provide a brief description of cell and organ printing. Describe how you would use cell and organ printing to bioengineer artificial tracheas.

12. In Chapter 5, we provided a process flow chart that can be used to determine the best strategy to induce vascularization in artificial tissue and organs. Using this process flow chart, develop a strategy to engineer vascularization in artificial tracheal tissue.

13. In Chapter 5, we discussed three strategies to induce vascularization in artificial tissue and organs: biologically replicated, biologically mediated, and

biologically inspired. Start by providing a brief description of each of these three strategies. Which one will you use for the vascularization of artificial tracheas and why?

14. Develop a process flow chart for the fabrication of artificial tracheas. Include a discussion of the following items: cell sourcing, biomaterial selection, scaffold cellularization, vascularization, and bioreactor conditioning.

15. What are some of the critical scientific and technological challenges in the field of tracheal tissue engineering? What can you do to overcome these challenges?

REFERENCES

1. Wailoo M, Emery JL. Structure of the membranous trachea in children. Acta Anat. (Basel) 1980;106(2):254–61.
2. Tandler B, Sherman JM, Boat TF, Wood RE. Surface architecture of the mucosal epithelium of the cat trachea: II. Structure and dynamics of the membranous portion. Am. J. Anat. 1983 Oct;168(2):133–44.
3. Baluk P, Fujiwara T, Matsuda S. The fine structure of the ganglia of the guinea-pig trachea. Cell Tissue Res. 1985;239(1):51–60.
4. Amiri MH, Gabella G. Structure of the guinea-pig trachea at rest and in contraction. Anat. Embryol. (Berl) 1988;178(5):389–97.
5. Breatnach E, Abbott GC, Fraser RG. Dimensions of the normal human trachea. AJR Am. J. Roentgenol. 1984 May;142(5):903–6.
6. Salassa JR, Pearson BW, Payne WS. Gross and microscopical blood supply of the trachea. Ann. Thorac. Surg. 1977 Aug;24(2):100–7.
7. Sturridge MF, Mueller MR, Treasure T. Blood supply of the trachea and proximal bronchi. Ann. Thorac. Surg. 2007 Aug;84(2):675.
8. Fisher AW. The Intrinsic Innervation of the Trachea. J. Anat. 1964 Jan;98:117–24. PMCID:PMC1261319.
9. Rikimaru A, Sudoh M. Innervation of the smooth muscle of the guinea-pig trachea. Nihon Heikatsukin. Gakkai Zasshi 1971 Mar;7(1):35–44.
10. Pack RJ, Al-Ugaily LH, Widdicombe JG. The innervation of the trachea and extrapulmonary bronchi of the mouse. Cell Tissue Res. 1984;238(1):61–8.
11. Baluk P, Gabella G. Innervation of the guinea pig trachea: a quantitative morphological study of intrinsic neurons and extrinsic nerves. J. Comp Neurol. 1989 Jul 1;285(1):117–32.
12. Phipps LM, Raymond JA, Angeletti TM. Congenital tracheal stenosis. Crit Care Nurse. 2006 Jun;26(3):60–9.
13. Loukanov T, Sebening C, Gorenflo M, Hagl S. Management of congenital tracheal stenosis in infancy. Eur. J. Cardiothorac. Surg. 2007 Feb;31(2):331.
14. Yang JH, Jun TG, Sung K, Choi JH, Lee YT, Park PW. Repair of long-segment congenital tracheal stenosis. J. Korean Med. Sci. 2007 Jun;22(3):491–6. PMCID:PMC2693643.

15. Herrera P, Caldarone C, Forte V, Campisi P, Holtby H, Chait P, Chiu P, Cox P, Yoo SJ, Manson D, et al. The current state of congenital tracheal stenosis. Pediatr. Surg. Int. 2007 Nov;23(11):1033–44.
16. Terada M, Hotoda K, Toma M, Hirobe S, Kamagata S. Surgical management of congenital tracheal stenosis. Gen. Thorac. Cardiovasc. Surg. 2009 Apr;57(4):175–83.
17. Wright CD. Treatment of congenital tracheal stenosis. Semin. Thorac. Cardiovasc. Surg. 2009;21(3):274–7.
18. Wijeweera O, Ng SB. Retrospective review of tracheoplasty for congenital tracheal stenosis. Singapore Med. J. 2011 Oct;52(10):726–9.
19. Hoffer ME, Tom LW, Wetmore RF, Handler SD, Potsic WP. Congenital tracheal stenosis. The otolaryngologist's perspective. Arch. Otolaryngol. Head Neck Surg. 1994 Apr;120(4):449–53.
20. Metzger RJ, Krasnow MA. Genetic control of branching morphogenesis. Science 1999 Jun 4;284(5420):1635–9.
21. Affolter M, Shilo BZ. Genetic control of branching morphogenesis during Drosophila tracheal development. Curr. Opin. Cell Biol. 2000 Dec;12(6):731–5.
22. Isaac DD, Andrew DJ. Tubulogenesis in Drosophila: a requirement for the trachealess gene product. Genes Dev. 1996 Jan 1;10(1):103–17.
23. Wilk R, Weizman I, Shilo BZ. trachealess encodes a bHLH-PAS protein that is an inducer of tracheal cell fates in Drosophila. Genes Dev. 1996 Jan 1;10(1):93–102.
24. Downward J, Leevers SJ. Trachealess–a new transcription factor target for PKB/Akt. Dev. Cell 2001 Dec;1(6):726–8.
25. Mortimer NT, Moberg KH. The Drosophila F-box protein Archipelago controls levels of the Trachealess transcription factor in the embryonic tracheal system. Dev. Biol. 2007 Dec 15;312(2):560–71. PMCID:PMC2170523.
26. Morozova T, Hackett J, Sedaghat Y, Sonnenfeld M. The Drosophila jing gene is a downstream target in the Trachealess/Tango tracheal pathway. Dev. Genes Evol. 2010 Dec;220(7–8):191–206.
27. Chung S, Chavez C, Andrew DJ. Trachealess (Trh) regulates all tracheal genes during Drosophila embryogenesis. Dev. Biol. 2011 Dec 1;360(1):160–72. PMCID: PMC3215829.
28. Isaac DD, Andrew DJ. Tubulogenesis in Drosophila: a requirement for the trachealess gene product. Genes Dev. 1996 Jan 1;10(1):103–17.
29. Wappner P, Gabay L, Shilo BZ. Interactions between the EGF receptor and DPP pathways establish distinct cell fates in the tracheal placodes. Development 1997 Nov;124(22):4707–16.
30. Llimargas M, Casanova J. EGF signalling regulates cell invagination as well as cell migration during formation of tracheal system in Drosophila. Dev. Genes Evol. 1999 Mar;209(3):174–9.
31. Sutherland D. Samakovlis C. Krasnow MA. branchless encodes a Drosophila FGF homolog that controls tracheal cell migration and the pattern of branching. Cell 1996 Dec 13;87(6):1091–101.
32. Kopecki Z, Ruzehaji N, Turner C, Iwata H, Ludwig RJ, Zillikens D, Murrell DF, Cowin AJ. Topically applied flightless I neutralizing antibodies improve healing of blistered skin in a murine model of epidermolysis bullosa acquisita. J. Invest Dermatol. 2013 Apr;133(4):1008–16.

33. Kontos GJ, Jr., Hedges CP, Rost MC, Nussbaum DK, Hanson JW. Postintubation tracheal stenosis: diagnosis and management. S.D. J. Med. 1993 Sep;46(9):323–5.
34. Grillo HC, Donahue DM, Mathisen DJ, Wain JC, Wright CD. Postintubation tracheal stenosis. Treatment and results. J. Thorac. Cardiovasc. Surg. 1995 Mar;109(3):486–92.
35. Grillo HC, Donahue DM. Postintubation tracheal stenosis. Chest Surg. Clin. N. Am. 1996 Nov;6(4):725–31.
36. Wain JC. Postintubation tracheal stenosis. Chest Surg. Clin. N. Am. 2003 May;13(2):231–46.
37. Kapidzic A, Alagic-Smailbegovic J, Sutalo K, Sarac E, Resic M. Postintubation tracheal stenosis. Med. Arh. 2004;58(6):384–5.
38. Wain JC, Jr. Postintubation tracheal stenosis. Semin. Thorac. Cardiovasc. Surg. 2009;21(3):284–9.
39. Zias N, Chroneou A, Tabba MK, Gonzalez AV, Gray AW, Lamb CR, Riker DR, Beamis JF, Jr. Post tracheostomy and post intubation tracheal stenosis: report of 31 cases and review of the literature. BMC. Pulm. Med. 2008;8:18. PMCID:PMC2556644.
40. Glover WJ. Artificial ventilation, prolonged endotracheal intubation and tracheostomy in paediatric surgery. Postgrad. Med. J. 1972 Aug;48(562):507–13. PMCID:PMC2495263.
41. Shaw EB. Endotracheal intubation and tracheostomy–clinical concepts. Dis. Mon. 1974 Mar;1–35.
42. Selecky PA. Tracheal damage and prolonged intubation with a cuffed endotracheal or tracheostomy tube. Heart Lung 1976 Sep;5(5):733.
43. Orringer MB. Endotracheal intubation and tracheostomy: indications, techniques, and complications. Surg. Clin. North Am. 1980 Dec;60(6):1447–64.
44. Berlauk JF. Prolonged endotracheal intubation vs. tracheostomy. Crit Care Med. 1986 Aug;14(8):742–5.
45. Scher N, Dobleman TJ, Panje WR. Endotracheal intubation as an alternative to tracheostomy after intraoral or oropharyngeal surgery. Head Neck 1989 Nov;11(6):500–4.
46. Cohen MD, Weber TR, Rao CC. Balloon dilatation of tracheal and bronchial stenosis. AJR Am. J. Roentgenol. 1984 Mar;142(3):477–8.
47. Philippart AI, Long JA, Greenholz SK. Balloon dilatation of postoperative tracheal stenosis. J. Pediatr. Surg. 1988 Dec;23(12):1178–9.
48. Betremieux P, Treguier C, Pladys P, Bourdiniere J, Leclech G, Lefrancois C. Tracheobronchography and balloon dilatation in acquired neonatal tracheal stenosis. Arch. Dis. Child Fetal Neonatal Ed 1995 Jan;72(1):F3–F7. PMCID:PMC2528410.
49. Maeda K, Yasufuku M, Yamamoto T. A new approach to the treatment of congenital tracheal stenosis: Balloon tracheoplasty and expandable metallic stenting. J. Pediatr. Surg. 2001 Nov;36(11):1646–9.
50. Lee WH, Kim JH, Park JH. Fluoroscopically Guided Balloon Dilation for Postintubation Tracheal Stenosis. Cardiovasc. Intervent. Radiol. 2013 Jan 26.
51. Interventional procedure overview of endoscopic balloon dilatation for subglottic or tracheal stenosis. NATIONAL INSTITUTE FOR HEALTH AND CLINICAL EXCELLENCE INTERVENTIONAL PROCEDURES PROGRAMME; 2013.
52. Nicolai T. Airway stents in children. Pediatric Pulmonology 2008 Apr;43(4):330–44.
53. Colt HG, Dumon JF. Airway Stents—Present and Future. Clinics in Chest Medicine 1995 Sep;16(3):465–78.

54. Lee P, Kupeli E, Mehta AC. Airway Stents. Clini.Chest Med. 2010 Mar;31(1):141–50.
55. Chin CS, Litle V, Yun J, Weiser T, Swanson SJ. Airway stents. Ann. Thorac. Surg. 2008 Feb;85(2):S792–S796.
56. Mehta AC, Dasgupta A. Airway stents. Clinics in Chest Medicine 1999 Mar;20(1):139–+.
57. Bolliger CT. Airway stents. Seminars in Respiratory and Critical Care Medicine 1997;18(6):563–70.
58. Liu YH, Ko PJ, Wu YC, Liu HP. Incorporated airway stent: a useful option for treating tracheal stenosis after metallic stenting. Interact. Cardiovasc. Thorac. Surg. 2004 Jun;3(2):254–6.
59. Mehta AC. AERO self-expanding hybrid stent for airway stenosis. Expert. Rev. Med. Devices 2008 Sep;5(5):553–7.
60. Parrington S, Tumber P, Wong D. Management of a patient with a large airway stent in situ. Can. J. Anaesth. 2009 Sep;56(9):712–3.
61. Liberman M, Wain JC. Balloon-guided, tapered, Polyflex stent guidance: an atraumatic technique for successful stent placement through tight, rigid airway stenoses. J. Thorac. Cardiovasc. Surg. 2010 Jul;140(1):248–9, 249.
62. Lin HC, Chou CL, Chen HC, Chung FT. Airway stent improves outcome in intubated oesophageal cancer patients. Eur. Respir. J. 2010 Jul;36(1):204–5.
63. Dutau H. Airway stenting for benign tracheal stenosis: what is really behind the choice of the stent? Eur. J. Cardiothorac. Surg. 2011 Oct;40(4):924–5.
64. Tanigawa N, Kariya S, Komemushi A, Nakatani M, Yagi R, Sawada S. Metallic stent placement for malignant airway stenosis. Minim. Invasive. Ther. Allied Technol. 2012 Mar;21(2):108–12.
65. Shepherd W. Endobronchial ultrasound for airway stent selection. J. Bronchology. Interv. Pulmonol. 2011 Jul;18(3):207–8.
66. Schmidt B, Olze H, Borges AC, John M, Liebers U, Kaschke O, Haake K, Witt C. Endotracheal balloon dilatation and stent implantation in benign stenoses. Ann. Thorac. Surg. 2001 May;71(5):1630–4.
67. LaMuraglia MV, Meister M, DiBona N. Tracheal resection and reconstruction: indications, surgical procedure, and postoperative care. Heart Lung. 1991 May;20(3):245–52.
68. Hedlund CS. Tracheal resection and reconstruction. Probl. Vet. Med. 1991 Jun;3(2):210–28.
69. Hannallah MS. The optimal breathing tube for tracheal resection and reconstruction. Anesthesiology 1995 Aug;83(2):419–21.
70. Sharpe DA, Moghissi K. Tracheal resection and reconstruction: a review of 82 patients. Eur. J. Cardiothorac. Surg. 1996;10(12):1040–5.
71. Pinsonneault C, Fortier J, Donati F. Tracheal resection and reconstruction. Can. J. Anaesth. 1999 May;46(5 Pt 1):439–55.
72. Sandberg W. Anesthesia and airway management for tracheal resection and reconstruction. Int. Anesthesiol. Clin. 2000;38(1):55–75.
73. Galetta D, Spaggiari L. Tracheal reconstruction for a long tracheal resection. Ann. Thorac. Surg. 2006 Nov;82(5):1953–4.
74. Hobai IA, Chhangani SV, Alfille PH. Anesthesia for tracheal resection and reconstruction. Anesthesiol. Clin. 2012 Dec;30(4):709–30.

75. Friedel G, Kyriss T, Leitenberger A, Toomes H. Long-term results after 110 tracheal resections. Ger Med. Sci. 2003;1:Doc10. PMCID:PMC2703230.
76. Jungebluth P, Moll G, Baiguera S, Macchiarini P. Tissue-engineered airway: a regenerative solution. Clin. Pharmacol. Ther. 2012 Jan;91(1):81–93.
77. Grillo HC. Tracheal replacement: a critical review. Ann. Thorac. Surg. 2002 Jun;73(6):1995–2004.
78. Komura M, Komura H, Kanamori Y, Tanaka Y, Suzuki K, Sugiyama M, Nakahara S, Kawashima H, Hatanaka A, Hoshi K, et al. An animal model study for tissue-engineered trachea fabricated from a biodegradable scaffold using chondrocytes to augment repair of tracheal stenosis. J. Pediatr. Surg. 2008 Dec;43(12):2141–6.
79. Naito H, Tojo T, Kimura M, Dohi Y, Zimmermann WH, Eschenhagen T, Taniguchi S. Engineering bioartificial tracheal tissue using hybrid fibroblast-mesenchymal stem cell cultures in collagen hydrogels. Interact. Cardiovasc. Thorac. Surg. 2011 Feb;12(2):156–61.
80. Kojima K, Bonassar LJ, Roy AK, Mizuno H, Cortiella J, Vacanti CA. A composite tissue-engineered trachea using sheep nasal chondrocyte and epithelial cells. Faseb Journal 2003 May;17(8):823–8.
81. Macchiarini P, Jungebluth P, Go T, Asnaghi MA, Rees LE, Cogan TA, Dodson A, Martorell J, Bellini S, Parnigotto PP, et al. Clinical transplantation of a tissue-engineered airway. Lancet 2008 Dec 13;372(9655):2023–30.
82. Elliott MJ, De CP, Speggiorin S, Roebuck D, Butler CR, Samuel E, Crowley C, McLaren C, Fierens A, Vondrys D, et al. Stem-cell-based, tissue engineered tracheal replacement in a child: a 2-year follow-up study. Lancet 2012 Sep 15;380(9846):994–1000.

8

BLADDER TISSUE ENGINEERING

Learning Objectives:

After completing this chapter, students should be able to:

1. Describe the structure, function, vascularization, and innervation of the urinary bladder.
2. Describe neurogenic bladder dysfunction and discuss complications associated with this condition and potential treatment modalities.
3. Describe surgical augmentation of the urinary bladder for patients with neurogenic bladder dysfunction.
4. Explain the development of the urinary bladder during embryogenesis.
5. List design considerations for bladder tissue engineering.
6. Describe the process of bioengineering artificial bladders using tissue engineering technology.
7. Discuss the use of cell sheet engineering for bladder tissue engineering.
8. Discuss the use of small intestine submucosa for bladder tissue engineering.
9. Discuss the use of poly(lactic-co-glycolic acid) (PLGA) for bladder tissue engineering.
10. Discuss the use of acellular scaffolds for bladder tissue engineering.
11. Describe organ models for bladder tissue engineering.
12. Describe the clinical use of artificial bladders.

Introduction to Tissue Engineering: Applications and Challenges, First Edition. Ravi Birla.
© 2014 The Institute of Electrical and Electronics Engineers, Inc. Published 2014 by John Wiley & Sons, Inc.

CHAPTER OVERVIEW

We begin this chapter by studying the structure and function of the urinary bladder and then look at one specific pathological condition: neurogenic bladder dysfunction. We then look at bladder augmentation as a surgical treatment for neurogenic bladder dysfunction. We also provide a brief introduction into the development of the urinary bladder during embryogenesis. We then proceed to identify the design considerations for bladder tissue engineering, and provide a general scheme for the fabrication of artificial bladders. In the rest of the chapter, we look at specific models that have been developed to bioengineer artificial bladder tissue or entire bladders. In particular, we look at cell sheet engineering to fabricate artificial bladder tissue. We also look at small intestine submucosa and PLGA as biomaterials for bladder tissue engineering. We discuss the role of acelullular grafts to support the fabrication of artificial bladders or bladder tissue. In the next section of this chapter, we look at organ models for urinary bladders. We end by discussing a seminal publication in the field that describes the use of bioengineered artificial bladders in patients.

8.1 BLADDER STRUCTURE AND FUNCTION

The bladder is part of the urinary system, which functions to regulate the composition of fluids, particularly water, in the body. The urinary system consists of the kidneys, ureters, the urinary bladder, and the urethra. In the urinary system, the kidneys are the major regulatory organ in which urine is produced and then transported through the ureters to the bladder, which serves as a storage vessel for the urine until it is removed from the body via the urethra. The kidneys, which do much of the heavy lifting in terms of fluid hemostasis, can be viewed as the primary functional and regulatory components of the urinary system. The ureters and urethra can be viewed as transport conduits to facilitate the movement of urine. And, the bladder is the primary storage location of urine until it exits the body. In this manner, the urinary system performs the critical function of fluid regulation. Any deficiencies in this process can lead to serious complications and adverse consequences.

As we have mentioned, the urinary bladder is an organ that stores urine until it exits the body. It is connected to the kidney by ureters and has a single output port, the urethra, which transports urine outside the body. From a structural and functional standpoint, the bladder is relatively simple, especially when compared to more complex organs like the kidneys. From a structural standpoint, the bladder can be viewed as a muscular container that acts to store and remove urine based on smooth muscle contraction; from a functional standpoint, the bladder can be viewed as storage vehicle for urine.

The urinary bladder consists of four layers, known as the mucosal layer, submucosal layer, muscular layer and finally, the adventitial layer (1–12). The luminal surface of the urinary bladder is known as the transitional epithelium and is followed by the lamina propria, which consists of connective tissue.Collectively, the transitional epithelium and the lamina propria are known as the mucosal layer. The transitional epithelium consists of specialized cells that can change the number of cell layers based on bladder distension. When the urinary bladder is empty, the transitional epithelium consists of four to five layers of cells; this number is reduced to two to three when the bladder is distended. The next layer of the bladder is known as the submucosal layer and consists of areolar tissue; it serves to connect the mucosal layer to the muscular smooth muscle layer. The third layer in the bladder is the smooth muscle layer, simply referred to as the muscular layer. The smooth muscle cells within the muscular layers are aligned in two different configurations: they can be aligned in a longitudinal manner or can be aligned in a circular fashion. The smooth muscle cells in the muscular layer can be divided into three regions, starting with cells that are longitudinally aligned, followed by cells that are aligned in a circular manner and finally, cells that are again longitudinally aligned. The final layer of the bladder, which is farthest away from the transitional epithelium, consists of connective tissue, is known as the adventitial layer, and is the site for tissue vascularization.

Blood is supplied to the urinary bladder primarily through the superior and inferior vesical arteries (13–16). The superior vesical artery receives its blood supply from the patent part of the umbilical artery, which in turn receives its blood supply from the common iliac artery through the internal iliac artery. The inferior vesical artery primarily feeds the base of the urinary bladder. The obturator and interior gluteal arteries provide smaller branches to the urinary bladder. In the female urinary bladder, additional blood supply is derived from the vaginal and uterine arteries.

The urinary bladder receives stimulation from both sympathetic and parasympathetic fibers, with sympathetic stimulation from the sacral nerves and parasympathetic stimulation from the pelvic splanchnic nerves (17–19).

8.2 NEUROGENIC BLADDER DYSFUNCTION

Introduction—There are several conditions that can adversely affect bladder function, some of which include trauma, infection, cancer, loss of neurological control, and congenital disorders and abnormalities in the formation of the bladder during development. In these cases, the bladder loses the ability to perform its primary function: the storage and removal of urine from the body. In this section, we will look at one specific example of a neurological disorder of the bladder: neurogenic bladder. In cases of neurogenic bladder, a person loses control of bladder functions due to damage and/or injury to the nervous system (20–33). In addition to loss of bladder function, neurogenic bladder increases the risk of kidney failure. Neurogenic bladder can be congenital, resulting from birth defects associated with the spinal cord formation; or can be acquired due to injury of the nerves that innervate the bladder, resulting from injury or trauma. These problems can lead to an overactive bladder, which refers to an increase in the frequency of urination, or an underactive bladder, in which case the bladder does not empty even when full and can leak.

Complications Associated with Neurogenic Bladders—There are several complications that occur as a result of neurogenic bladders, including hydronephrosis, renal failure, urinary tract infections, calculus diseases, and cancer of the bladder. Hydronephrosis is a condition in which water accumulates in the kidneys due to inefficient removal from the body resulting from a poorly functioning bladder. This condition is a very severe and serious consequence of neurogenic bladder; it can lead to loss of renal function and eventually lead to patient mortality. In addition to hydronephrosis, renal failure can also occur due to pyelonenephritis, which is a urinary tract infection affecting the kidney; it can lead to kidney failure and can prove to be fatal. Urinary tract infections are associated with the accumulation of urine and the inability of the bladder to remove the accumulated urine from the body. Another complication steming from neurogenic bladders is calculus disease, which refers to the formation of stones in the urinary tract due to the accumulation and/or immobility of the urine. The risk of bladder cancer is also significantly higher in cases of neurogenic bladders due to infections and bladder stones.

Catheterization Management of Symptoms Relating to Neurogenic Bladders—Catheterization is the standard procedure for the management of neurogenic bladder dysfunction. In this procedure, a catheter is inserted directly in the urinary bladder by passing it through the urethra. The catheter can be used to drain urine from the bladder by way of gravity flow. Catheterization can be for short durations, in which case it is referred to as intermittent catheterization (IC), or can be long-term, in which case the catheter is retained for 30 days or longer. In the case of IC, the patient is typically able to insert the catheter and can do so on a regular basis, typically 4–6 times per day. This process is designed to simulate normal urinary function. In the case of long-term catheterization, the catheter is retained for an extended period of time. The Foley catheter, manufactured by Barb Medical, is an example of a commercially available catheter used for extended time periods. While long-term catheterization offers the advantage of continuous draining, it also significantly increases the risk of urinary tract infection due to the direct route from the external environment to the urinary tract.

Pharmacological Treatment of Neurogenic Bladders—Anticholinergic drugs are effective in the management of overactive bladders. They act by relaxing by the detrusor muscle in the bladder wall. Examples of anticholinergic drugs that are commonly used include oxybutynin, trospium chloride, and propiverine. Anticholinergic agents work by blocking the activity of acetylcholine, which is a neurotransmitter that activates the detrusor muscle in the bladder wall leading to muscle contraction and drainage of urine form the bladder. Inactivation of acetylcholine will prevent stimulation and, therefore, contraction of the detrusor muscle and will act to reduce the frequency of contraction in overactive bladders.

Neuromodulation as a Therapeutic Strategy for Neurogenic Bladders—The term neuromodulation refers to electrical stimulation of the nerves that innervate the bladder. As we have seen before, neurogenic bladder is a condition that occurs due to the loss of nerve stimulation. Therefore, it has been hypothesized that utilization of electrical stimulation can regulate bladder function in a way that is similar to the nerves that innervate it. In other words, neuromodulation provides a mechanism to restore lost nerve innervation to the bladder (34–39). Electrical stimulation has been used extensively as a therapeutic strategy for many medical conditions and has been discussed in Chapter 6. In addition, electrical stimulation has been shown to correlate with the development and maturation of bioengineered artificial tissue, including artificial heart muscle. this material was presented in Chapter 6. Here, we discuss another application of electrical stimulation: as a therapeutic modality for the treatment of neurogenic bladders. In two separate studies, electrical stimulation of the sacral nerve has been conducted as a treatment for neurogenic bladder dysfunction in adult and pediatric patients (40,41). In the study with adult patients, neuromodulation was accomplished by electrodes that were inserted into one or both dorsal foramina of the S3 segment (41). While there were variations in the results of this study, the symptoms were reduced by at least 50% in 30% of the patients (41). In the second study, sacral nerve stimulation was evaluated in pediatric patients; as in the case of adult patients, there were variations in the results, although the general trend was toward functional improvement in a large percentage of the patients (40).

8.3 SURGICAL BLADDER AUGMENTATION

A surgical procedure known as bladder augmentation is an option for patients with neurogenic bladder disorders or other abnormalities of the lower urinary tract. Bladder augmentation refers to the process by which the urinary bladder is expanded by suturing autologous segments of the ileum, small or large bowel, or segments of the ureters (42–53). Surgical intervention is only considered when other treatment strategies have failed to provide the desired functional benefit. Patients with neurogenic bladder disorders will first be treated with pharmacological agents like anticholinergic compounds, while urinary incontinence is managed by use of catheters. Surgical intervention for bladder augmentation can be considered when these procedures fail to provide significant functional benefits to the patient. Once the decision to proceed with bladder augmentation has been made, one of the most important decisions is regarding the choice of tissue specimen to be used. Some of the choices include gastric segments, small and large bowel segments, segments of the ureter, and segments of the ileum. The use of ileac segments for bladder reconstruction is known as ileocystoplasty and will be described here.

The primary objective of bladder augmentation is to increase the volume of the bladder and increase total muscle mass by suturing autologous tissue specimens, as muscular tissue is used as the graft. An increase in bladder pressure is known to adversely affect renal tissue function, and by decreasing this pressure, renal function can be preserved. The increase in muscle mass serves to increase the ability of the bladder to remove urine to the external environment. Therefore, bladder augmentation surgery can increase the functional performance of the urinary bladder by reducing internal pressure, preserving renal function, and increasing muscle mass and contractile activity.

As we have mentioned before, ileocystoplasty is the surgical augmentation of the bladder using a segment of the ileum. An ileac segment with a length of about 30 cm is excised from the patient, and the remaining portions of the ileum are reconnected. The excised portion of the ileum is flushed and opened by making an incision at the midline, and a pouch is created by anastomosis of the opposite ends of the excised ileum. This process leads to the fabrication of an autologous tissue graft, which can then be used for bladder augmentation. An incision is made in the bladder, the tissue graft is sutured on the open bladder, and the open bladder with the sutured ileum segment is closed by anastomosis of the open ends. This process expands the bladder by increasing its internal volume and increases the muscular component of the bladder; it preserves renal function and supports urine discharge from the bladder.

8.4 DEVELOPMENT OF THE URINARY BLADDER

During embryogenesis, the cloaca, Wolffian ducts, and the caudal nephric duct are important structures that give rise to the urinary bladder (54–56). The cloaca is a sac that contains endodermal cells. The lower region of the sac is referred to as the urogenital sinus, which contains the cells which give rise to the urinary bladder and the urethra. The location of the urogenital sinus during embryonic development is aligned with the location of the fully formed bladder and urethra. The Wolffian ducts (WDs), also known as the mesonephric ducts, and the nephric duct consist of two tubes connecting the primordial kidney to the cloaca, eventually leading to the formation of the trigone region of the urinary bladder. During one of the earlier steps of embryogenesis, the Wolffian duct inserts within the cloaca, and later on, the ureter develops and migrates from the primordial kidney toward the cloaca. As the ureter approaches the cloaca, it forms a common tubular connection with the Wolffian duct, and the region of shared connection is known as the common nephric duct (CND), also referred to as the caudal nephric duct. During the course of development, the cells of the CND undergo apoptosis, allowing the ureter to form an insertion with the cloaca, which is independent of the Wolffian duct. Using this mechanism, the two ureters form and are inserted within the cloaca; subsequent maturation of the cloaca leads to formation of the bladder and the urethra.

8.5 DESIGN CONSIDERATIONS FOR BLADDER TISSUE ENGINEERING

In the previous chapter, we looked at the design considerations for tracheal tissue engineering; many of the design considerations that were presented for tracheal tissue engineering also apply for the fabrication of artificial bladders. In the previous chapter, we also presented an overarching design statement for tracheal tissue engineering, which also applies for bladder tissue engineering: *bioengineered bladders should be similar in form and function to mammalian bladders.*

The specific design considerations for bladder tissue engineering are: 1) bioengineered bladders must serve several important functions, including the storage of urine and the timely and efficient removal of the urine from the body, 2) biocompatibility, 3) nonimmunogenic and minimal inflammatory response, 4) nontoxic and noncarcinogenic, 5) avoidance of collapse by reasonable strength, 6) support cell engraftment, 7) support neovascularization, 8) possibility of growth, 9) resistance to fibroblastic and bacterial invasion, 10) standardized easy and short fabrication, 11) customizable and low cost, 12) easy surgical handling, 13) provide physiological environment such as ECM, 14) minimal necessity of donors and accessibility, 15) the results of engraftment are predictably successful, 16) provide or support epithelial resurfacing, 17) must not dislocate or erode over time, and 18) durability.

8.6 PROCESS OF BIOENGINEERING ARTIFICIAL BLADDERS

Introduction—In Chapter 1, we presented a general scheme to bioengineer artificial tissue, and in Chapter 7 we presented a scheme to fabricate artificial tracheas. There were similarities between the two processes, although there were significant differences, particularly related to the tissue fabrication and cellularization strategies. In this section, we present a general process flow sheet for the fabrication of artificial bladders, along with a discussion of cell sourcing, biomaterials, and tissue fabrication technology (Figure 8.1). The process flow sheet is generic and does not represent any specific technology from any research laboratory. In subsequent sections, we will present specific examples of strategies that have been used to bioengineer artificial bladders.

Tissue Graft Development—There are three tissue engineering strategies that can be used for the fabrication of artificial bladder tissue and/or entire bladders: 1) bioengineer an entire artificial bladder, 2) bioengineer a patch or graft that can be sutured with the host graft, similar to tissue grafts used in bladder augmentation surgery and 3) bioengineer partial bladders that can be sutured onto host tissue to support and/or improve bladder function. These three bioengineering strategies are presented in Figure 8.1. In the first case, an entire bioartificial bladder can be fabricated in the laboratory and used for organ transplantation. This strategy involves fabrication of an entire organ and therefore is associated with significant scientific and technology challenges that must first be overcome. The second strategy is designed to fabricate tissue patches that can be used in bladder augmentation. During our discussion of bladder augmentation, we saw that tissue grafts are often derived from gastric segments, segments of small and large bowel, segments of the ureter, and segments of the ileum. This requires incision of tissue grafts from the host followed by reanastomosis with host bladder tissue. This is a very invasive surgery that can be prevented by the use of bioengineered tissue grafts, which can be directly sutured with the host bladder. The third strategy is based on the ability to bioengineer partial bladders. While this is similar to the development of tissue grafts, it is focused on the fabrication of hollow chambers rather than planar scaffolds. This strategy is based on the fabrication of segments of hollow chambers

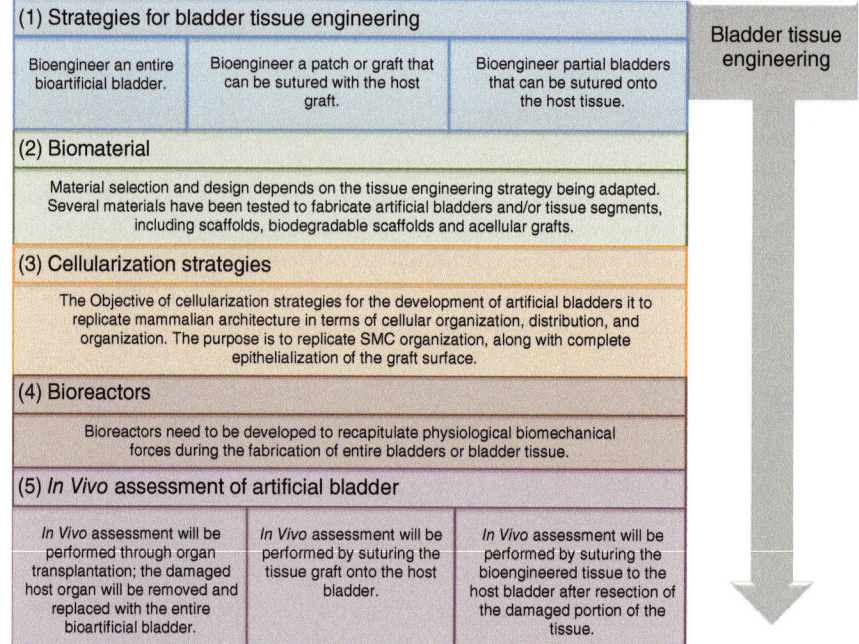

Figure 8.1 Process of Bioengineering Artificial Bladders—Three strategies for bladder tissue engineering are presented: fabrication of entire bladders, partial bladders, or bladder tissue segments. Many different biomaterial platforms have been used, including acellular grafts, biodegradable hydrogels, and polymeric scaffolds. Cellularization strategies have been focused on epithelization of the luminal surface with smooth muscle cells within the interior. Bioreactors need to be developed to simulate natural bladder function. Finally, several models have been developed to test the effectiveness of artificial bladders and/or artificial bladder tissue *in vivo*.

which can be sutured with the host bladder; compared to planar tissue grafts, hollow structures can provide greater functionality.

Biomaterial Design and Fabrication—Several materials have been tested to fabricate artificial bladders and/or tissue segments, including synthetic scaffolds, biodegradable scaffolds, and acellular grafts, with varying degrees of success. Material selection and design depends on the tissue engineering strategy being adapted, with greater mechanical strength and stability required when fabricating hollow grafts and entire bioartificial bladders than when performing planar tissue grafts. As was the case for tracheal tissue engineering, acellular scaffolds have been proven to be successful in artificial bladders that have been used clinically in patients.

Cellularization for the Fabrication of Artificial Bladders—The objective of cellularization strategies for the development of artificial bladders is to replicate mammalian architecture in terms of cellular organization, distribution, and organization.

As we have seen earlier in this chapter, the urinary bladder is composed of epithelial cells on the outer surface and smooth muscle cells within the interior. In addition, the smooth muscle cells (SMCs) were organized in three distinct layers consisting of longitudinal cells, circular cells, and longitudinal cells. Tissue engineering strategies are focused on replicating this complex organization of SMCs, along with complete epithelialization of the graft surface.

Bioreactors—Interestingly, bioreactors have not played a prominent role in supporting the fabrication and/or maturation of artificial bladder tissue. While the bladder is not exposed to extensive hemodynamic forces like the heart or blood vessels are exposed to, it is exposed to external forces on a frequent basis. For example, the bladder is constantly exposed to changes in the volume of urine, which is accompanied by changes in the stress fields on the walls of the urinary bladder. Similarly, removal of urine from the bladder is due to smooth muscle contraction, which results in significant changes in the pressure-volume relationship. These biomechanical forces have an impact on cell phenotype and tissue function, as the effects of stress on endothelial cells and smooth muscle cells are well documented. Therefore, there is a need to develop bioreactors to recapitulate physiological biomechanical forces during the fabrication of entire bladders or bladder tissue. However, bioreactor design and fabrication has not been a high-priority area of research; there are very few publications in this area, as we will see in subsequent sections.

In Vivo Functional Assessment of Artificial Bladders or Tissue Equivalents—The final step in the process scheme is the development of *in vivo* models to assess the functionality of bioengineered tissue grafts and/or entire artificial bladders. In the first case, which involves the fabrication of an entire bladder, *in vivo* assessment will be performed through organ transplantation. The damaged or diseased host organ will be removed and replaced with the bioartificial bladder, ensuring connectivity to the ureters and the urethra. In the second case, which involves the fabrication of planar grafts, *in vivo* assessment will be similar to what has been described for bladder augmentation surgery, and the tissue graft will be sutured onto the host bladder. In this case, an incision will be made to open the bladder, and the bioengineered graft will be sutured onto the host bladder. In the third case, which involves the fabrication of partial bladder components, the bioengineered tissue can be sutured to the host bladder after resection of the damaged or diseased portion of the tissue.

8.7 CELL SHEET ENGINEERING FOR BLADDER TISSUE ENGINEERING

Introduction—During our discussion of tissue fabrication technologies in Chapter 4, we looked at cell sheet engineering. As a reminder, cell sheet engineering is a scaffold-free technology, which relies on ECM produced by the cells to support 3D tissue formation and function. Isolated cells are plated on a temperature-sensitive culture surface, and cell culture conditions are optimized to support the formation of a cohesive cell monolayer; the temperature of the tissue culture plate is then reduced, resulting in detachment of the cohesive cell monolayer. Individual cell sheets are fabricated using these temperature-sensitive surfaces, and individual sheets can be stacked together to form thicker multilayer tissue. The primary advantage of using cell sheet engineering to fabricate artificial tissue is that ECM is produced by the cells and contains the right composition of proteins and other ECM components to support 3D tissue formation. ECM that is fabricated by the cells is very well-suited to support cell attachment, viability, and functionality; the ECM fabricated by the cells leads to the fabrication of functional 3D artificial tissue. Cell sheet engineering has been used for the fabrication of artificial bladders, as will be presented in the remainder of this section (57,58).

Cell Sourcing—This study was conducted using epithelial cells that were isolated from bladder tissue of dogs (57). However, the study did not take into account the smooth muscle cell layer of the bladder tissue, and did not incorporate a muscular layer in the artificial bladder. The primary epithelial cells were isolated from dog bladder, which was cut into small pieces, and the epithelial layer was separated from the underlying tissue. This epithelial tissue was suspended in an enzymatic solution to separate the cells from the tissue.

Biomaterials—Cell sheet engineering technology is based on scaffold-free methods, which rely on the ECM generated by the cells. The cells fabricate ECM and then use the newly formed ECM to fabricate 3D artificial tissue. Tissue fabrication does not require any external scaffolding.

Tissue Fabrication Technology—Primary cells were cultured on temperature-sensitive surfaces for a period of three weeks. During the culture period, the tissue culture plates were maintained in a cell culture incubator at 37°C, with frequent media changes. A three week culture period was necessary to support the formation of a cohesive cell monolayer and to promote ECM production by the cells. At the end of this culture period, the temperature was reduced from 37°C to 20°C, which resulted in detachment of the cohesive cell monolayer from the underlying surface. After detachment, the cells had formed a cohesive cell monolayer that can be referred to as a cell sheet.

In Vivo Testing—In order to test the cell sheets during *in vivo* implantation, a novel strategy was developed for autologous cell transplantation (58). However, in order to support tissue fabrication and implantation of artificial tissue, autologous cells were not used, as complications can result from using a single bladder to harvest primary cells and for implantation of artificial tissue. In order to overcome this problem, primary cells were isolated from the oral cavity and used to bioengineer cell sheets using the technology described for epithelial cells isolated from bladder tissue (58). The cell sheets were used in conjunction with a gastric flap as a graft for bladder reconstruction, as the cell sheets alone do not have the mechanical strength to support bladder function. The cell sheets, fabricated with epithelial cells from the oral cavity, were placed on top of the gastric flap and used for bladder construction. A section of the host bladder was resected and replaced with the gastric flap with the cell sheet attached to it.

Discussion of Cell Sheet Engineering for Bladder Tissue Engineering—The use of cell sheet engineering for bladder reconstruction was used in conjunction with bladder augmentation. Gastric flaps were coupled with cell sheets and then used for transplantation. The primary functional component was provided by the gastric flap, while the cell sheets served to provide a supporting role. This is somewhat expected, as cell sheets do not possess the mechanical properties to support the fabrication of complex hollow organs like the bladder. Significant advancement of the cell sheet technology will be required prior to development of complex organs like the urinary bladder. However, cell sheet engineering remains a novel technology that has been used extensively for many tissue fabrication applications.

8.8 SMALL INTESTINAL SUBMUCOSA (SIS) FOR BLADDER TISSUE ENGINEERING

Introduction—Small intestinal submucosa (SIS) has been extensively used for bladder tissue engineering (Figure 8.2).

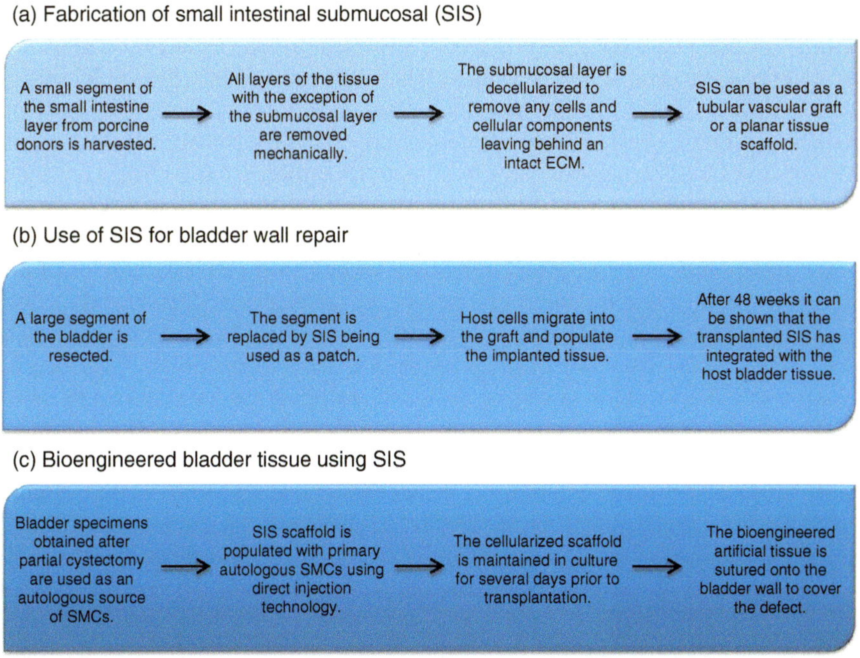

(a) Fabrication of small intestinal submucosal (SIS)

A small segment of the small intestine layer from porcine donors is harvested. → All layers of the tissue with the exception of the submucosal layer are removed mechanically. → The submucosal layer is decellularized to remove any cells and cellular components leaving behind an intact ECM. → SIS can be used as a tubular vascular graft or a planar tissue scaffold.

(b) Use of SIS for bladder wall repair

A large segment of the bladder is resected. → The segment is replaced by SIS being used as a patch. → Host cells migrate into the graft and populate the implanted tissue. → After 48 weeks it can be shown that the transplanted SIS has integrated with the host bladder tissue.

(c) Bioengineered bladder tissue using SIS

Bladder specimens obtained after partial cystectomy are used as an autologous source of SMCs. → SIS scaffold is populated with primary autologous SMCs using direct injection technology. → The cellularized scaffold is maintained in culture for several days prior to transplantation. → The bioengineered artificial tissue is sutured onto the bladder wall to cover the defect.

Figure 8.2 SIS for Bladder Tissue Engineering—(a) Fabrication of SIS—SIS is fabricated from a segment of the small intestine after removal of all layers of the tissue except the submucosal layer. **(b) Use of SIS for Bladder Wall Repair**—SIS is used, in the absence of cells, for bladder reconstruction. **(c) Bioengineered Bladder Tissue using SIS**—SIS is populated with cells using direct cell injection and the artificial tissue graft is used for bladder augmentation.

SIS is a very interesting biomaterial that has been used for many different tissue engineering applications (59–68), and one that we have not discussed before. We begin this section with an overview of SIS as a biomaterial and then provide specific examples of the use of SIS for bladder reconstruction and bladder tissue engineering. We will look at two examples, one of which describes the utilization of SIS as a biomaterial (in the absence of cells) for bladder repair and the second example, in which artificial bladder tissue is fabricated by cellularization of the SIS biomaterial, which is then used as a graft for bladder tissue repair.

Small Intestinal Submucosa as a Biomaterial—SIS is obtained from the submucosal layer of a small intestine segment that has been harvested from porcine donors (64). The small intestine consists of several distinct tissue layers with an epithelial layer on the luminal surface, followed by goblet cells, the submucosal layer, and longitudinally and circularly aligned muscle cells. During the preparation of SIS, a segment of the small intestine layer is harvested, commonly from pigs, and all layers of the tissue, with the exception of the submucosal layer, are removed mechanically. The submucosal layer is next subjected to a decellularization protocol to remove any cells and cellular components, leaving behind an intact ECM. This two-step process that consists of mechanically removing layers of porcine small intestine followed by decellularization results in the production of SIS. SIS can be used as a tubular vascular graft or can be transformed to form a planar tissue scaffold by making an incision along the length of the tubular graft. Since SIS is biologically derived, it contains several ECM components, including collagen, proteoglycans, GAGs, glycoproteins, and growth factors; these components make SIS an attractive biomaterial for tissue engineering applications to support cell attachment, remodeling, and subsequently, tissue fabrication. The primary advantage of SIS is the biologically derived ECM, which contains proteins and other components of the ECM that support cell attachment and spreading and support cell-matrix interactions; these functions in turn modulate cell phenotype and support 3D tissue fabrication.

SIS as a Xenogeneic Material—It should be noted that SIS is a xenogeneic material, which means that it is isolated from animal donors, and after being processed using specific conditions, it can be designed and used for transplantation. In the case of xenogeneic transplantation, cells, tissue, or entire organs obtained from one species, commonly pigs, are used for transplantation in a different species, typically humans. This can be compared to autologous and allogeneic sources, which were discussed in Chapter 2 in the context of cell sourcing. In the case of autologous transplantation, the donor and recipient of cells or biomaterials are the same, while in the case of allogeneic transplantation, the donor and recipient are different but from the same species. The primary advantage of autologous transplantation is immune acceptance by the recipient. Immune rejection is a concern with allogeneic and xenogeneic transplantation. However, SIS is nonimmunogenic; cell surface antigens are responsible for host immunogenic response, and xenogeneic materials like SIS are nonimmunogenic due to the complete removal of the cellular components.

SIS as a Bladder Wall Substitute—There have been several studies demonstrating the utility of SIS for bladder reconstruction using tissue engineering as well as alternative strategies (69,70). In one study conducted on rats, SIS was used as a patch for bladder augmentation surgery (70). A large segment of the bladder was resected and replaced with SIS, which was sutured onto the remaining bladder wall (70). After a period of 48 weeks, it was shown that the transplanted SIS had integrated with the host bladder tissue, and the two were indistinguishable (70). In addition, the implanted graft had formed three layers of bladder tissue: the epithelium, the smooth muscle layer, and the serosa. The SIS was not cellularized at the time of implantation, but host cells migrated into the graft and populated the implanted tissue, supporting the formation of bladder tissue.

This was a very simple application of SIS biomaterial for bladder wall substitution. Since there were no cells involved in the study, this strategy is not considered to be tissue engineering. However, there are several examples in the recent literature where biomaterials (without cells) have been used for the repair of damaged and/or injured tissue. This study clearly demonstrated the integration of SIS with bladder tissue, and the results of this study suggest that SIS can be used as a tissue graft for bladder augmentation surgery. In addition, the results of this study suggest that SIS may have the potential to be bioengineered into artificial bladders, and the coupling of the biomaterial with cells may provide additional functional benefit. The ability of SIS to functionally integrate with host tissue over a period of 48 weeks and withstand the constant changes in the bladder pressure clearly demonstrates the potential of the biomaterial in bladder reconstruction and bladder tissue engineering.

Cell-Seeded SIS Graft for Bladder Reconstruction—In the previous example, we looked at the use of SIS as a biomaterial for bladder reconstruction without the incorporation of cells. However, the utility of SIS to support engraftment of cells and then use the cell-seeded graft for bladder reconstruction has also been evaluated. In a recent study, SIS grafts were fabricated from small intestine tissue after mechanical removal of all tissue layers except the submucosal layer followed by decellularization of the submucosal layer to remove any remaining cells (69). New Zealand white rabbits were used for bladder augmentation studies and were subjected to a partial cystectomy to create an injury model. The tissue specimens that were obtained after the partial cystectomy were used to isolate SMCs, which were cultured and expanded on tissue culture plates. After partial cystectomy, the animals were housed for one month and used as recipients for tissue engineered grafts. Using this strategy, the researchers were able to use an autologous cell source to bioengineer artificial bladder tissue; SMCs were isolated from New Zealand White rabbits and used to bioengineer tissue grafts, which were later implanted into the same animal. In order to support the formation of artificial bladder tissue, the SIS scaffold was populated with primary autologous SMCs using direct injection technology. The cellularized scaffolds were maintained in culture for several days prior to transplantation and then used as a tissue graft for bladder augmentation. The bioengineered artificial tissue was sutured onto the bladder wall to cover the defect, and the host bladder was evaluated at several time intervals after transplantation of the tissue graft.

Discussion—In this section, we have seen two contrasting strategies for bladder tissue repair; one relies upon the use of a biomaterial for bladder tissue augmentation, while the second one uses bioengineered artificial tissue for bladder repair. Stated another way, the first strategy is based on the use of biomaterials for tissue repair, while the second strategy is based on biomaterials and cells for tissue repair. When a biomaterial is used without cells, the strategy is based on the hypothesis that host cells will migrate into the implanted biomaterial to support the formation of functional tissue. While both strategies continue to be aggressively researched and developed, we believe that the latter of the two offers greater benefit due to the presence of cells at the time of graft implantation.

We will end our discussion in this section by relating some of the principles of tissue engineering presented in earlier chapters with the models that have been discussed here. It should be noted that in the tissue engineering study (the second of the two presented), primary cells were used and were isolated from an autologous source. The cells were maintained and expanded in culture using standard cell culture procedures. The SIS biomaterial used in the study was obtained using a deceullarization strategy and populated with cells using direct injection technology. Many of the terms used here (primary cells, autologous, decellularization, cell culture, and direct cell injection) have been discussed in earlier chapters, and the reader is encouraged to relate these fundamental principles of tissue engineering to the specific examples of bladder tissue engineering.

8.9 PLGA AS A BIOMATERIAL FOR BLADDER TISSUE ENGINEERING

Introduction—In the previous section, we looked at SIS as a biomaterial for bladder augmentation both with and without cells. In this section, we will study another such example. In this section, we will study the use of the polymeric scaffold poly (lactic-co-glycolic acid), PLGA, to support bladder repair, and as in the previous example, we will look at the use of PLGA with and without cells (Figure 8.3). PLGA has been used extensively as a biomaterial for tissue engineering along with many other medical applications (71–79). We begin this section with a discussion of the properties of PLGA and then move onto our discussion of the use of this material for bladder repair and augmentation.

PLGA as a Biomaterial—PLGA has been extensively used as a vehicle for controlled delivery of therapeutic agents and as a porous scaffold for tissue engineering applications. PLGA is a degradable copolymer of lactic acid and glycolic acid; it is often described in terms of the relative percentage of these two monomers; PLGA 25:75 refers to PLGA with a composition of 25% lactic acid and 75% glycolic acid (79). One of the main advantages of PLGA is the nontoxicity of its degradation products; PLGA undergoes hydrolysis, and the degradation products of this reaction are the monomers lactic acid and glycolic acid, both of which are easily metabolized by the body (79). The degradation kinetics of PLGA can be modulated over a period of weeks or months by changing the relative proportions of monomer units. The ability to control the degradation kinetics of PLGA has made

(a) PLGA as a biomaterial

PLGA is a degradable copolymer of lactic acid and glycolic acid. When PLGA undergoes hydrolysis, the degradation products of this reaction are the monomers lactic acid and glycolic acid, both of which are easily metabolized by the body. The degradation kinetics of PLGA can be modulated over a period of weeks or months providing a mechanism to regulate the rate of release of an embedded therapeutic agent. Additionally PLGA allows the fabrication of scaffolds with varying pore size and porosity to satisfy the design requirements for any tissue fabrication application.

(b) Use of PLGA for bladder wall repair

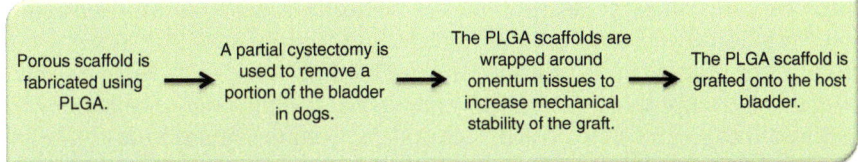

(c) Bioengineered bladder tissue using PLGA

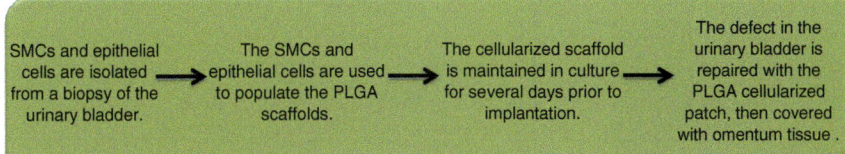

Figure 8.3 **PLGA for Bladder Tissue Engineering**—(a) PLGA as a Biomaterial–PLGA is a copolymer of lactic acid and glycolic acid. (b) **Use of PLGA for Bladder Wall Repair**—3D scaffolds were fabricated using PLGA and used for bladder augmentation without the use of any cells in the scaffold. (c) **Bioengineered Bladder Tissue Using PLGA**—3D scaffolds were populated with cells prior to use for bladder augmentation.

it a suitable biomaterial for controlled release strategies. Regulating the degradation kinetics of PLGA provides a mechanism to regulate the rate of release of an embedded therapeutic agent.

PLGA has also been widely used as a porous scaffold for various tissue engineering applications (79–96). There have been several techniques described for the fabrication of PLGA into porous scaffolds, and the properties of the scaffolds, like the pore size and porosity, can be modulated by varying the fabrication process; this modulation provides the ability to fine-tune the properties of the porous scaffold to satisfy the design requirements for any tissue fabrication application.

PLGA Scaffolds for Bladder Reconstruction—In one particular study, porous scaffolds were fabricated using PLGA, though the details of the fabrication technology were not presented in the published manuscript. The PLGA scaffolds were implanted in dogs after a partial cystectomy to remove a portion of the host bladder (97). The PLGA scaffolds were not used alone, but rather wrapped around omentum tissue at the time of implantation in order to increase the mechanical stability of the graft. Bladder function was assessed at various time intervals after implantation of the tissue grafts. Functional performance of the bladder was assessed by measuring the bladder capacity and bladder compliance, both of which showed improvement after implantation of the PLGA scaffold.

Cellularized PLGA Scaffolds for Bladder Reconstruction—PLGA scaffolds were cellularized using autologous SMCs and epithelial cells that were isolated from a biopsy of the urinary bladder (98). The cells were maintained in culture and subpassaged to increase the number of cells. SMCs and epithelial cells were used to populate the PLGA scaffolds. The details of the cellularization protocol were not provided in the study, but it can be assumed that direct injection technology was used for scaffold cellularization. The cellularized scaffolds were maintained in culture for several days prior to implantation. The implantation model for the cellularized scaffolds was similar to that for the PLGA scaffolds without cells. A partial cystectomy was performed to remove part of the urinary bladder in dogs, repaired with a cellularized patch, then covered with omentum tissue. As before, biological and functional performance metrics were evaluated. It was found that the cellularized PLGA patch performed better than the PLGA patch without cells both in terms of functional performance and in terms of histological data.

Discussion of the Model—In this example, we have seen the use of PLGA, a biodegradable biomaterial, to support the functional repair of bladder tissue. The PLGA was used with and without cells in a side-by-side comparison, and the cellularized scaffold performed better than the scaffold without cells. This study serves to demonstrate the clear advantages of tissue engineering strategies over other methods and showcase the ability of cell-seeded scaffold to support functional recovery of bladder tissue. In addition, the study also illustrates many important principles of tissue engineering that we have discussed in earlier chapters: autologous cells, cell culture, biodegradable scaffold, and direct injection technology.

8.10 ACELLULAR GRAFTS FOR BLADDER TISSUE ENGINEERING

Introduction—In this section, we will look at the use of acellular grafts for bladder tissue engineering; there have been several examples in the literature demonstrating the feasibility of acellular grafts to support the formation of artificial bladder tissue (Figure 8.4). We provide three specific examples in this section: 1) use of acellular grafts without the incorporation of cells, 2) use of acellular grafts with the incorporation of cells, and 3) the use of acellular grafts with the incorporation of adipose-derived stem cells.

Acellular Grafts for Bladder Tissue Engineering—We have been introduced to acellular grafts as a scaffold for tissue engineering applications during our discussion of biomaterials. Acellular scaffolds are fabricated by complete removal of cellular components from tissue/organ specimens that have been harvested from animal or human sources. After removal of all the cellular components, an intact ECM is retained. The retained ECM is rich is ECM proteins and other components of the ECM. This acellular graft has binding sites for cells and supports cell-matrix interactions, leading to the formation of functional artificial tissue. These properties

(a) Acellular bladder grafts

Acellular scaffolds are fabricated by complete removal of cellular components from tissue/organ specimens that have been harvested from animal or human sources. After removal of all the cellular components, an intact ECM is retained. The retained ECM is rich is ECM proteins and other components of the ECM. This acellular graft has binding sites for cells and supports cell-matrix interactions, leading to the formation of functional artificial tissue.

(b) Use of acellular graft for bladder wall repair

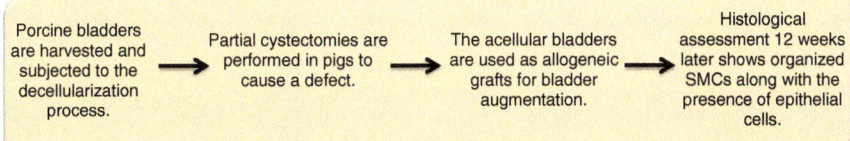

(c) Bioengineered bladder tissue using acellular graft

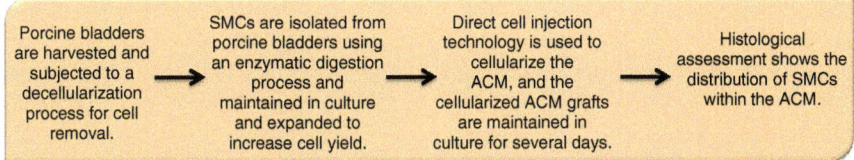

Figure 8.4 Acellular Scaffolds for Bladder Tissue Engineering—(a) Acellular Bladder Grafts—Acellular grafts are fabricated by the complete removal of cells, leaving behind an intact extracellular matrix. **(b) Use of Acellular Grafts for Bladder Wall Repair**—In this case, acellular grafts were used in the absence of any cells. **(c) Bioengineered Bladder Tissue using Acellular Grafts**—Acellular grafts were populated with cells and then used for bladder augmentation.

of acellular grafts have proven to be important in the fabrication of artificial tissue for many different tissue and organ systems, including the urinary bladder, as we will see in this section.

Use of Bladder Acellular Matrix Allograft (BAMA) for Bladder Augmentation— In this study, porcine bladders were harvested and subjected to a detergent- and enzyme-based decellularization process to remove all the cells (99,100). Partial cystectomies were conducted in pigs, and the acellular bladders were used as allogeneic grafts for bladder augmentation. During an initial short-term study, animals were sacrificed after 30 days; histological assessment showed that scattered SMCs had infiltrated the acellular bladder grafts. In a follow-up study, animals were sacrificed after 12 weeks, and histological assessment showed organized SMCs along with the presence of epithelial cells. Combined, these two studies demonstrated the feasibility of using acelullar bladder grafts to support bladder repair (99,100). It should be noted that the acellular grafts were not populated with cells at the time of transplantation; rather, this strategy relies upon infiltration of host cells to populate the transplanted graft. This is similar to the strategy that was presented for SIS earlier in this chapter.

Cell-Seeded Acellular Matrix (ACM) for Bladder Reconstruction—In a recent study, porcine bladders were harvested and subjected to a decellularization process for cell removal (101). SMCs were isolated from porcine bladders using an enzymatic digestion process. The SMCs were maintained in culture and expanded to increase cell yield. Allogeneic SMCs were used to populate the ACM; this study was designed for *in vitro* testing and not for *in vivo* bladder augmentation. These studies were designed for model development and assessment. Therefore, the use of allogeneic cells was completely justifiable. Once a sufficient number of cells were obtained, the cells were used to populate the ACM. Direct cell injection technology was used to cellularize the ACM, and the cellularized ACM grafts were maintained in culture for several days. Histological evidence showed the distribution of SMCs within the ACM and provided early evidence for the formation of artificial bladder tissue.

Use of Adipose-Derived Stem Cells for Bladder Tissue Engineering—A very interesting study was published that makes use of adipose-derived stem cells to populate acellular bladder grafts and uses cellularized artificial tissue for bladder repair (102). In this study, bladders were harvested from New Zealand White rabbits and subjected to a decellularization protocol for cell removal. A tissue biopsy was obtained from the backs of the rabbits and subjected to a collagenase digestion protocol for the isolation of adipose-derived stem cells. These cells were maintained and expanded in culture and used to populate segments of an acellular bladder; autologous cells were used for this study. The cellularized tissue was maintained in culture for several days and then used as a graft for bladder reconstruction. A defect was created in the bladder wall, and the bioengineered bladder tissue was sutured on the site of the defect. The grafts were implanted for a period of 24 weeks, after which the animals showed improvement in bladder function when compared to controls (acellular grafts without any cells). Perhaps more interesting was the differentiation of adipose-derived stem cells to form SMCs and epithelial cells, as demonstrated by positive staining for cell specific markers; the differentiation of the stem cells was likely mediated by cues within the *in vivo* environment.

Discussion of Acellular Grafts for Bladder Tissue Engineering—Acellular grafts have proven to be successful for bladder augmentation either alone or in combination with cells. When acellular grafts were used in the absence of cells, infiltration of cells from the host tissue leads to cellularization of the acellular grafts. However, this process is difficult to regulate, and the spatial distribution of the cells cannot be modulated. This limitation can be overcome by the use of cell-seeded scaffolds in which the distribution and organization of the cells can be controlled. In the three models that have been discussed, we have seen the use of allogeneic SMCs for studies designed for *in vitro* testing and have seen the use of autologous adipose-derived stem cells for *in vivo* bladder augmentation. In this section, we have seen many of the concepts that were described in earlier chapters, some of which include acellullar scaffolds, autologous and allogeneic cells, direct injection technology, and adipose-derived stem cells.

8.11 ORGAN MODELS FOR BLADDER TISSUE ENGINEERING

Introduction—We have looked at several tissue engineering models for fabricating artificial bladder tissue. However, in all of the examples, the focus has been on the development of artificial bladder tissue and not entire hollow chamber organs. The development of artificial bladder tissue is important, as it provides valuable information about the organization and the remodeling of isolated cells to form functional 3D artificial tissue. In addition, artificial bladder tissue can also be used as a patch to support and/or augment function in injured or diseased bladders. Undoubtedly, tissue models play a vital role in the development of tissue engineering therapies to support bladder function. In addition to the development of tissue models, several research groups have developed organ models that more closely replicate organ-level architecture of the urinary bladder. In this section, we will look at two examples of organ models for bladder tissue engineering, one of which was developed using a novel *in vivo* approach and one of which was developed using a novel *in vitro* approach.

In Vivo Organ Model for Bladder Tissue Engineering—A very interesting strategy was developed for the fabrication of hollow chamber bladders by implantation of acellular grafts. In this study, entire bladders where harvested from Sprague-Darley rats and subjected to a decellularization protocol to remove all the cells (103). The acellular bladders were referred to as BAMGs—bladder acellular matrix grafts. Using a second set of rats, a partial cystectomy was created to remove greater than 50% of the host bladder, and the BAMG was sutured onto the host bladder (103). The purpose of this surgery was not to support functional recovery of the host bladder. Instead, the objective of the study was to use the *in vivo* environment to fabricate artificial bladders. The host environment provides a range of physiological cues, growth factors for tissue development and maturation, and many other signals essential for tissue/organ function. In addition, the host tissue provides a potential source of cells that can infiltrate into the implanted BAMG and support scaffold cellularization. The BAMG was explanted after a period of several months, and the biological and functional performance metrics were evaluated and shown to be comparable to that of mammalian bladders. At the time of explanation, the tissue graft can be viewed as a bioengineered artificial bladder.

During the course of this book, we have studied several steps in the tissue fabrication process, including cell sourcing, scaffold cellularization, and bioreactor conditioning. In most of the examples that we have looked at thus far, these steps are conducted during controlled *in vitro* conditions; indeed, this is the most popular strategy to support the development of artificial tissue. However, in the current example, we see a novel application of the *in vivo* host environment to provide many of the cues required to support tissue development. This is advantageous, as the host environment has all the necessary signals to guide tissue development. However, the major limitation with this strategy is that the researcher does not have any control over the processing variables and cannot change any of the input variables.

In Vitro Organ Model for Bladder Tissue Engineering—Another very interesting study described the fabrication of hollow chamber artificial bladders using an *in vitro* strategy. PGA was used as the biomaterial, and a large segment of PGA was used to create hollow chamber structures that resembled the urinary bladder. In order to increase the mechanical strength of the bioengineered structure, the PGA was coated with a liquefied copolymer that consisted of PLGA and methylene chloride (98). Cells were isolated from tissue specimens harvested from urinary bladders of dogs. The tissue specimens were subjected to a digestion protocol to isolate the SMCs and the epithelial cells, which were maintained in cell culture and expanded to increase the cell yield, as a very large number of cells were required to populate the PGA graft. The SMCs were seeded onto the exterior surface of the bioengineered graft, while the epithelial cells were seeded onto the luminal surface of the graft. The tissue graft was maintained in culture and then implanted *in vivo* in a dog model. A subtotal cystectomy was performed, sparing the trigone region, and the bioengineered bladder was implanted in recipient dogs. The functional performance of the implanted artificial bladder was assessed over a period of eleven months and compared with controls (grafts without any cells). The functional performance of the implanted artificial bladder was significantly greater than that of the controls' functional performance.

This study was clearly novel: it demonstrated the fabrication of artificial bladders using tissue engineering strategies. Many of the principles that we have covered in previous chapters were used in the development of this technology: biomaterial design and development, tissue fabrication, cell sourcing, and cell culture and scaffold cellularization.

Discussion of Organ Models for Bladder Tissue Engineering—In this section, we have looked at two examples for the fabrication of artificial bladders, one using an *in vivo* strategy and another using an *in vitro* method. These two examples serve to demonstrate the progression of research in the field of bladder tissue engineering from planar tissue specimens to complex artificial organs. The *in vivo* strategy is a novel method that makes use of internal physiological conditions to direct the growth and development of artificial bladders. The *in vitro* strategy is novel due to the ability to combine all the elements necessary for tissue fabrication toward the development of artificial bladders; this strategy demonstrates the utility of many of the tissue engineering principles that we have seen in earlier chapters and is clearly a seminal study in the field.

8.12 CLINICAL STUDY FOR BLADDER TISSUE ENGINEERING

There has been one reported clinical study using artificial bladders for the treatment of patients (104). This study is one of the seminal publications in the field of tissue engineering and proves the feasibility of bioengineered tissue (104). The current study provides significant impetus for the field and defines a path for the development of other artificial tissue and organ systems.

The clinical study was conducted by the same research group that developed the *in vitro* organ model for bladder tissue engineering that was described in the previous section. Therefore, there are several similarities between the work described here and the work described in the previous section. The method used for the fabrication of scaffolds for the artificial bladder was similar to what was described in the previous section; the biomaterial was molded to form a hollow chamber structure which resembled the urinary bladder. In this study, different biomaterials were used. There were seven patients enrolled in this clinical study, and out of these seven patients, a collagen matrix was used in three patients, a collagen matrix wrapped with omentum was used in one patient, and a hybrid matrix of collagen and PGA wrapped with omentum was used in the final three patients. Three different scaffolding materials were used to bioengineer artificial bladders; this was a result of the progress made in understanding and optimizing fabrication technology for artificial bladders. The seven patients were recruited for this study over time rather than being recruited at the start of the clinical study. During this time, the researchers continued their efforts to optimize and improve artificial bladder fabrication and incorporated the refinements into the clinical study. It should be noted that the clinical study was not designed to compare different scaffolding materials for artificial bladder fabrication.

Autologous cells were obtained from a tissue biopsy from the patient's bladder and the SMCs and epithelial cells were isolated using a digestion protocol. The cells were maintained in culture and expanded to increase the cell yield. The SMCs were seeded on the exterior surface of the scaffold, and the epithelial cells were seeded on the luminal surface of the scaffold. The cell-seeded scaffold was maintained in a cell culture incubator for a period of 3–4 days prior to implantation. For surgical implantation in patients, the artificial bladders were anastomosed to the native bladders, and the patients were monitored for several months after the surgery. The results of the study demonstrated positive results both in terms of the functional performance of the implanted bladder and in terms of the histological assessment showing the distribution and alignment of the cells and ECM.

This study was a remarkable accomplishment and a cornerstone for the field of bladder tissue engineering and tissue engineering as a whole. In this study, many of the principles of tissue engineering were adapted, including biomaterials, tissue fabrication, cell isolation, culture and expansion, and scaffold cellularization. The principles of tissue engineering were used to fabricate artificial organs that were then implanted to help several patients and significantly improve their quality of life.

SUMMARY

Current State of the Art—Bladder tissue engineering is one of the more mature branches of the field; it has experienced proven clinical success. The clinical study remains a seminal publication and a cornerstone in the field of tissue engineering. There have been three primary sources of cells used to support the fabrication of artificial bladders. First, primary autologous cells have been used as a source of SMCs and epithelial cells. Second, autologous adipose-derived stem cells have been used, with the *in vivo* microenvironment supporting differentiation of the stem cells to SMCs and epithelial cells. Third, infiltration of host cells into implanted acellullar grafts has also been used as a potential source of cells for bladder tissue engineering. Several biomaterials have been used to support the fabrication of artificial bladder tissue or artificial bladders, some of which include SIS, acellular grafts, and PLGA. In addition, cell sheeting engineering has been tested as a scaffold-free platform to support the development of artificial bladder tissue. Finally, we have looked at several tissue fabrication technologies that have been used to bioengineer artificial bladder tissue and entire organs. The transition has been from monolayer cell sheets to multilayer tissue grafts all the way to artificial bladders. Cell sheets have been used to coat the surface of gastric flaps prior to implantation, tissue patches have been engineered and shown to be effective in bladder augmentation, and entire bladders have been fabricated using *in vivo* and *in vitro* tissue engineering approaches.

Thoughts for Future Research—Considerable progress has been made in the field of bladder tissue engineering. However, there are several areas that have not been explored in great detail. The use of stem cells, including hES cells, iPS cells, and mesenchymal cells has not be explored in great detail; in addition, the signals that drive the differentiation of these cells toward bladder SMCs and epithelial cells need to be explored. There have not been many studies that look at the controlled vascularization and innervation of artificial bladder tissue under controlled *in vitro* culture conditions. Another area of research that has not received significant attention is the development of bioreactors to support the development and maturation of artificial bladder tissue. The normal function of the urinary bladder relies on SMC contraction, which results in changes in the pressure-volume relationship. There is a need to develop bioreactors that simulate these normal physiological conditions during the culture of artificial bladder tissue/organs.

PRACTICE QUESTIONS

1. Describe the structure of the urinary bladder, including the different cell types and the functions of these cells.

2. Discuss neurogenic bladder dysfunction as it relates to urinary bladder function. What are the clinical manifestations of neurogenic bladder dysfunction? What are some potential treatment modalities for this condition?

3. Describe the process of surgical bladder augmentation.

4. The process of surgical bladder augmentation relies upon autologous tissue grafts to expand the size of the urinary bladder. How can tissue engineering strategies be developed to improve this technique?

5. Describe the development of the urinary bladder during embryogenesis.

6. What can we learn about the development of the urinary bladder during embryogenesis that can be applied toward the fabrication of 3D artificial bladders?

7. During the course of this chapter, we provided a general description of the process of bioengineering artificial bladders and/or bladder tissue. Describe this process. If you had to add one additional step to the generic process, what would it be and why?

8. We have now studied two process schemes, one for the fabrication of artificial tracheas (Chapter 7) and one for the fabrication of artificial bladders (Chapter 8). What are some of the similarities between these two process schemes? What are some of the differences between these two process schemes?

9. Discuss the role of cell sheet engineering in bladder tissue engineering. Cell sheets have been used in conjunction with gastric segments for bladder reconstruction surgery; discuss the relative merits and weaknesses of this strategy.

10. Describe how cell sheet engineering can be used to fabricate complete artificial bladders.

11. In Chapter 4, we studied self-organization strategies for the fabrication of artificial tissue. Develop a strategy to fabricate artificial bladders or artificial bladder tissue using self-organization strategies.

12. Describe the use of SIS as a biomaterial. How is SIS fabricated? What are some of the properties of SIS that make it an attractive biomaterial? SIS is obtained from xenogenic sources; does this lead to immune rejection problems?

13. During our discussion of SIS in bladder tissue engineering, we described the use of SIS as a biomaterial for bladder augmentation and as a scaffold to support the fabrication of artificial bladder tissue. Discuss the relative advantages and disadvantages of each of the two strategies.

14. Develop a process in which SIS can be used to fabricate an entire artificial bladder.

15. Discuss the use of PLGA as a biomaterial in tissue engineering. What are some of the properties of the material that make it useful for tissue engineering? What fabrication technologies can be used for the fabrication of 3D scaffolds using PLGA?

16. Discuss the use of PLGA for bladder augmentation with and without cells. What are the relative advantages of each of the two methods?

17. Develop a process in which PLGA can be used to fabricate 3D artificial bladder tissue.

18. Discuss the relative advantages and disadvantages of acellular grafts for bladder tissue engineering.

19. What are some potential uses of bioreactors during the fabrication of artificial bladders? Design a bioreactor to support the fabrication and/or culture of artificial bladders.

20. What is the potential role of perfusion systems during the culture of artificial bladders? Design a perfusion system to support the culture of artificial bladders.

REFERENCES

1. Preusser S, Diener PA, Schmid HP, Leippold T. Submucosal endocervicosis of the bladder: an ectopic, glandular structure of Mullerian origin. Scand. J. Urol. Nephrol. 2008;42(1):88–90.
2. Dorschner W, Stolzenburg JU, Neuhaus J. Structure and function of the bladder neck. Adv. Anat. Embryol. Cell Biol. 2001;159:III-109.
3. Gabella G. Structure of the intramural nerves of the rat bladder. J. Neurocytol. 1999 Aug;28(8):615–37.
4. Gosling JA. The structure of the bladder neck, urethra and pelvic floor in relation to female urinary continence. Int. Urogynecol. J. Pelvic. Floor. Dysfunct. 1996;7(4): 177–8.
5. Gabella G, Uvelius B. Urinary bladder of rat: fine structure of normal and hypertrophic musculature. Cell Tissue Res. 1990 Oct;262(1):67–79.
6. Holm-Bentzen M, Ammitzboll T. Structure and function of glycosaminoglycans in the bladder. Ann. Urol. (Paris) 1989;23(2):167–8.
7. Wade RH, Brisson A. Three-dimensional structure of bladder membrane protein. Ultramicroscopy 1984;13(1–2):47–56.
8. Gosling J. The structure of the bladder and urethra in relation to function. Urol. Clin. North Am. 1979 Feb;6(1):31–8.
9. Gosling JA, Dixon JS, Dunn M. The structure of the rabbit urinary bladder after experimental distension. Invest Urol. 1977 Mar;14(5):386–9.
10. Ellis DJ. The bladder neck: a theory of its structure and function. Br. J. Urol. 1972 Dec;44(6):727–8.
11. Hoyes AD, Ramus NI, Martin BG. Fine structure of the epithelium of the human fetal bladder. J. Anat. 1972 Apr;111(Pt 3):415–25. PMCID:PMC1271131.
12. Bro-Rasmussen F, Sorensen AH, Bredahl E, Kelstrup A. The Structure and Function of the Urinary Bladder. Urol. Int. 1965;19:280–95.
13. Brading AF, Greenland JE, Mills IW, McMurray G, Symes S. Blood supply to the bladder during filling. Scand. J. Urol. Nephrol. Suppl 1999;201:25–31.
14. Hossler FE, Monson FC. Structure and blood supply of intrinsic lymph nodes in the wall of the rabbit urinary bladder–studies with light microscopy, electron microscopy, and vascular corrosion casting. Anat. Rec. 1998 Nov;252(3):477–84.
15. Irwin P, Galloway NT. Impaired bladder perfusion in interstitial cystitis: a study of blood supply using laser Doppler flowmetry. J. Urol. 1993 Apr;149(4):890–2.
16. Michels NA. Variations in the blood-supply of the liver, gall bladder, stomach, duodenum, pancreas and spleen; 200 dissections. Am. J. Med. Sci. 1948 Jul;216(1):115.

17. Clemens JQ. Basic bladder neurophysiology. Urol. Clin. North Am. 2010 Nov;37(4): 487–94.
18. Campioni P, Goletti S, Palladino F, Nanni M, Napoli M, Valentini AL. The neurogenic bladder: anatomy and neurophysiology. Rays 2002 Apr;27(2):107–14.
19. Yeates WK. Neurophysiology of the bladder. Paraplegia 1974 Aug;12(2):73–82.
20. Agarwal SK, Bagli DJ. Neurogenic bladder. Indian J. Pediatr. 1997 May;64(3): 313–26.
21. Madersbacher HG. Neurogenic bladder dysfunction. Curr. Opin. Urol. 1999 Jul;9(4): 303–7.
22. Nijman RJ. Neurogenic and non-neurogenic bladder dysfunction. Curr. Opin. Urol. 2001 Nov;11(6):577–83.
23. Salvaggio E, Arces L, Rendeli C. Clinical patterns of neurogenic bladder. Rays 2002 Apr;27(2):115–20.
24. Verpoorten C, Buyse GM. The neurogenic bladder: medical treatment. Pediatr. Nephrol. 2008 May;23(5):717–25. PMCID:PMC2275777.
25. Bauer SB. Neurogenic bladder: etiology and assessment. Pediatr. Nephrol. 2008 Apr;23(4):541–51. PMCID:PMC2259256.
26. MacLellan DL. Management of pediatric neurogenic bladder. Curr. Opin. Urol. 2009 Jul;19(4):407–11.
27. 27.Cameron AP. Pharmacologic therapy for the neurogenic bladder. Urol. Clin. North Am. 2010 Nov;37(4):495–506.
28. Westney OL. The neurogenic bladder and incontinent urinary diversion. Urol. Clin. North Am. 2010 Nov;37(4):581–92.
29. Klausner AP, Steers WD. The neurogenic bladder: an update with management strategies for primary care physicians. Med. Clin. North Am. 2011 Jan;95(1):111–20.
30. McGuire EJ. Urodynamics of the neurogenic bladder. Urol. Clin. North Am. 2010 Nov;37(4):507–16.
31. Gormley EA. Urologic complications of the neurogenic bladder. Urol. Clin. North Am. 2010 Nov;37(4):601–7.
32. Dorsher PT, McIntosh PM. Neurogenic bladder. Adv. Urol. 2012;2012:816274. PMCID:PMC3287034.
33. Yeates WK. Neurophysiology of the bladder. Paraplegia 1974 Aug;12(2):73–82.
34. Halverstadt DB, Leadbetter WF. Electrical stimulation of the human bladder: experience in three patients with hypotonic neurogenic bladder dysfunction. Br. J. Urol. 1968 Apr;40(2):175–82.
35. Hald T. Neurogenic dysfunction of the urinary bladder. An experimental and clinical study with special reference to the ability of electrical stimulation to establish voluntary micturition. Dan. Med. Bull. 1969 Jun;16:Suppl.
36. Tanagho EA, Schmidt RA. Electrical stimulation in the clinical management of the neurogenic bladder. J. Urol. 1988 Dec;140(6):1331–9.
37. Madias JE. Electrocardiographic artifact induced by an electrical stimulator implanted for management of neurogenic bladder. J. Electrocardiol. 2008 Sep;41(5):401–3.
38. Tellenbach M, Schneider M, Mordasini L, Thalmann GN, Kessler TM. Transcutaneous electrical nerve stimulation: an effective treatment for refractory non-neurogenic overactive bladder syndrome? Urol: World J; 2012 May 24.

39. Radziszewski K. Outcomes of electrical stimulation of the neurogenic bladder: Results of a two-year follow-up study. NeuroRehabilitation 2013 Jan 1;32(4):867–73.
40. Guys JM, Haddad M, Planche D, Torre M, Louis-Borrione C, Breaud J. Sacral neuromodulation for neurogenic bladder dysfunction in children. J. Urol. 2004 Oct;172(4 Pt 2):1673–6.
41. Hohenfellner M, Humke J, Hampel C, Dahms S, Matzel K, Roth S, Thuroff JW, Schultz-Lampel D. Chronic sacral neuromodulation for treatment of neurogenic bladder dysfunction: long-term results with unilateral implants. Urology 2001 Dec;58(6):887–92.
42. Gurocak S, Nuininga J, Ure I, De Gier RP, Tan MO, Feitz W. Bladder augmentation: Review of the literature and recent advances. Indian J. Urol. 2007 Oct;23(4):452–7. PMCID:PMC2721579.
43. Babu R, Ragoori D. Bladder augmentation: Distal ureterocystoplasty with proximal ureteric reimplantation: A novel technique. J. Indian Assoc. Pediatr. Surg. 2012 Oct;17(4):165–7. PMCID:PMC3518995.
44. Metcalfe PD, Rink RC. Bladder augmentation: complications in the pediatric population. Curr. Urol. Rep. 2007 Mar;8(2):152–6.
45. Gurocak S, De Gier RP, Feitz W. Bladder augmentation without integration of intact bowel segments: critical review and future perspectives. J. Urol. 2007 Mar;177(3):839–44.
46. Schaefer M, Kaiser A, Stehr M, Beyer HJ. Bladder augmentation with small intestinal submucosa leads to unsatisfactory long-term results. Urol: J. Pediatr. 2013 Jan 15.
47. Escudero RM, Patino GE, Fernandez ER, Gil MJ, Garcia EL, Alonso AH, Pinies GO, Sanchez JP, Fernandez CH. Bladder augmentation using the gastrointestinal tract. Indication, follow up and complications. Arch. Esp. Urol. 2011 Dec;64(10):953–9.
48. Stein R, Kamal MM, Rubenwolf P, Ziesel C, Schroder A, Thuroff JW. Bladder augmentation using bowel segments (enterocystoplasty). BJU. Int. 2012 Oct;110(7):1078–94.
49. Quiroz-Guerrero J, Badillo M, Munoz N, Anaya J, Rico G, Maldonado-Valadez R. Bladder augmentation in a young adult female exstrophy patient with associated omphalocele: an extremely unusual case. J. Pediatr. Urol. 2009 Aug;5(4):330–2.
50. Daher P, Zeidan S, Riachy E, Iskandarani F. Bladder augmentation and/or continent urinary diversion: 10-year experience. Eur. J. Pediatr. Surg. 2007 Apr;17(2):119–23.
51. Stein R, Schroder A, Thuroff JW. Bladder augmentation and urinary diversion in patients with neurogenic bladder: non-surgical considerations. J. Pediatr. Urol. 2012 Apr;8(2):145–52.
52. Sajadi KP, Goldman HB. Bladder augmentation and urinary diversion for neurogenic LUTS: current indications. Curr. Urol. Rep. 2012 Oct;13(5):389–93.
53. Rodo JS, Caceres FA, Lerena JR, Rossy E. Bladder augmentation and artificial sphincter implantation: urodynamic behavior and effects on continence. J. Pediatr. Urol. 2008 Feb;4(1):8–13.
54. Hannema SE, Hughes IA. Regulation of Wolffian duct development. Horm. Res. 2007;67(3):142–51.
55. Staack A, Donjacour AA, Brody J, Cunha GR, Carroll P. Mouse urogenital development: a practical approach. Differentiation 2003 Sep;71(7):402–13.

56. Cuckow PM, Nyirady P, Winyard PJ. Normal and abnormal development of the urogenital tract. Prenat. Diagn. 2001 Nov;21(11):908–16.
57. Shiroyanagi Y, Yamato M, Yamazaki Y, Toma H, Okano T. Transplantable urothelial cell sheets harvested noninvasively from temperature-responsive culture surfaces by reducing temperature. Tissue Eng. 2003 Oct;9(5):1005–12.
58. Watanabe E, Yamato M, Shiroyanagi Y, Tanabe K, Okano T. Bladder augmentation using tissue-engineered autologous oral mucosal epithelial cell sheets grafted on demucosalized gastric flaps. Transplantation 2011 Apr 15;91(7):700–6.
59. Kang KN, Lee JY, Kim dY, Lee BN, Ahn HH, Lee B, Khang G, Park SR, Min BH, Kim JH, et al. Regeneration of completely transected spinal cord using scaffold of poly(D,L-lactide-co-glycolide)/small intestinal submucosa seeded with rat bone marrow stem cells. Tissue Eng. Part A 2011 Sep;17(17–18):2143–52.
60. Schnoeller TJ, de PR, Hefty R, Jentzmik F, Waalkes S, Zengerling F, Schrader M, Schrader AJ. Partial nephrectomy using porcine small intestinal submucosa. World J. Surg. Oncol. 2011;9:126. PMCID:PMC3233505.
61. Palminteri E, Berdondini E, Fusco F, De NC, Salonia A. Long-term results of small intestinal submucosa graft in bulbar urethral reconstruction. Urology 2012 Mar;79(3):695–701.
62. Ma L, Yang Y, Sikka SC, Kadowitz PJ, Ignarro LJ, Abdel-Mageed AB, Hellstrom WJ. Adipose tissue-derived stem cell-seeded small intestinal submucosa for tunica albuginea grafting and reconstruction. Proc. Natl. Acad. Sci. U.S.A 2012 Feb 7; 109(6):2090–5. PMCID:PMC3277542.
63. Lee AJ, Chung WH, Kim DH, Lee KP, Suh HJ, Do SH, Eom KD, Kim HY. Use of canine small intestinal submucosa allograft for treating perineal hernias in two dogs. J. Vet. Sci. 2012 Sep;13(3):327–30. PMCID:PMC3467410.
64. Andree B, Bar A, Haverich A, Hilfiker A. Small intestinal submucosa segments as matrix for tissue engineering: review. Tissue Eng. Part B Rev 2013 Aug;19(4):279–91.
65. Witt RG, Raff G, Van GJ, Rodgers-Ohlau M, Si MS. Short-term experience of porcine small intestinal submucosa patches in paediatric cardiovascular surgery. Eur. J. Cardiothorac. Surg. 2013 Jul;44(1):72–6.
66. Song Z, Peng Z, Liu Z, Yang J, Tang R, Gu Y. Reconstruction of abdominal wall musculofascial defects with small intestinal submucosa scaffolds seeded with tenocytes in rats. Tissue Eng. Part A 2013 Jul;19(13–14):1543–53. PMCID:PMC3665322.
67. Yi JS, Lee HJ, Lee HJ, Lee IW, Yang JH. Rat peripheral nerve regeneration using nerve guidance channel by porcine small intestinal submucosa. J. Korean Neurosurg. Soc. 2013 Feb;53(2):65–71. PMCID:PMC3611061.
68. Rossetto VJ, da Mota LS, Rocha NS, Miot HA, Grandi F, Brandao CV. Grafts of porcine small intestinal submucosa seeded with cultured homologous smooth muscle cells for bladder repair in dogs. Acta Vet. Scand. 2013; 55:39. PMCID:PMC3663814.
69. Lai JY, Chang PY, Lin JN. Bladder autoaugmentation using various biodegradable scaffolds seeded with autologous smooth muscle cells in a rabbit model. J. Pediatr. Surg. 2005 Dec;40(12):1869–73.
70. Kropp BP, Eppley BL, Prevel CD, Rippy MK, Harruff RC, Badylak SF, Adams MC, Rink RC, Keating MA. Experimental assessment of small intestinal submucosa as a bladder wall substitute. Urology 1995 Sep;46(3):396–400.

71. Hawley AE, Illum L, Davis SS. Preparation of biodegradable, surface engineered PLGA nanospheres with enhanced lymphatic drainage and lymph node uptake. Pharm. Res. 1997 May;14(5):657–61.
72. Nakanishi Y, Chen G, Komuro H, Ushida T, Kaneko S, Tateishi T, Kaneko M. Tissue-engineered urinary bladder wall using PLGA mesh-collagen hybrid scaffolds: a comparison study of collagen sponge and gel as a scaffold. J. Pediatr. Surg. 2003 Dec;38(12):1781–4.
73. Baek CH, Ko YJ. Characteristics of tissue-engineered cartilage on macroporous biodegradable PLGA scaffold. Laryngoscope 2006 Oct;116(10):1829–34.
74. Jeong SI, Kim SY, Cho SK, Chong MS, Kim KS, Kim H, Lee SB, Lee YM. Tissue-engineered vascular grafts composed of marine collagen and PLGA fibers using pulsatile perfusion bioreactors. Biomaterials 2007 Feb;28(6):1115–22.
75. Holgado MA, Cozar-Bernal MJ, Salas S, Arias JL, Alvarez-Fuentes J, Fernandez-Arevalo M. Protein-loaded PLGA microparticles engineered by flow focusing: physicochemical characterization and protein detection by reversed-phase HPLC. Int. J. Pharm. 2009 Oct 1;380(1–2):147–54.
76. Kang SW, Lee SJ, Kim JS, Choi EH, Cha BH, Shim JH, Cho DW, Lee SH. Effect of a scaffold fabricated thermally from acetylated PLGA on the formation of engineered cartilage. Macromol. Biosci. 2011 Feb 11;11(2):267–74.
77. Enlow EM, Luft JC, Napier ME, DeSimone JM. Potent engineered PLGA nanoparticles by virtue of exceptionally high chemotherapeutic loadings. Nano. Lett. 2011 Feb 9;11(2):808–13. PMCID:PMC3122105.
78. Jain AK, Das M, Swarnakar NK, Jain S. Engineered PLGA nanoparticles: an emerging delivery tool in cancer therapeutics. Crit Rev. Ther. Drug Carrier Syst. 2011;28(1):1–45.
79. Ungaro F, d'Angelo I, Miro A, La Rotonda MI, Quaglia F. Engineered PLGA nano- and micro-carriers for pulmonary delivery: challenges and promises. J. Pharm. Pharmacol 2012 Sep;64(9):1217–35.
80. Patrick CW Jr, Chauvin PB, Hobley J, Reece GP. Preadipocyte seeded PLGA scaffolds for adipose tissue engineering. Tissue Eng. 1999 Apr;5(2):139–51.
81. Holy CE, Cheng C, Davies JE, Shoichet MS. Optimizing the sterilization of PLGA scaffolds for use in tissue engineering. Biomaterials 2001 Jan;22(1):25–31.
82. Koegler WS, Griffith LG. Osteoblast response to PLGA tissue engineering scaffolds with PEO modified surface chemistries and demonstration of patterned cell response. Biomaterials 2004 Jun;25(14):2819–30.
83. Hwang CM, Khademhosseini A, Park Y, Sun K, Lee SH. Microfluidic chip-based fabrication of PLGA microfiber scaffolds for tissue engineering. Langmuir 2008 Jun 1;24(13):6845–51.
84. Tan H, Wu J, Lao L, Gao C. Gelatin/chitosan/hyaluronan scaffold integrated with PLGA microspheres for cartilage tissue engineering. Acta Biomater. 2009 Jan;5(1):328–37.
85. Jose MV, Thomas V, Johnson KT, Dean DR, Nyairo E. Aligned PLGA/HA nanofibrous nanocomposite scaffolds for bone tissue engineering. Acta Biomater. 2009 Jan;5(1):305–15.
86. Baker SC, Rohman G, Southgate J, Cameron NR. The relationship between the mechanical properties and cell behaviour on PLGA and PCL scaffolds for bladder tissue engineering. Biomaterials 2009 Mar;30(7):1321–8.

87. van EF, Saris DB, Fedorovich NE, Kruyt MC, Willems WJ, Verbout AJ, Martens AC, Dhert WJ, Creemers L. In vivo matrix production by bone marrow stromal cells seeded on PLGA scaffolds for ligament tissue engineering. Tissue Eng. Part A 2009 Oct;15(10):3109–17.
88. Ngiam M, Liao S, Patil AJ, Cheng Z, Chan CK, Ramakrishna S. The fabrication of nano-hydroxyapatite on PLGA and PLGA/collagen nanofibrous composite scaffolds and their effects in osteoblastic behavior for bone tissue engineering. Bone 2009 Jul;45(1):4–16.
89. Sahoo S, Toh SL, Goh JC. PLGA nanofiber-coated silk microfibrous scaffold for connective tissue engineering. J. Biomed. Mater. Res. B Appl. Biomater. 2010 Oct;95(1):19–28.
90. Andreas K, Zehbe R, Kazubek M, Grzeschik K, Sternberg N, Baumler H, Schubert H, Sittinger M, Ringe J. Biodegradable insulin-loaded PLGA microspheres fabricated by three different emulsification techniques: investigation for cartilage tissue engineering. Acta Biomater. 2011 Apr;7(4):1485–95.
91. Han J, Lazarovici P, Pomerantz C, Chen X, Wei Y, Lelkes PI. Co-electrospun blends of PLGA, gelatin, and elastin as potential nonthrombogenic scaffolds for vascular tissue engineering. Biomacromolecules 2011 Feb 14;12(2):399–408.
92. Go DP, Palmer JA, Gras SL, O'Connor AJ. Coating and release of an anti-inflammatory hormone from PLGA microspheres for tissue engineering. J. Biomed. Mater. Res. A 2011 Nov 29.
93. Takechi M, Ohta K, Ninomiya Y, Tada M, Minami M, Takamoto M, Ohta A, Nakagawa T, Fukui A, Miyamoto Y, et al. 3-dimensional composite scaffolds consisting of apatite-PLGA-atelocollagen for bone tissue engineering. Dent. Mater. J. 2012;31(3):465–71.
94. Chang NJ, Jhung YR, Yao CK, Yeh ML. Hydrophilic gelatin and hyaluronic acid-treated PLGA scaffolds for cartilage tissue engineering. J. Appl. Biomater. Funct. Mater. 2013;11(1):e45–e52.
95. Horst M, Madduri S, Milleret V, Sulser T, Gobet R, Eberli D. A bilayered hybrid microfibrous PLGA–acellular matrix scaffold for hollow organ tissue engineering. Biomaterials 2013 Feb; 34(5):1537–45.
96. Wang PY, Wu TH, Tsai WB, Kuo WH, Wang MJ. Grooved PLGA films incorporated with RGD/YIGSR peptides for potential application on skeletal muscle tissue engineering. Colloids Surf. B Biointerfaces. 2013 Oct 1;110:88–95.
97. Jayo MJ, Jain D, Wagner BJ, Bertram TA. Early cellular and stromal responses in regeneration versus repair of a mammalian bladder using autologous cell and biodegradable scaffold technologies. J. Urol. 2008 Jul;180(1):392–7.
98. Oberpenning F, Meng J, Yoo JJ, Atala A. De novo reconstitution of a functional mammalian urinary bladder by tissue engineering. Nat. Biotechnol. 1999 Feb;17(2):149–55.
99. Reddy PP, Barrieras DJ, Wilson G, Bagli DJ, McLorie GA, Khoury AE, Merguerian PA. Regeneration of functional bladder substitutes using large segment acellular matrix allografts in a porcine model. J. Urol. 2000 Sep;164(3 Pt 2):936–41.
100. Merguerian PA, Reddy PP, Barrieras DJ, Wilson GJ, Woodhouse K, Bagli DJ, McLorie GA, Khoury AE. Acellular bladder matrix allografts in the regeneration of functional bladders: evaluation of large-segment (>24 cm) substitution in a porcine model. BJU. Int. 2000 May;85(7):894–8.

101. Cheng HL, Loai Y, Farhat WA. Monitoring tissue development in acellular matrix-based regeneration for bladder tissue engineering: multiexponential diffusion and T2* for improved specificity. NMR Biomed. 2012 Mar;25(3):418–26.
102. Zhu WD, Xu YM, Feng C, Fu Q, Song LJ, Cui L. Bladder reconstruction with adipose-derived stem cell-seeded bladder acellular matrix grafts improve morphology composition. World J. Urol. 2010 Aug;28(4):493–8.
103. Piechota HJ, Dahms SE, Nunes LS, Dahiya R, Lue TF, Tanagho EA. In vitro functional properties of the rat bladder regenerated by the bladder acellular matrix graft. J. Urol. 1998 May;159(5):1717–24.
104. Atala A, Bauer SB, Soker S, Yoo JJ, Retik AB. Tissue-engineered autologous bladders for patients needing cystoplasty. Lancet 2006 Apr 15;367(9518):1241–6.

9

LIVER TISSUE ENGINEERING

Learning Objectives

After completing this chapter, students should be able to:

1. Describe the structure and function of the liver.
2. Describe symptoms associated with acute liver failure, along with potential treatment options.
3. Discuss orthotopic and partial liver transplantation.
4. Briefly describe the molecular and cellular events that take place during liver regeneration after partial hepatectomy.
5. Briefly discuss the process of liver development during human development.
6. Discuss design considerations for liver tissue engineering.
7. Describe the process scheme to bioengineer artificial liver tissue.
8. Describe the differentiation of hES cells, iPS cells, bone marrow MSCs, and hepatic stem cells to form hepatocytes.
9. Discuss robotic protein printing, photo-responsive culture surfaces, and PDMS stencils for spatial control of hepatocytes.
10. Describe different biomaterial platforms that have been used for the fabrication of artificial liver tissue.
11. Describe strategies for the fabrication of artificial liver tissue.

Introduction to Tissue Engineering: Applications and Challenges, First Edition. Ravi Birla.
© 2014 The Institute of Electrical and Electronics Engineers, Inc. Published 2014 by John Wiley & Sons, Inc.

12. Describe strategies to support vascularization of artificial liver tissue.
13. Describe perfusion systems that have been developed to support the culture of artificial liver tissue.
14. Discuss the role of spheroid culture in liver tissue engineering.

CHAPTER OVERVIEW

We begin this chapter with a discussion of the structure and function of the mammalian liver. We then look at some liver disorders, with particular attention to acute liver failure. This is followed by a discussion on liver transplantation, and we proceed to describe the chronic shortage of donor livers. We next provide a brief description of liver regeneration and development. This is followed by a list of design criteria for the fabrication of artificial liver tissue. We next provide a general process flow sheet for liver tissue engineering. We provide a discussion of the use of stem cells for liver tissue engineering, including strategies that have been used to drive the differentiation fate of stem cells to form hepatocytes. We next look at surface patterning technologies and biomaterial platforms as applied to the field of liver tissue engineering. This is followed by a discussion on strategies that have been used to support the fabrication of 3D artificial liver tissue. We follow this up with a discussion of vascularization of artificial liver tissue and bioreactors to support the culture of artificial liver tissue. We end this chapter with a discussion of spheroid culture of hepatocytes and a comparison of this technology with tissue engineering.

9.1 STRUCTURE AND FUNCTION OF THE LIVER

Introduction—The liver is a component of the digestive system and performs many functions necessary for digestion. The liver is the largest internal organ in humans, weighing an average of three pounds. In the previous two chapters, we studied the trachea and the bladder that are hollow organs; in comparison, the liver is a solid organ. Hepatocytes are the primary functional cell type found in the liver; they are responsible for most of the functional properties of the liver. Under normal physiological conditions, hepatocytes have a slow rate of turnaround and are not very proliferative. However, in response to injury, the liver has a remarkable regenerative capacity, and in rats, the loss of liver tissue by partial hepatectomy to remove two-thirds of the organ is compensated for by an increase in the rate of hepatocyte proliferation within two weeks.

Liver Function—The liver has several functions related to digestion. Hepatocytes in the liver make a fluid known as bile, which contains cholesterol, bile acids, and bilirubin and aides in the digestion process by breaking down fats to fatty acids (1–5). Bile is made in the liver by hepatocytes and is transported to the gallbladder for storage via the bile canaliculi and the hepatic ducts. The liver is also the primary site for the storage of glycogen, vitamins, and minerals; and is responsible for metabolism of fats, proteins, and carbohydrates, and excretion of metabolic waste products. The liver also plays an important role in the synthesis of components of the blood, including plasma proteins and clotting agents.

Liver Structure—Anatomically, the liver consists of four lobes that are known as the left, right, caudate, and quadrate lobes (6). The functional unit of the liver is known as the lobule and consists of hepatocytes, blood vessels, nerves, and bile ducts—all of which are uniformly arranged. A central vein is located at the center of the each lobule; the central veins come together to form the hepatic veins, which in turn feed into the inferior vena cava. The blood that exits the liver follows the aforementioned path. The portal triad is located at the ends of lobules and consists of three primary structures: the hepatic bile duct, hepatic portal vein, and hepatic artery. In addition to these three primary structures, the portal triad also contains nerve tissue and lymphatic vessels.

Liver Blood Supply—The liver has a dual blood supply from the hepatic portal vein and the hepatic artery (7–13). The hepatic artery delivers oxygenated blood from the circulatory system, while the hepatic portal vein delivers blood from the small intestines containing nutrients. About three-quarters of the blood entering the liver is from the hepatic portal vein, while the remaining amount enters from the hepatic artery. Blood exits the liver via the central vein that is located in the middle of the liver lobule and feeds into the hepatic veins.

Liver Innervation—Sympathetic nervous stimulation to the liver is from the thoracic nerves T7-T12 of the spinal cord, while parasympathetic nervous stimulation is from the dorsal motor nucleus of the vagus nerve, which is located in the dorsal brainstem (14,15).

Hepatocytes—Hepatocytes are the primary functional cells of the liver and perform many of the functions that have been described (16,17). As expected, hepatocytes have been the focus of much research in tissue engineering, as we will study in subsequent sections of this chapter. The structure of hepatocytes is consistent with their role in energy metabolism; they contain a single round nucleus and numerous mitochondria. Consistent with the role of hepatocytes in protein synthesis, hepatocytes contain large amounts of rough endoplasmic reticulum, which in turn has large amounts of ribosomes that are necessary for protein synthesis. Hepatocytes are also functionally coupled to other hepatocytes via an extensive network of gap junctions that occupy as much as three percent of the cell membrane surface. The gap junctions allow intracellular cellular communication between neighboring cells and support the growth, proliferation, and function of hepatocytes and hence, liver function. The predominant gap junction proteins are connexin32 and connexin26.

9.2 ACUTE LIVER FAILURE

Definition of Acute Liver Failure—Acute liver failure (ALF) is defined as the onset of hepatocellular dysfunction in the absence of pre-existing liver disease characterized by coagulopathy and encephalopathy within 8 weeks of the hepatic insult (18–24). The onset of ALF is sudden without any pre-existing liver conditions, and progression of the disease over time can lead to multiorgan failure, which can result in patient mortality. Acute liver failure (ALF) is a rare disease, as classified by the NIH Office of Rare Disease Research, with an incidence rate of one to six cases per year for every one million people in the developed world. The incidence for ALF in developing countries is more difficult to estimate, though it is expected to be significantly higher. Although the incidence of ALF is low, the consequences are significant and the mortality rate is very high, often exceeding 50%.

ALF Symptoms—Jaundice is one of the earlier symptoms of ALF, and can be easily recognized by the skin becoming yellowish. This change in color is due to an accumulation of bilirubin, which is a byproduct of red blood cells' deterioration. Other symptoms of ALF are linked to a loss of liver function and include a loss of metabolic function, decreased gluconeogenesis (formation of glucose from molecules other than carbohydrates) leading to hypoglycemia (reduction in blood glucose concentration), and decreased ammonia clearance leading to hyperammonemia (increase in blood ammonia concentration). While ALF primarily affects the liver, the manifestation and progression of ALF can lead to dysfunction in many organs, including the heart, lungs, kidneys, and brain. ALF can also lead to coagulopathy, which impairs the mechanism responsible for blood clotting, and encephalopathy, which refers to damage of brain function.

Classification of ALF—There are three classification schemes that are used to describe ALF (25–27): hyperacute liver failure, acute liver failure, and subacute liver failure. In each of these three cases, there are differences in the time from jaundice to encephalopathy, the severity of jaundice and coagulopathy, and the survival rate without emergency liver transplantation. As we progress from hyperacute liver failure to subacute, there is an increase in the time from jaundice encephalopathy, an increase in the severity of the jaundice, and a decrease in the severity of coagulopathy. There is also a decrease in the survival rate of the patients without emergency liver transplantation as the disease progresses from hyperacute to subacute liver failure.

Causes of ALF—ALF is caused by a significant loss of liver function, brought about by apoptosis of hepatocytes, the primary functional cells of the liver. The liver is responsible for numerous critical functions, most of which are carried out by the hepatocytes. Therefore, any loss in the number of hepatocytes will directly correlate with a decrease in liver function. Apoptosis of hepatocytes during ALF can be brought about by many different agents, some of which include viral infection, primarily by the hepatitis virus, and drug induced injury, particularly by nonprescription acetaminophen. ALF by viral infection is more predominant in developing nations, while ALF by drug-induced injury is more prevalent in developed countries.

Treatment Strategies for Patients with ALF—The primary treatment options for patients with ALF include pharmacological intervention, organ transplantation, and the use of mechanical support devices. High dose N-acetylcysteine (NAC) is used for the treatment of ALF that has been induced by an overdose of acetaminophen, and has shown to be effective when used during the early stages of ALF. Later stage ALF may require liver transplantation, and since the progression of ALF is rapid, emergency transplantation is usually required. While effective, emergency liver transplantation has a lower rate of success compared to elective liver transplantation. Mechanical liver support devices are also used for the treatment and management of ALF, although such devices are primarily used as a bridge to transplantation.

9.3 LIVER TRANSPLANTATION

Introduction—Liver transplantation is the standard of care for adult patients with end stage liver failure (28–35). Due to the regenerative capacity of the liver, many different transplantation strategies have been developed. The most common method has been orthotopic liver transplantation (OLT), in which case donor livers harvested from cadavers are transplanted to patients after complete removal of the damaged or diseased liver. In addition to OLT, other methods have been developed for liver transplantation. Some of these methods include the use of partial liver grafts obtained from living donors (living donor liver transplantation, LDLT) or the use of a single liver divided into two allografts that are transplanted into two patients (split-liver transplantation, SLT).

Liver Transplantation Statistics for Adult Patients—The most recent year for which transplantation data was available from the US Department of Health and Human Services Scientific Registry of Transplant Patients was 2011 (36–38). This data shows that close to 12,000 adult patients were on the waiting list for a liver transplant (Figure 9.1a), with around 6000 patients receiving a transplant (Figure 9.1b). The data further shows a mortality rate of just over 10% for patients on the waiting list for a liver transplant (Figure 9.1c). However, in patients who are able to receive a liver transplant, the survival rate has been high, with a one-year survival rate of greater than 80%, a three-year survival rate of greater than 75%, and a five-year survival rate just under 70% (Figure 9.1d). Therefore, while liver transplantation has saved numerous lives, there remains a chronic shortage of donor livers with a high mortality rate of patients on the waiting list.

Liver Transplantation using Partial Liver Segments—Livers for use in adult liver transplantations primarily originate from deceased donors, while a very small fraction are from living donors, as can been seen in Figure 9.1b. Living donor liver transplantation (LDLT) is performed in pediatric patients or in adults with a small size. The regenerative capacity of the liver allows transplantation from living donors; a portion of the liver is removed from the donor and transplanted into the recipient, from whom the diseased and/or damaged liver has been completely removed. In both cases, the liver is able to regenerate and provide functional support for both the donor and recipient of the liver. Within a time period of 8–12 weeks, normal liver volume is restored in both the donor and the recipient, as a result of the regenerative capacity of the liver. This is particularly beneficial for pediatric patients, for whom size-matched livers are often difficult to source; the use of partial liver grafts provides an option for these patients. In addition to LDLT, another strategy that has been used for partial liver transplantation is

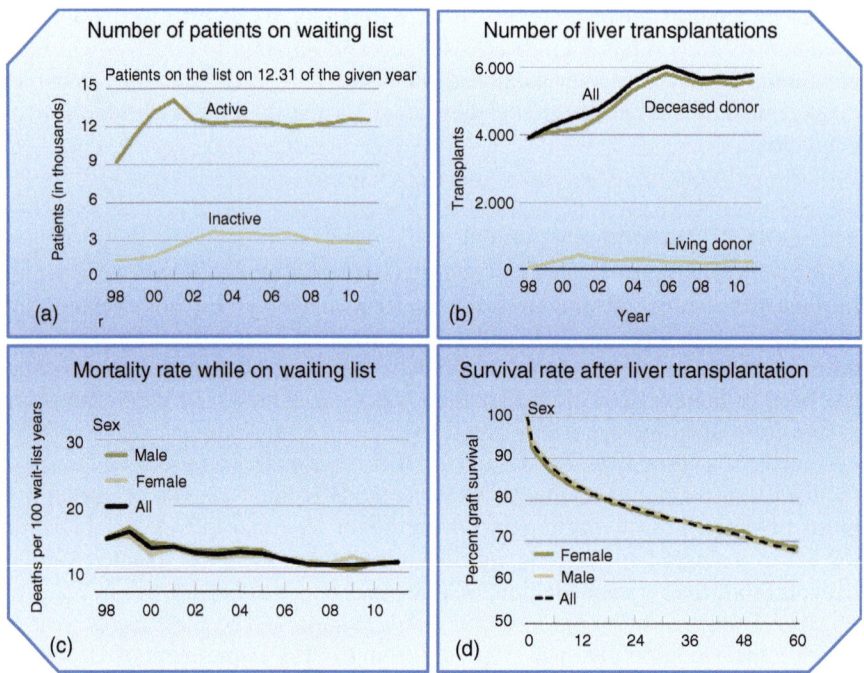

Figure 9.1 Statistics for Liver Transplantation—(a) **Number of Patients on Waiting List**—There are about 12,000 patients currently on the waiting list for a liver transplant. (b) **Number of Liver Transplantations**—Just under 6000 patients are able to receive a liver transplant. (c) **Mortality Rate While on Waiting List**—The mortality rate of patients while waiting for a liver transplantation is in excess of 10%. (d) **Survival Rate after Liver Transplantation**—For patients who do receive a liver transplantation, the survival rate is high: reported to be about 75% at the three-year time point. *Note*–The data presented here has been obtained from the Organ Procurement and Transplantation Network (OPTN) and Scientific Registry of Transplant Recipients (SRTR). The data and analyses reported in the 2011 Annual Data Report of the Organ Procurement and Transplantation Network and the US Scientific Registry of Transplant Recipients have been supplied by the Minneapolis Medical Research Foundation and UNOS under contract with HHS/HRSA. The authors alone are responsible for reporting and interpreting these data; the views expressed herein are those of the authors and not necessarily those of the US Government.

split-liver transplantation (SLT), in which a single adult cadaveric liver is divided or split into two, and the two pieces serve as transplantation grafts, one for an adult patient and one for a pediatric patient. As in the case for LDLT, regeneration of the transplanted grafts results in restoration of normal liver function for both recipients.

Indications for Adult Liver Transplantation—In the previous section, we looked at ALF, which can lead to end stage liver failure and require liver transplantation. ALF is the primary cause of 5%–6% of all liver transplants in the US. Other diseases which can require liver transplantation include chronic liver failure due to cirrhosis caused by hepatitis C virus (HCV), hepatocellular carcinoma (HCC), and liver diseases related to alcohol consumption, which accounted for one in six transplants in 2011.

Quality of Life after Liver Transplantation—After undergoing liver transplantation, the patient undoubtedly extends his/her life. Cases of patients who have lived for more than 30 years post-liver transplantation have been reported. However, there is a decrease in the quality of life for patients after liver transplantation. There is a tendency for liver transplant patients to have a reduced social life and a decrease in physical activity due to excessive fatigue and poor sleep quality. An increase in the rate of depression for patients after liver transplantation has also been observed. Complications resulting from immunosuppression therapy are also present, some of which include hypertension, new-onset diabetes mellitus, and dyslipidemia. Obesity, renal disease, and an increase in the risk of cancer are additional factors that may affect the quality of life of liver transplant patients.

9.4 LIVER REGENERATION

The mammalian liver has a remarkable regenerative capacity, and after partial hepatectomy to remove 70% of the liver tissue, it is able to completely recover lost functionality (39–48). The increase in tissue mass is primarily due to an increase in the number of mature hepatocytes. Recruitment of stem cells does not appear to play a significant role in the regenerative capacity of the liver. This unique characteristic of the mammalian liver has allowed the development of liver transplantation techniques using liver segments, as we have seen in the previous section. In addition, understanding the molecular mechanisms leading to the regenerative capacity of the liver has far-reaching implications in tissue engineering and can be exploited to support the tissue fabrication process.

The regenerative capacity of the mammalian liver is due to the proliferative capacity of hepatocytes. After partial hepatectomy to remove 70% of the liver tissue, cells in the remaining 30% of the tissue undergo one round of cell division, doubling the number of cells and resulting in an increase of tissue mass. This process increases the tissue mass of the liver to about 60% of the original mass; a subset of the cells undergo a second round of cell division, which allows recovery of the entire tissue mass lost during partial hepatectomy. In addition to supporting functional recovery after partial hepatectomy, the hepatocytes are able to maintain normal function of the liver during the recovery phase. This is indeed a remarkable characteristic of the liver.

There are changes in liver hemodynamics after partial hepatectomy due to changes in blood flow regimes, and these hemodynamic changes are important in initiating liver regeneration. This is followed by a complex cascade of molecular and cellular changes along with activation of several intracellular signaling pathways, which orchestrate the functional recovery of the liver. These changes affect both the cellular and extracellular components of the liver tissue. One of the earlier events of liver regeneration is the breakdown and remodeling of the extracellular matrix, which is important to release the cells to support cell proliferation and increase in cell number. The breakdown of the liver extracellular matrix is brought about by several matrix metalloproteinases, which are upregulated during the regeneration process. The breakdown of extracellular matrix components result in release of several growth factors, which are stored locally, with hepatocyte growth factor (HGF) being one such example. HGF has proliferative effects and acts by binding to the cell surface receptor *cMet*.

In addition to changes in the extracellular matrix, there are significant intracellular molecular events that take place during the liver regeneration process. Partial hepatectomy leads to an increase in the expression of more than 100 genes; this increase orchestrates the proliferative response of hepatocytes. Shortly after partial hepatectomy, there is an increase in the expression of Stat3 and NFkB, which are important signaling molecules that trigger the proliferative response of the hepatocytes. The purpose of the regenerative response of the liver is to restore normal functionality after partial hepatectomy, and this is achieved in part by an increase in the rate of proliferation of hepatocytes. However, as this process continues, there is also a need to stop the proliferation of hepatocytes once functional recovery has been accomplished. Termination of the regenerative response makes use of a complex feedback system between growth factors, extracellular matrix components, and the cells. There are several compounds that participate in the process, with transforming growth factor-β1 (TGF-β1) being one such example. TGF-β1 production is increased by stellate cells in response to HGF. However, proliferating hepatocytes become resistant to TGF-β1, which plays a part in the termination of the regenerative process.

9.5 LIVER DEVELOPMENT

The liver develops from cells in the endoderm, and one of the early events of liver development is the expression of albumin, transthyretin, and α-fetoprotein; the expression of these proteins is also used as a marker for liver function during tissue engineering studies (49–51). During the early stages of liver development, hepatoblasts, which are early stem cells giving rise to hepatocytes, undergo a complex series of steps which include proliferation, cell migration, and loss of adhesion. Several transcription factors regulate this process, including Hex, Prox-1, Tbx3, HNF-6, and OC-2. In addition, BMP and FGF signaling is important for cell proliferation. Once the endoderm cells have been specified, a liver diverticulum forms at day 22 in humans; the endoderm cells are known as hepatoblasts at this stage. The hepatoblasts give rise to a pseudostratified epithelium and proliferate to form a tissue bud, which is delineated by a basement membrane which contains laminin, collagen IV, nidogen, fibronectin, and heparan sulfate proteoglycan. The hepatocytes then migrate away from the epithelial lining of the endoderm, travel through the basement membrane, and invade the septum transversum. During later stages of liver development, the hepatoblasts give rise to mature hepatocytes under a careful gene expression pattern.

9.6 DESIGN CONSIDERATIONS FOR LIVER TISSUE ENGINEERING

In the previous two chapters, we looked at the design considerations for tracheal and bladder tissue engineering. Many of the design considerations that were presented for tracheal and bladder tissue engineering also apply for the fabrication of artificial liver tissue. In the previous two chapters, we also presented an overarching design statement for tracheal and tissue engineering, which also applies for liver tissue engineering: *"bioengineered liver tissue should be similar in form and functional to a mammalian liver."*

The specific design considerations for liver tissue engineering are: 1) functionality: albumin synthesis is used as an early indicator of artificial liver function, 2) biocompatibility, 3) nonimmunogenic and minimal inflammatory response, 4) nontoxic and noncarcinogenic, 5) avoidance of collapse by reasonable strength, 6) support cell engraftment, 7) support neovascularization, 8) possibility of growth, 9) resistance to fibroblastic and bacterial invasion, 10) standardized easy and short fabrication, 11) customizable and low cost, 12) easy surgical handling, 13) provide physiological environment such as ECM, 14) minimal necessity of donors and accessibility, 15) the results of engraftment are predictably successful, 16) provide or support epithelial resurfacing, 17) must not dislocate or erode over time, and 18) permanent constructions.

9.7 PROCESS OF BIOENGINEERING ARTIFICIAL LIVER TISSUE

Introduction—In Chapter 1, we presented a general scheme to bioengineer artificial tissue, and in the previous two chapters we applied this scheme for tracheal and bladder tissue engineering. In this section, we will adopt the general scheme for tissue engineering toward the fabrication of artificial liver tissue (Figure 9.2). Our discussion will focus on general points based on what we have learned about liver structure and function in the previous few sections. We will structure our discussion to answer one question: *based on what we know about liver structure and function, what strategies can be implemented to fabricate artificial liver tissue?* In subsequent sections, we will look at specific examples from the recent literature of different methods that have been adopted to bioengineer artificial tissue.

Cell Sourcing for Liver Tissue Engineering—The liver has a remarkable regenerative capacity, which is due to hepatocyte proliferation in response to partial hepatectomy. This provides a unique option for cell sourcing that may not be available for other tissue systems. A tissue biopsy from the patient can be used to isolate and expand primary hepatocytes during culture, and these cells then can be used to bioengineer artificial tissue; this strategy will provide an autologous cell source for the fabrication of artificial liver tissue. Other sources of cells for liver tissue engineering are from the differentiation of stem cells, including hES and iPS cells. Stem cells have not been used extensively in tracheal and bladder tissue engineering, but have been used for the fabrication of artificial liver tissue; we look at some specific examples in a subsequent section.

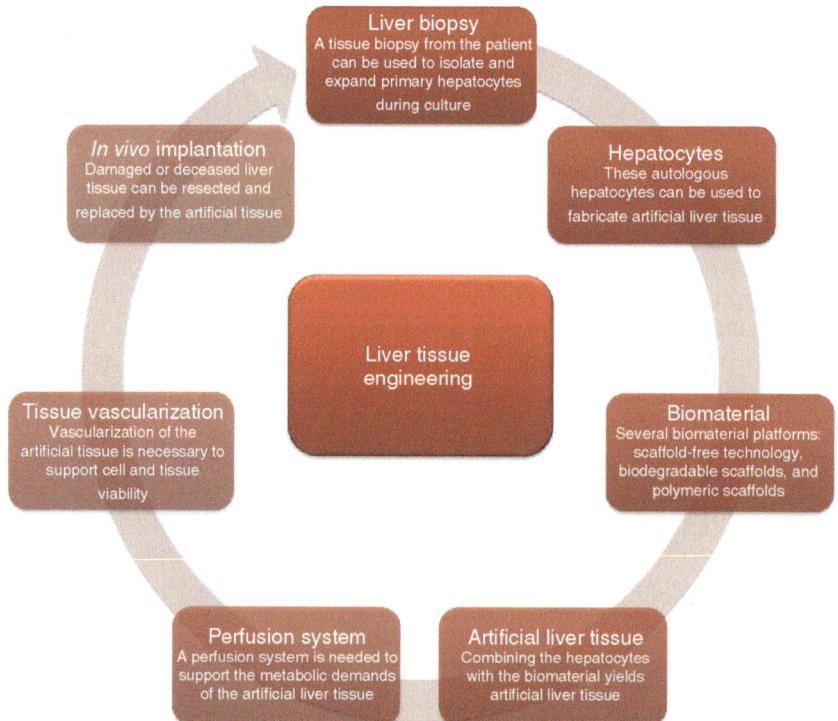

Figure 9.2 Overview of Liver Tissue Engineering—Primary hepatocytes are isolated from a liver biopsy and cultured and expanded using controlled *in vitro* conditions. These cells are then cultured within a 3D scaffold to support the formation of artificial liver tissue. Vascularization of artificial tissue is required to support metabolic activity, and the newly formed blood vessels are perfused using custom bioreactors. Artificial liver tissue fabricated using this process can be implanted *in vivo* to support the functional activity of damaged or diseased livers.

Biomaterials for Liver Tissue Engineering—In the previous two chapters, we looked at biomaterials that have been used for tracheal and bladder tissue engineering; as we have seen, acellular scaffolds have been a preferred biomaterial to support artificial tissue development. The trachea and the bladder are hollow structures and therefore require materials that have a high mechanical strength. However, for liver tissue engineering, the requirements for mechanical strength are not as high. Therefore, many other biomaterials platforms have been tested: scaffold-free technologies, biodegradable scaffolds, and polymeric scaffolds have been used to support the fabrication of 3D artificial liver tissue.

Bioreactors for Liver Tissue Engineering—In the previous two chapters, there was an evident lack of interest in the development of bioreactors for tracheal and bladder tissue engineering. However, this is not the case for liver tissue engineering, in which bioreactor technology has been an integral part of the development process. The liver has a very high metabolic activity to support synthesis of thousands of proteins. Therefore, it is self-evident that perfusion systems are necessary to support the metabolic demands of artificial liver tissue. Electrical stimulation and mechanical stretch are not essential for the development of liver tissue. Liver tissue is not excitable and therefore does require electrical stimulation to maintain function. Similarly, the liver is not exposed to large hemodynamic loads during normal tissue function, hence, mechanical stretch is not required during the tissue fabrication process.

Vascularization of Artificial Liver Tissue—Since liver tissue is known to be highly metabolic, vascularization of the artificial tissue will be necessary to support cell and tissue viability. During our discussion of vascularization strategies in an earlier chapter, we looked at several *in vitro* and *in vivo* strategies being developed for tissue engineering. Many of these strategies, including additional novel methods, have been developed to support the neovascularization of liver tissue, and we will study these in a subsequent section.

In Vivo Implantation—Development of patches of liver tissue can be used directly as a graft for implantation in the host. Damaged or diseased liver tissue can be resected from the patient and replaced by tissue segments that have been fabricated in the laboratory. Coupling between the implanted and the host liver tissue can lead to recovery of lost tissue functionality and restore normal liver function.

The discussion in this section has been designed to provide a general scheme to bioengineer artificial liver tissue. As we study specific examples, we will see how these elements come together, along with many other novel methods, to bioengineer artificial tissue.

9.8 STEM CELLS FOR LIVER TISSUE ENGINEERING

In Chapter 2, we studied stem cell engineering and looked at several sources for stem cells and strategies to regulate the differentiation fate of these cells toward specific cell lineages. In this section, we will study specific examples that have been implemented to drive the differentiation fate of stem cells toward a hepatic lineage. Specially, we will study human embryonic stem cells, induced pluripotent stem cells, hepatic stem cells and mesenchymal stem cells. There have been several strategies published in the recent literature describing different strategies to drive the differentiation fate of these stem cells toward a hepatic lineage. For illustrative purposes, we will describe one strategy for each of the four stem cell types.

Embryonic Stem Cells for Liver Tissue Engineering—The strategy to drive the differentiation of ES cells toward a hepatic lineage has been to mimic the process as it happens during embryogenesis (52). It may be recalled that during embryogenesis, early stem cells are first differentiated toward the three germ layers—the ectoderm, the mesoderm, and the endoderm—which then give rise to different organ and tissue systems. It may be further recalled that the liver is derived from endodermal cells. Therefore, the strategy for liver tissue engineering has been to first drive the differentiation ES cells toward endodermal cells, followed by differentiation of endodermal cells toward a hepatic lineage. In one specific example, a three-step process was used to drive the differentiation of mouse ES cells to form hepatocytes. The first two steps were designed to differentiate mouse ES cells to endodermal cells and then to form hepatocytes; the third step was designed to support functional maturation of hepatocytes. ES cells were cultured in the presence of Activin A for three days, followed by culture in the presence of aFGF and sodium butyrate for an additional five days; these steps were designed to drive the differentiation fate of ES cells to endodermal cells and then to hepatocytes, respectively. Hepatocyte maturation was accomplished by culturing the differentiated cells in culture media containing HGF for five days, followed by OSM and Dex for an additional five days.

Induced Pluripotent Stem Cells for Liver Tissue Engineering—The concept of iPS cells may be recalled from our earlier discussion of stem cell engineering in Chapter 2. Mature somatic cells like fibroblasts are transformed to an embryonic lineage, which can be differentiated to multiple cell types. As iPS cells resemble ES cells, the differentiation strategy to form hepatocytes from iPS cells has been similar to that for ES cells. In a recent study, a series of chemical conditioning steps were used to drive the differentiation of iPS cells to form hepatocytes (53). iPS cells were first differentiated to form definitive endodermal cells and then progressively differentiated to form more specialized cells. Definitive endodermal cells refer to early endodermal cells that differentiate to form all tissue/organs derived from the endoderm, while anterior definite endodermal (ADE) cells are more specialized and only give rise to the liver, pancreas, lungs, and thyroid. Using this strategy, various chemical compounds were used to differentiate the iPS cells to form early endodermal cells, ADE cells, and then hepatic progenitor stem cells. This was followed by differentiation of the hepatic progenitor stem cells to form immature hepatocytes and then mature hepatocytes.

Hepatic Stem Cells for Liver Tissue Engineering—Hepatic stem cells are present in the human liver and have the potential to differentiate into hepatocytes to support liver function in cases of injury and disease. Hepatic stem cells can be isolated and maintained in culture, and several strategies have been developed to support the differentiation of these cells to form hepatocytes. Hepatic stem cells are isolated from liver tissue specimens using an enzymatic digestion process and are cultured on the surface of a feeder layer of fibroblast cells. Chemical conditioning was used to drive the differentiation of hepatic stem cells to form hepatocytes using a two-step process (54). In the first step, the stem cells were conditioned with epidermal growth factor to form immature hepatocytes. In the second step, the immature hepatocytes were conditioned with HGF to support the formation of mature hepatocytes.

There are significant differences in the strategies that have been used to drive the differentiation of ES cells and iPS cells when compared with differentiation strategies for hepatic stem cells. In the case of ES and iPS cells, conditions were optimized to drive the differentiation of the stem cells to form endodermal cells; this was followed by differentiation of the endodermal stem cells to form hepatocytes. The differentiation strategy was designed to mimic embryogenesis. However, when compared with ES and iPS cells, hepatic stem cells have limited differentiation potential and are committed to forming either hepatocytes or bile duct epithelium. Therefore, chemical conditioning can be used to drive the differentiation of these cells toward a hepatic lineage without the intermediate step to generate endodermal stem cells.

Bone Marrow-Derived Mesenchymal Stem Cells for Liver Tissue Engineering— We have studied bone marrow MSCs for tracheal and bladder tissue engineering in the previous two chapters. Due to the many advantages these cells offer, they have also been used for applications in liver tissue engineering. In one specific example, bone marrow MSCs were isolated from human donors and maintained and expanded under controlled *in vitro* conditions (55). The MSCs were cultured in a 3D nanofiber scaffold that was fabricated using PCL as the polymer; electrospinning was used as the biomaterial fabrication technology. As controls in the study, MSCs were also cultured on the surface of monolayer 2D tissue culture plates. Differentiation of MSCs to form hepatocytes was accomplished using a two-step chemical conditioning process, as in the previous examples. In the first step, HGF and DEX were used to drive the differentiation of MSCs to form immature hepatocytes; this was followed by chemical conditioning using OSM to support the formation mature hepatocytes. The hepatocytes formed during 3D culture of MSCs were shown to have higher functional performance when compared with hepatocytes formed during 2D culture of the MSCs. This study showcased a novel application of tissue engineering technology to regulate the differentiation of stem cells to a specific cell lineage, and serves to highlights the advantages of 3D culture when compared to monolayer 2D culture.

9.9 SURFACE PATTERNING TECHNOLOGY FOR LIVER TISSUE ENGINEERING

Introduction—During our discussion of tissue fabrication technology in Chapter 4, we looked at strategies to regulate the spatial distribution of cells. Cell and organ printing, soft lithography, and surface patterning were some examples of techniques that have been used to control the spatial distribution of cells. The primary advantage of these technologies is the ability to regulate the placement of different cell types and extracellular matrix components during the tissue fabrication process; this in turn results in artificial tissue that is closer in form to mammalian tissue. In this section, we will study three examples that have been used to regulate the spatial distribution of hepatocytes relative to fibroblasts.

Robotic Protein Printing—A new method was developed to control the placement of extracellular matrix components on the culture surface (56). This method, known as robotic protein printing, allows 2D spatial control of the placement of proteins; cells then bind to the protein using cell-surface integrins (56). This process not only promotes the spatial alignment of cells but also elicits very specific cell-matrix interactions.

In one study, robotic protein printing was used to develop a novel co-culture system using hepatocytes and fibroblasts (56). Collagen was first printed in a specific configuration on the surface of glass slides that had been modified and prepared to support robotic protein printing. A cell suspension of hepatocytes was placed on the surface of the printed culture surface; the hepatocytes attach to regions at which collagen was printed. A cell suspension containing fibroblasts was then added to the glass slide with the hepatocytes attached. The fibroblasts attached to regions of the culture surface where hepatocytes did not attach. Using this method, the 2D spatial distribution of the hepatocytes and fibroblasts was controlled by placement of collagen. In addition to regulating spatial placement of multiple cell types, this process offers many other advantages, including the formation of a multi-cellular culture system and promoting cell-cell and cell-matrix interactions.

Photoresponsive Culture Surface—Another very interesting way to regulate spatial distribution of cells is via photoresponsive culture surfaces. Using this technology, the culture surface is coated with a photoresponsive polymer, which under normal conditions does not support cell attachment (57). At the start of the process, the entire culture surface is coated with photoresponsive polymer; this process results in a culture surface that does not support cell adhesion. Specific regions of the culture surface are then exposed to UV radiation, which causes a change in the configuration of the photoresponsive polymer; this in turn results in the culture surface switching to "cell friendly."

In one study, hepatocytes were added to the culture surface and preferentially attached to the regions that were coated with the photoresponsive polymer PEG and treated with UV light (57). Selective areas of the culture surface that did not have hepatocytes attached were exposed to a second round of UV treatment. A second cell type, which in this case was fibroblasts, was added to the treated regions of the tissue culture plate. Using this method, the placement of different cell types was regulated on a 2D culture surface. As in the previous example, this process offers advantages of promoting cell-cell and cell-matrix interactions.

PDMS Stencils for Spatial Positioning of Hepatocytes—Another method to spatially regulate hepatocytes and fibroblasts has been via the use of PDMS stencils (58). Using this technology, a stencil is fabricated on PDMS using soft lithography; the stencil can be generated in any pattern or configuration. The PDMS stencil is then placed on the culture surface, and cells can be added to the culture surface; the cells attach to regions of the culture surface not protected by the pattern on the stencil; the PDMS stencil serves to guide the placement of cells on the culture surface.

In one example, the PDMS stencil was used to generate a specific pattern, and hepatocytes were added to the culture surface (58). The hepatocytes attached to regions of the culture surface not protected by the stencil. The second cell type, which in this case was fibroblasts, was added to the culture surface. As before, the fibroblasts attached to regions of the tissue culture surface not protected by the PDMS stencil. In order words, the fibroblasts attach directly on top of the hepatocytes, thereby creating a bilayer structure. The PDMS stencil was removed, leaving a controlled pattern of hepatocytes and fibroblasts. The spatial positioning of the cells is regulated by the pattern of the stencil, and as can be envisioned, many different configurations can be created.

9.10 BIOMATERIAL PLATFORMS FOR LIVER TISSUE ENGINEERING

During our discussion of biomaterial platforms in Chapter 3, we looked at four strategies: scaffold-free methods, polymeric scaffolds, biodegradable hydrogels, and acellular tissue grafts. We also looked at the relative advantages and disadvantages of each strategy. During our discussion of tracheal and bladder tissue engineering in the previous two chapters, we saw that a subset of these biomaterial platforms have been tested to support the fabrication of artificial tracheas and/or bladders. The rigid design constraints placed by the hollow structures of the mammalian trachea and the mammalian urinary bladder, has limited the application of some of these technologies. For example, scaffold-free methods have not matured to the point at which they can meet the design requirements of mechanical strength and stability to support artificial bladders. The case of liver tissue is different, as the requirements for mechanical strength are not as high. However, mechanical stability is still a critical design requirement, as the biomaterial needs to support hepatocyte culture and remodeling. Therefore, it comes as no surprise that there has been a very high degree of interest in testing many different biomaterial platforms to support the fabrication of artificial liver tissue. All four biomaterial platforms that we have studied earlier—scaffold-free technology, polymeric scaffolds, biodegradable hydrogels, and acellular tissue grafts—have been used to support liver tissue fabrication. In the next section, we will look at specific examples of these four biomaterial platforms to support the fabrication of 3D artificial liver tissue.

9.11 FABRICATION OF 3D ARTIFICIAL LIVER TISSUE

Introduction—There have been numerous strategies implemented for the fabrication of artificial liver tissue, and during the course of this discussion, we will have an opportunity to apply many of the principles we learned in earlier chapters. Many of the technologies we have discussed have been tested in liver tissue engineering, including scaffold-free technologies, polymeric scaffolds, biodegradable hydrogels, and acellular matrix. In this section, we will look at four examples of methods that have been used to bioengineer 3D artificial liver tissue: cell sheet engineering, alginate scaffolds, poly(l-lactic acid) (PLLA) scaffolds, and acellular matrices.

Cell Sheet Engineering to Bioengineer Artificial Liver Tissue—As we have seen before, the primary advantage of cell sheet engineering is that external scaffolding is not required; rather, the cells make their own extracellular matrix, which is used to support 3D tissue formation and remodeling. We have discussed the method of tissue fabrication before, and the same method has been used to support the fabrication of artificial liver tissue (59). Primary hepatocytes were plated on a temperature-sensitive culture surface. When the cells are maintained at 37°C, the properties of the culture surface support cell adhesion; at this stage, the culture surface is "cell-friendly." This promotes the formation of a cohesive cell monolayer, which is referred to as a cell sheet. A decrease in culture temperature changes the properties of the culture surface, making it "cell-unfriendly." The cohesive cell monolayer detaches from the culture surface and remains intact, resulting in the formation of artificial liver tissue.

Porous Alginate Scaffolds for Liver Tissue Engineering—During our discussion of biomaterials for tissue engineering in Chapter 3, we studied biodegradable hydrogels and their use to support tissue fabrication. Collagen, fibrin, matrigel, and alginate are examples of biodegradable hydrogels that have been used extensively in tissue engineering. These are all naturally occurring compounds and have extracellular matrix components similar to that of mammalian tissue; this similarity supports cell-matrix interaction. In this particular study, matrigel was used as the biomaterial; alginate is the monovalent salt of alginic acid and is a block polymer of β-D-mannuronic acid and α-L-guluronic acid (60). The properties of alginate can be carefully controlled based on processing conditions; resulting in control of the porosity, 3D scaffold architecture, degradation kinetics, and material properties. In this example, porous sponges with an average pore size of 100–150 µm were fabricated using alginate, and primary hepatocytes were seeded into the scaffold using direct injection technology. The cellularized scaffolds were maintained in culture for two weeks and shown to support the formation of artificial liver tissue.

Porous PLLA Scaffolds for Liver Tissue Engineering—During our discussion of biomaterials for tissue engineering, we compared porous polymeric scaffolds to biodegradable hydrogels. As we discussed earlier, the primary advantage of polymer scaffolds was the ability to control the properties of the scaffold by changing processing conditions, polymer composition, and/or addition of cross-linking agents, to name a few. The primary drawback of polymeric scaffolds was the lack of specific binding sites for integrin-mediated cell binding. In this study, PLLA scaffolds were used to support the fabrication of artificial liver tissue (61). The strategy used in this study was similar to the one described for alginate scaffolds. Porous scaffolds were fabricated using PLLA and cellularized with primary hepatocytes. Direct cell injection was used as the cellularization technology; over time in culture, this strategy resulted in the formation of artificial liver tissue.

Acellular Scaffolds for Liver Tissue Engineering—In the previous two chapters, we have looked at several examples using acellular scaffolds to support the fabrication of artificial tracheal and bladder tissue. Acellular scaffolds are fabricated by the removal of cells from naturally occurring tissue; after removal of all cellular components, an intact extracellular matrix is left behind. This ECM has the right composition and distribution of proteins to support artificial tissue fabrication. Due to these advantages, acellular scaffolds have found extensive applications in the tissue engineering literature. In this particular study, acellular scaffolds were fabricated by decellularization of porcine liver specimens (62). Cellularization of the acellular scaffolds was performed via direct injection of primary hepatocytes. As in the previous cases, over time in culture, this method resulted in the fabrication of artificial liver tissue.

9.12 VASCULARIZATION FOR LIVER TISSUE ENGINEERING

Introduction—Vascularization is important to support cell viability in any given tissue system, and in Chapter 5, we studied several strategies to induce vascularization in bioengineered artificial tissue. In this section, we will look at the applications of these principles to support the fabrication of vascularized artificial liver tissue. We will look at two specific examples from the literature, both using *in vivo* vascularization methods, coupled with controlled release of angiogenic growth factors.

Prevascularized Scaffolds for Liver Tissue Engineering—In one study, several elements of neovascularization were incorporated into fabricating vascularized liver tissue. A novel scaffold was fabricated using alginate and was embedded with microspheres designed for the controlled release of vascular endothelial growth factor (VEGF) (63). The alginate scaffold was designed to support neovascularization and the growth and functionality of hepatocytes. The objective of this study was to use the alginate scaffold as the biomaterial during the tissue fabrication process. The purpose of the microspheres was to provide a mechanism for the controlled release of VEGF into the local culture environment; VEGF is known to increase the rate of neovascularization in mammalian tissue. VEGF was encapsulated in custom microspheres with known degradation kinetics; as the microspheres degrade, VEGF was gradually released into the local culture environment. The rate of release of VEGF was reported to be 8–10 ng/day over a two-week period. The purpose of the VEGF was to promote neovascularization of the alginate scaffold. Therefore, the novelty in scaffold design should be appreciated, as it serves two critical functions—supporting hepatocyte culture and activity (a function performed by the alginate) and promoting neovascularization (a function performed by the controlled release of VEGF).

The next step in the process was prevascularization of the scaffold. This was achieved by implantation of the scaffold, without seeding any hepatocytes, onto the liver lobes of recipient rats. In this particular study, *in vivo* vascularization strategies were implemented, as discussed in Chapter 5. This means that neovascularization of the implanted tissue was a result of the host response and not due to any external user intervention. The presence of VEGF enhanced neovascularization of the implanted tissue, as measured by the capillary density. It should be noted that the prevascularization of the scaffold was undertaken prior to implantation of the hepatocytes. Through this method, when the hepatocytes were implanted, a vascular bed was in place to support the metabolic activity of the cells.

After a seven-day implantation period, primary hepatocytes were injected into the prevascularized scaffold while the scaffold remained implanted at the original site. The scaffold was not explanted before hepatocytes were injected; instead, the hepatocytes were injected into the scaffold while it was still implanted onto the liver lobe of the recipient rats. At various intervals after scaffold cellularization, the scaffolds were recovered and processed for histological assessment. As controls, hepatocytes were also injected into scaffolds that were prevascularized in the absence of the controlled release of VEGF. As expected, the functional performance of the prevascularized scaffold in the presence of VEGF was superior to the controls' performance.

In Vivo Vascularization of Scaffolds for Liver Tissue Engineering—In this study, polylactic acid (PLA) was used as the biomaterial and was engineered to form porous discs designed to support the culture and viability of hepatocytes (64). The polymer was first coated with bFGF, an angiogenic agent known to support neovascularization in mammalian tissue. The PLA scaffold was designed to support controlled release of bFGF into the local environment, as was the case in the previous example; however, the release kinetics were different, and most of the bFGF was released within the first 72 hours. Hepatocytes were seeded onto the scaffold and then implanted *in vivo*; as in the previous example, controlled release of the bFGF was designed to support neovascularization of the liver patch during development and maturation. This was indeed the case, as two weeks after implantation, vascularized liver tissue was formed. During the course of this study, neovascularization progressed in parallel with liver tissue formation and maturation, meaning that both happened at the same time. In the previous example, a prevascularized scaffold was used for liver tissue formation. These studies were not conducted side-by- side, and therefore, a direct comparison cannot be made between the two. However, the differences between the two methods should be appreciated, along with the novelty associated with each one.

9.13 BIOREACTORS FOR LIVER TISSUE ENGINEERING

Bioreactors are critical to support the tissue fabrication process. We studied bioreactor design in great detail in Chapter 6. During our discussion of tracheal and bladder tissue engineering, we noted a significant absence of bioreactor technology during the tissue fabrication process. While bioreactors have not been extensively used in the development of mammalian tracheal and bladder tissue, they have been used during the fabrication of artificial livers. This provides our first opportunity to study the application of bioreactor technology during the tissue fabrication process. Liver tissue is highly metabolic due to the number of functions performed; as such, perfusion is important to support the high metabolic activity of hepatocytes. Electrical stimulation is not important and, in fact, not required for the fabrication and/or culture of artificial liver tissue due to the non-excitable nature of the tissue. Similarly, mechanical stretch is not important, as liver tissue is not constantly exposed to changes in the hemodynamic environment (as is the case for the cardiovascular system). In this section, we will look at two specific bioreactor systems that have been developed to support the culture of artificial liver tissue.

Perfusion Culture of Artificial Liver Tissue—As one example, a perfusion system was developed to support the culture of artificial liver tissue. The liver tissue was fabricated by direct injection of primary hepatocytes into a porous PLGA scaffold (65). The perfusion system consisted of a reservoir to hold the artificial tissue constructs, a reservoir to hold the cell culture media, and a peristaltic pump. Cell culture media was perfused to the artificial liver tissue, and spent media was recirculated through the system. The perfusion system was housed in a cell culture incubator to regulate temperature and pH. As expected, culture of artificial liver tissue in the presence of continuous media flow significantly enhanced functional performance when compared with controls that were maintained under static culture conditions.

Dual-Compartment Perfusion System for Liver Tissue Engineering—A second perfusion system was developed to support the culture of artificial liver tissue. In this system, artificial liver tissue was fabricated by direct injection of primary hepatocytes within a porous hybrid scaffold fabricated using collagen and PLGA (66). As in the previous example, the system consisted of culture vessels to accommodate artificial liver tissue and cell culture media. A peristaltic pump was used for perfusion of cell culture media to the tissue specimens, and spent media was recirculated through the system. The tissue specimens were maintained in a cell culture environment for temperature and pH regulation. The novelty of this perfusion system was the development of a dual-chamber compartment to accommodate two different samples. A porous membrane separated the two compartments, which allowed the flow of soluble factors between the two compartments. Artificial liver tissue and stellate cells, also cultured on the 3D scaffolds, were maintained in each of the two compartments; it was hypothesized that the release of soluble factors from stellate cells would enhance the functional performance of the artificial liver tissue. It was demonstrated that the presence of the stellate cells did, in fact, have a significant impact on the functional performance of the artificial liver tissue.

9.14 SPHEROID CULTURE FOR LIVER TISSUE ENGINEERING

During the course of our discussion of tissue engineering for the development of artificial liver tissue, we have looked at the isolation and culture of primary hepatocytes using monolayer 2D culture. Using monolayer cell culture techniques described in Chapter 2, primary cells are isolated after enzymatic digestion of tissue specimens, and these cells are maintained on the surface of tissue culture plates. The culture conditions are optimized to support the proliferation and expansion of primary cells and are subpassaged to increase cell yield. This strategy for monolayer 2D cell culture has been used extensively for tissue engineering studies, as we have seen earlier in this chapter and during our discussions for the fabrication of artificial trachea and bladder tissue. In addition to the examples presented in this book, monolayer cell culture has been used extensively for almost all tissue fabrication efforts, and is a core technology required to bioengineer 3D artificial tissue.

During our discussion of 2D monolayer cell culture, we looked at some limitations of this technique. Under normal physiological conditions, cells are maintained in 3D, and the 3D culture environment is important to maintain cell/tissue function and support 3D tissue architecture. During 2D monolayer culture, cells maintain partial functionality due to lack of complete cell-cell and cell-matrix interactions seen during normal mammalian tissue function. This functionality is one major advantage of developing 3D tissue engineering models that mimic many of the physiological cues seen during normal mammalian tissue function.

In order to address this limitation, culture techniques have been developed to support the culture of primary cells in 3D aggregates known as spheroids; this technique has been referred to as spheroid culture. In this technique, the culture surface is modified with agents that change the properties of the surface, making them "cell-unfriendly." When the cells are cultured on these "unfriendly" surfaces, they do not attach to the culture surface; rather, the cells remain in suspension and form aggregates or spheroids. As the cells are maintained in a 3D environment during spheroid culture, they exhibit a higher degree of functionality than cells that are maintained during monolayer 2D culture.

Spheroid culture is a technique used to culture cells in spheroids or aggregates to support cell functionality. There are similarities between spheroid culture and tissue engineering, as both require 3D culture of cells. However, spheroid culture techniques are primarily used to maintain cell phenotype and function during culture and are not often used to design therapeutic strategies or to support artificial tissue or organ fabrication. Therefore, spheroid culture should be considered a specialized method of cell culture designed to increase cell phenotype and functionality.

Spheroid culture has been used extensively to support the culture of hepatocytes. Several methods have been described in the literature to support the formation of spheroids using hepatocytes. In one example, the culture surface was coated with poly (2-hydroxyethyl methacrylate), which prevents cell adhesion and promotes the formation of spheroids of hepatocytes (67). In another study, the culture surface was also coated with poly (2-hydroxyethyl methacrylate) to prevent cell adhesion; however, this strategy was coupled with gentle mixing to further prevent the hepatocytes from attaching to the culture surface (68). In another example, spheroid culture of hepatocytes was promoted by maintaining cells in a custom bioreactor that was designed to provide continuous rotation of cells, thereby preventing attachment to a culture surface. In this example, primary hepatocytes were maintained in a bioreactor designed to culture the cells in the presence of an oscillatory frequency, which was shown to support spheroid formation. It was also demonstrated that spheroid culture of hepatocytes was associated with a decrease in cell death and increase in cell function when compared to hepatocytes that are maintained under standard monolayer 2D conditions (69). In another example of spheroid culture of hepatocytes, the primary cells were cultured by entrapment within a synthetic thermoreversible extracellular matrix. The functional performance of primary hepatocytes was enhanced when maintained in spheroid culture, as compared to monolayer 2D culture of the cells (70).

SUMMARY

Current State of the Art—The field of liver tissue engineering is considerably mature; it is more mature than the fields of tracheal and bladder tissue engineering. Many resources have been invested in developing strategies to fabricate artificial liver tissue, as we have seen throughout the course of this chapter. There has been a large amount of research invested in driving the differentiation fate of stem cells to form hepatocytes. Induced pluripotent stem cells, embryonic stem cells, adipose-derived mesenchymal stem cells, and hepatic stem cells have all been evaluated, with chemical conditioning being the preferred differentiation strategy. Similarly, considerable resources have been invested in evaluating different biomaterial platforms to test suitability for applications in liver tissue engineering. Some of the biomaterials platforms used to fabricate artificial liver tissue include scaffold-free methods, polymeric scaffolds, biodegradable hydrogels, and acellular grafts. There have been several strategies for vascularization of artificial tissue that involve implantation of artificial liver tissue coupled with angiogenic growth factors to enhance blood vessel formation. Bioreactor technology has also been developed to support the culture of artificial liver tissue and has primarily been focused on fabrication of perfusion systems to support the metabolic activity of hepatocytes.

Thoughts for Future Research—While there has been considerable progress in the field of liver tissue engineering, there are several areas that could benefit from additional research. A large amount of research has been published in the recent literature describing the differentiation of stem cells to form hepatocytes, particularly using chemical compounds. Due to this extensive knowledge base, we now have an understanding of the drivers of stem cell differentiation to form hepatocytes; this knowledge base can be expanded, and chemical factors can be coupled to bioreactors to deliver controlled perfusion protocols. The combined use of chemical compounds and fluid stresses resulting from media perfusion can increase differentiation efficiency and/or support the formation of mature hepatocytes. Another area of research that requires attention is the development of perfusion systems to support the metabolic activity of artificial liver tissue. Most of the systems developed thus far require the use of a cell culture incubator for regulation of processing variables. In addition, the current generation of perfusion systems does not have the capacity to monitor the functional performance of the artificial liver tissue in real- time or make use of noninvasive monitoring technology. The development of perfusion systems that operate independently of a cell culture incubator, with inline monitoring of processing variables, provide greater control over the system. In addition, the incorporation of real-time, noninvasive monitoring of the functional performance of artificial liver tissue will provide significant information about 3D tissue remodeling and functional reorganization in response to media perfusion. This information can be used to regulate perfusion variables. For example, as the metabolic activity of the artificial liver tissue increases or decreases, this information can be monitored in real- time and can be used to increase or decrease the flow rate of the culture media.

PRACTICE QUESTIONS

1. Describe the structure and function of the mammalian liver.
2. Describe the causes, symptoms, and treatment options for acute liver failure.

3. Discuss various strategies that have been used for liver transplantation. What are the relative advantages and disadvantages of each of these strategies?

4. Describe the molecular mechanism of liver regeneration.

5. Discuss development of the liver during embryogenesis.

6. The liver has a remarkable regenerative capacity. However, this capability has not been widely used to support the fabrication of artificial liver tissue. Based on your understanding of liver regeneration and liver tissue engineering, develop a strategy that makes use of liver regeneration to bioengineer artificial liver tissue.

7. In this chapter, we described a general process scheme to bioengineer artificial liver tissue. Develop a process scheme to bioengineer artificial livers, starting with a tissue biopsy and working up to the fabrication of artificial tissue. Provide a description for cell sourcing, material selection, and scaffold cellularization, and justify and explain your choice for each of the three.

8. During our discussion of stem cells for liver tissue engineering, we looked at several strategies to drive the differentiation fate of stem cells to form hepatocytes. However, all strategies were based on the use of growth factors or other chemical compounds. Develop a strategy that does not rely solely on the use of growth factors or other chemical compounds to drive the differentiation fate of induced pluripotent stem cells to form hepatocytes.

9. During our discussion of biomaterial platforms for liver tissue engineering, we looked at scaffold-free methods, polymeric scaffolds, and biodegradable hydrogels to support the fabrication of artificial liver tissue. Which one of these biomaterial platforms is more suited to support the fabrication of artificial liver tissue and why?

10. We looked at several models of artificial liver tissue, all of which used direct cell injection to populate the 3D scaffold with cells. Discuss the relative advantages and disadvantages of direct cell injection as a strategy to support scaffold cellularization. Develop an alternative strategy for scaffold cellularization to support the fabrication of artificial liver tissue. Describe the strategy and explain the relative advantages and disadvantages of the selected scaffold cellularization strategy when compared with direct cell injection.

11. In Chapter 4, we looked at cell and organ printing as strategies that have been used to support the fabrication of artificial tissue. Discuss the feasibility of using cell and organ printing to support the fabrication of artificial liver tissue. Describe the advantages and disadvantages of using this technology in liver tissue engineering. Develop a strategy to implement cell and organ printing to support artificial liver tissue fabrication.

12. During our discussion of bioreactors for liver tissue engineering, we looked at two examples of perfusion systems that were developed to support the metabolic activity of artificial liver tissue. Both of these perfusion systems were designed to function inside of a cell culture incubator, which provided temperature and pH regulation. Another strategy in the development of perfusion systems is to engineer sensors and feedback loops for regulation of processing variables within the perfusion system. Discuss the relative advantages and disadvantages of each of these two strategies. Design a perfusion system with embedded sensors and feedback loops.

13. During our discussion of vascularization strategies for liver tissue engineering, we looked at two examples that utilized *in vivo* methods to support liver vascularization. It may be recalled from our discussion in Chapter 5, that *in vivo* and *in vitro* strategies can be used to support the neovascularization of artificial tissue. Discuss the relative advantages and disadvantages of *in vivo* and *in vitro* strategies for the vascularization of liver tissue. Develop an *in vitro* strategy to support the vascularization of artificial liver tissue.

14. During our discussion of tissue fabrication strategies for liver tissue engineering, we looked at one example of scaffold-free technology. In this example, cell sheet engineering was used to support the fabrication of artificial liver tissue. It may be recalled from our discussion in Chapter 4 that self-organization strategies have also been used to support the fabrication of artificial tissue. Compare cell sheet engineering and self-organization strategies as technologies to support the fabrication of artificial liver tissue. Develop a method to bioengineer artificial liver tissue using self-organization technology.

15. We discussed spheroid culture of primary hepatocytes. This method involves the culture of primary hepatocytes in aggregates, thereby supporting 3D culture of the cells. There are similarities and differences between spheroid culture and 2D monolayer culture. There are also similarities and differences between spheroid culture and tissue engineering. Discuss the similarities and differences between spheroid culture and monolayer cell culture. Also discuss the similarities and differences between spheroid culture and tissue engineering.

REFERENCES

1. Oppe TE. Liver Function in the Newborn Infant. Biochem. Clin. 1964;3:5–11.
2. Hoe CM, Wilkinson JS. Liver function: a review. Aust. Vet. J. 1973 Mar;49(3):163–9.
3. Lathe GH. Liver function in the newborn. Med. Chir Dig. 1974;3(3):215–20.
4. Reichen J, Paumgartner G. Excretory function of the liver. Int. Rev. Physiol. 1980;21:103–50.
5. Corless JK, Middleton HM, III. Normal liver function. A basis for understanding hepatic disease. Arch. Intern. Med. 1983 Dec;143(12):2291–4.

6. Israel J, London WT. Liver structure, function, and anatomy: effects of hepatitis B virus. Curr. Top. Microbiol. Immunol. 1991;168:1–20.
7. Michels NA. Variations in blood supply of liver, gallbladder, stomach, duodenum and pancreas; summary based on one hundred Dissections. J. Int. Coll. Surg. 1945 Nov;8:502–4.
8. Michels NA. Variations in blood supply of liver, gallbladder, stomach, duodenum and pancreas. Anat. Rec. 1946 Mar;94:481.
9. Wakim KG. The blood supply of the normal liver. Proc. Staff. Meet. Mayo Clin. 1953 Apr 22;28(8):218–27.
10. Popper HL, Jefferson NC, Wulkan E, Necheles H. Bile secretion and blood supply of the liver. Am. J. Physiol. 1955 May;181(2):435–8.
11. Michels NA. The ever varied blood supply of the liver and its collateral circulation. J. Int. Coll. Surg. 1957 Jan;27(1):1–17.
12. Woodburne RT. Segmental anatomy of the liver: blood supply and collateral circulation. Med. Bull. (Ann. Arbor) 1962 Jul;28:189–99.
13. Sleight DR, Thomford NR. Gross anatomy of the blood supply and biliary drainage of the canine liver. Anat. Rec. 1970 Feb;166(2):153–60.
14. McCuskey RS. Anatomy of efferent hepatic nerves. Anat. Rec. A Discov. Mol. Cell Evol. Biol. 2004 Sep;280(1):821–6.
15. Berthoud HR. Anatomy and function of sensory hepatic nerves. Anat. Rec. A Discov. Mol. Cell Evol. Biol. 2004 Sep;280(1):827–35.
16. Klover PJ, Mooney RA. Hepatocytes: critical for glucose homeostasis. Int. J. Biochem. Cell Biol. 2004 May;36(5):753–8.
17. Sell S. The hepatocyte: heterogeneity and plasticity of liver cells. Int. J. Biochem. Cell Biol. 2003 Mar;35(3):267–71.
18. Herrera JL. Management of acute liver failure. Dig. Dis. 1998 Sep;16(5):274–83.
19. Berry P, Wendon J. Recognition and early management of acute liver failure. Acute. Med. 2005;4(3):92–8.
20. O'Grady J. Modern management of acute liver failure. Clin. Liver Dis. 2007 May;11(2):291–303.
21. Williams R. Acute liver failure–practical management. Acta Gastroenterol. Belg. 2007 Apr;70(2):210–3.
22. Stravitz RT, Kramer DJ. Management of acute liver failure. Nat. Rev. Gastroenterol. Hepatol. 2009 Sep;6(9):542–53.
23. Craig DG, Lee A, Hayes PC, Simpson KJ. Review article: the current management of acute liver failure. Aliment. Pharmacol. Ther. 2010 Feb 1;31(3):345–58.
24. D'Agostino D, Diaz S, Sanchez MC, Boldrini G. Management and prognosis of acute liver failure in children. Curr. Gastroenterol. Rep. 2012 Jun;14(3):262–9.
25. O'Grady JG, Williams R. Classification of acute liver failure. Lancet 1993 Sep 18;342(8873):743.
26. Williams R. Classification, etiology, and considerations of outcome in acute liver failure. Semin. Liver Dis. 1996 Nov;16(4):343–8.
27. Rosmawati M. Aetiology and classification of acute liver failure. Med. J. Malaysia 2005 Jul;60 Suppl B:125–6.

28. Saidi RF. Current status of liver transplantation. Arch. Iran Med. 2012 Dec;15(12):772–6.
29. Carlisle EM, Testa G. Adult to adult living related liver transplantation: where do we currently stand? World J. Gastroenterol. 2012 Dec 14;18(46):6729–36. PMCID:PMC3520161.
30. Yang X, Gong J, Gong J. The value of living donor liver transplantation. Ann. Transplant. 2012 Dec 31;17(4):120–4.
31. Bodenheimer HC, Jr., Okun JM, Tajik W, Obadia J, Icitovic N, Friedmann P, Marquez E, Goldstein MJ. The impact of race on organ donation authorization discussed in the context of liver transplantation. Trans. Am. Clin. Climatol. Assoc. 2012;123:64–77. PMCID:PMC3540608.
32. O'Mahony CA, Goss JA. The future of liver transplantation. Tex. Heart Inst. J. 2012;39(6):874–5. PMCID:PMC3528242.
33. Zazula CP. Liver transplantation: the illusion of choice. Virtual. Mentor. 2012 Mar;14(3):269–71.
34. Wakade VA, Mathur SK. Donor safety in live-related liver transplantation. Indian J. Surg. 2012 Feb;74(1):118–26. PMCID:PMC3259172.
35. Lobritto S, Kato T, Emond J. Living-donor liver transplantation: current perspective. Semin. Liver Dis. 2012 Nov;32(4):333–40.
36. Organ Procurement and Transplantation Network and Scientific Registry of Transplant Recipients 2010 Data Report. American Journal of Transplantation 2012 Jan;12:1–154.
37. Issue S. Organ Procurement and Transplantation Network and Scientific Registry of Transplant Recipients 2011 Data Report introduction. American Journal of Transplantation 2013 Jan;13:8–10.
38. Issue S. Organ Procurement and Transplantation Network and Scientific Registry of Transplant Recipients 2011 Data Report preface. American Journal of Transplantation 2013 Jan;13:5–7.
39. Tarla MR, Ramalho FS, Ramalho LN, Silva TC, Brandao DF, Ferreira J, Silva OC, Zucoloto S. A molecular view of liver regeneration. Acta Cir. Bras. 2006;21 Suppl 1:58–62.
40. Tarla MR, Ramalho F, Ramalho LN, Silva TC, Brandao DF, Ferreira J, Silva OC, Zucoloto S. Cellular aspects of liver regeneration. Acta Cir. Bras. 2006;21 Suppl 1:63–6.
41. Tanimizu N, Miyajima A. Molecular mechanism of liver development and regeneration. Int. Rev. Cytol. 2007;259:1–48.
42. Michalopoulos GK. Liver regeneration. J. Cell Physiol. 2007 Nov;213(2):286–300. PMCID:PMC2701258.
43. Michalopoulos GK. Liver regeneration: alternative epithelial pathways. Int. J. Biochem. Cell Biol. 2011 Feb;43(2):173–9. PMCID:PMC2888836.
44. Jia C. Advances in the regulation of liver regeneration. Expert. Rev. Gastroenterol. Hepatol. 2011 Feb;5(1):105–21.
45. Gilgenkrantz H, Collin de lA. New insights into liver regeneration. Clin. Res. Hepatol. Gastroenterol. 2011 Oct;35(10):623–9.
46. Papadimas GK, Tzirogiannis KN, Mykoniatis MG, Grypioti AD, Manta GA, Panoutsopoulos GI. The emerging role of serotonin in liver regeneration. Swiss. Med. Wkly. 2012;142:w13548.

47. Karkampouna S, Ten DP, Dooley S, Julio MK. TGFbeta signaling in liver regeneration. Curr. Pharm. Des. 2012;18(27):4103–13.
48. Carnovale CE, Ronco MT. Role of nitric oxide in liver regeneration. Ann. Hepatol. 2012 Sep;11(5):636–47.
49. Shin D, Monga SP. Cellular and molecular basis of liver development. Compr. Physiol. 2013 Apr;3(2):799–815.
50. Zong Y, Stanger BZ. Molecular mechanisms of liver and bile duct development. Wiley. Interdiscip. Rev. Dev. Biol. 2012 Sep;1(5):643–55.
51. Wilkins BJ, Pack M. Zebrafish models of human liver development and disease. Compr. Physiol. 2013 Jul 1;3(3):1213–30.
52. Zhou M, Li P, Tan L, Qu S, Ying QL, Song H. Differentiation of mouse embryonic stem cells into hepatocytes induced by a combination of cytokines and sodium butyrate. J. Cell Biochem. 2010 Feb 15;109(3):606–14.
53. Hannan NR, Segeritz CP, Touboul T, Vallier L. Production of hepatocyte-like cells from human pluripotent stem cells. Nat. Protoc. 2013 Feb;8(2):430–7. PMCID:PMC3673228.
54. He ZP, Tan WQ, Tang YF, Feng MF. Differentiation of putative hepatic stem cells derived from adult rats into mature hepatocytes in the presence of epidermal growth factor and hepatocyte growth factor. Differentiation 2003 Jun;71(4–5):281–90.
55. Kazemnejad S, Allameh A, Soleimani M, Gharehbaghian A, Mohammadi Y, Amirizadeh N, Jazayery M. Biochemical and molecular characterization of hepatocyte-like cells derived from human bone marrow mesenchymal stem cells on a novel three-dimensional biocompatible nanofibrous scaffold. J. Gastroenterol. Hepatol. 2009 Feb;24(2):278–87.
56. Revzin A, Rajagopalan P, Tilles AW, Berthiaume F, Yarmush ML, Toner M. Designing a hepatocellular microenvironment with protein microarraying and poly(ethylene glycol) photolithography. Langmuir 2004 Apr 13;20(8):2999–3005.
57. Kikuchi K, Sumaru K, Edahiro J, Ooshima Y, Sugiura S, Takagi T, Kanamori T. Stepwise assembly of micropatterned co-cultures using photoresponsive culture surfaces and its application to hepatic tissue arrays. Biotechnol. Bioeng. 2009 Jun 15;103(3):552–61.
58. Cho CH, Park J, Tilles AW, Berthiaume F, Toner M, Yarmush ML. Layered patterning of hepatocytes in co-culture systems using microfabricated stencils. Biotechniques 2010 Jan;48(1):47–52. PMCID:PMC3147300.
59. Ohashi K, Yokoyama T, Yamato M, Kuge H, Kanehiro H, Tsutsumi M, Amanuma T, Iwata H, Yang J, Okano T, et al. Engineering functional two- and three-dimensional liver systems in vivo using hepatic tissue sheets. Nat. Med. 2007 Jul;13(7):880–5.
60. Glicklis R, Shapiro L, Agbaria R, Merchuk JC, Cohen S. Hepatocyte behavior within three-dimensional porous alginate scaffolds. Biotechnol. Bioeng. 2000 Feb 5;67(3):344–53.
61. Torok E, Lutgehetmann M, Bierwolf J, Melbeck S, Dullmann J, Nashan B, Ma PX, Pollok JM. Primary human hepatocytes on biodegradable poly(l-lactic acid) matrices: a promising model for improving transplantation efficiency with tissue engineering. Liver Transpl. 2011 Feb;17(2):104–14.
62. Lang R, Stern MM, Smith L, Liu Y, Bharadwaj S, Liu G, Baptista PM, Bergman CR, Soker S, Yoo JJ, et al. Three-dimensional culture of hepatocytes on porcine liver tissue-derived extracellular matrix. Biomaterials 2011 Oct;32(29):7042–52.

63. Kedem A, Perets A, Gamlieli-Bonshtein I, Dvir-Ginzberg M, Mizrahi S, Cohen S. Vascular endothelial growth factor-releasing scaffolds enhance vascularization and engraftment of hepatocytes transplanted on liver lobes. Tissue Eng. 2005 May; 11(5-6):715-22.
64. Lee H, Cusick RA, Browne F, Ho KT, Ma PX, Utsunomiya H, Langer R, Vacanti JP. Local delivery of basic fibroblast growth factor increases both angiogenesis and engraftment of hepatocytes in tissue-engineered polymer devices. Transplantation 2002 May 27;73(10):1589-93.
65. Fiegel HC, Havers J, Kneser U, Smith MK, Moeller T, Kluth D, Mooney DJ, Rogiers X, Kaufmann PM. Influence of flow conditions and matrix coatings on growth and differentiation of three-dimensionally cultured rat hepatocytes. Tissue Eng. 2004 Jan;10(1-2):165-74.
66. Wen F, Chang S, Toh YC, Arooz T, Zhuo L, Teoh SH, Yu H. Development of dual-compartment perfusion bioreactor for serial coculture of hepatocytes and stellate cells in poly(lactic-co-glycolic acid)-collagen scaffolds. J. Biomed. Mater. Res. B Appl. Biomater. 2008 Oct;87(1):154-62.
67. Landry J, Bernier D, Ouellet C, Goyette R, Marceau N. Spheroidal aggregate culture of rat liver cells: histotypic reorganization, biomatrix deposition, and maintenance of functional activities. J. Cell Biol. 1985 Sep;101(3):914-23. PMCID:PMC2113699.
68. Wang YJ, Li MD, Wang YM, Ding J, Nie QH. Simplified isolation and spheroidal aggregate culture of rat hepatocytes. World J.Gastroenterol. 1998 Feb;4(1):74-6.
69. Nyberg SL, Hardin J, Amiot B, Argikar UA, Remmel RP, Rinaldo P. Rapid, large-scale formation of porcine hepatocyte spheroids in a novel spheroid reservoir bioartificial liver. Liver Transpl. 2005 Aug;11(8):901-10.
70. Park KH, Song SC. Morphology of spheroidal hepatocytes within injectable, biodegradable, and thermosensitive poly(organophosphazene) hydrogel as cell delivery vehicle. J. Biosci. Bioeng. 2006 Mar;101(3):238-42.

INDEX

Actin filaments, 46
Acute inflammation, 102
Acute liver failure, 297
Adherens junctions, 51
Adipocytes, 72
Adult stem cells, 56, 72
Adventitia, 185, 239, 266
Airway stents, 245
Allogeneic cells, 53, 282
Anchorage-dependent, 62
Angiogenesis, 29, 30, 60, 157, 159, 160, 161, 163, 164, 166, 167, 168, 169, 170, 171, 172, 173, 177, 179
Apple, 24
Arteriologenesis, 160, 161
Artificial heart muscle, 4
Artificial hearts, 4
Atala, A, 7
Autocrine signaling, 49
Autologous cells, 53, 54, 251, 256, 274
Autophagy, 46

Bell, E, 5
Bile, 296
Biocompatibility, 6, 85, 87, 89, 90, 100, 101, 102, 116, 119, 271, 303

Bioink, 140, 141, 142
Biological metrics, 17
Biologically inspired, 157, 174, 177, 178, 184, 185
Biologically mediated, 157, 177, 178, 181, 182, 186
Biologically replicated, 157, 177, 178, 179, 180, 186
Biomaterial development, 88
Biomaterial synthesis, 11
Biomaterials, 8, 11, 14, 248, 251, 274, 304
Biomechanical forces, 194
Biomimetic, 23, 89, 104, 105, 106, 112, 113, 116, 119
Biopaper, 140, 141, 142
Bioprinting, 140, 141, 142
Bioreactor classification, 197
Bioreactor design, 200
Bioreactors, 11, 14, 24, 26, 61, 139, 195, 196, 197, 198, 199
Black, J, 86
Blistered, 243
Blood coagulation, 104
Blood islands, 162
Blood vessels, 3, 4, 11, 29, 104, 132
Bone marrow-derived mesenchymal cells, 166

Branchless, 242
Breaking strength, 94
Brittle, 94, 95
Building blocks, 14

Cardiac pacemakers, 222
Cardiovascular tissue engineering, 4
Cartilaginous rings, 239
Cataracts, 91
Cell attachment, 62
Cell based cardiac pumps, 4
Cell culture, 61
Cell encapsulation, 28, 32, 33
Cell isolation, 62
Cell lines, 54
Cell patterning, 131, 132, 133, 140, 145, 146, 147, 148
Cell potency, 69
Cell printing, 133, 140, 141, 185
Cell proliferation, 63
Cell sheet engineering, 132, 136, 266, 273, 274, 309
Cell signaling, 48
Cell sourcing, 9, 23, 52, 139, 248, 253, 256, 274, 303
Cell spreading, 63
Cell therapy, 28, 30, 55
Cell transplantation, 28, 33, 34, 41, 42, 43, 55
Cell viability, 63
Cellerity™, 207
Cell-matrix interactions, 47
Cells, 14, 54, 55, 58, 59, 61, 62
Cellular junctions, 50
Central dogma, 66
Centrifugal force, 214
Ceramics, 107
Chemical degradation, 99
Chondrocytes, 47, 54, 72, 248
Chondroitin sulfate, 48
Chromatin, 43
Chromosomes, 43
Chronic inflammation, 102
Ciliary escalator, 238
Clinical applications, 20, 248, 253
Cloaca, 270
Coagulopathy, 297
Collagen, 48, 63, 104, 107, 110, 139, 226, 248
Common nephric duct, 270
Complement activation, 103
Confluency, 63
Congenital tracheal stenosis, 240
Connexin, 51
Contact inhibition, 63
Contractile force, 17, 68, 115

Controlled release, 31, 251, 279, 311
Core technologies, 22
Cornea, 91
Cristae, 45
Crosshead extension indicator, 93
Crystalline lens, 91
Cytoskeleton, 46

Decellularized matrices, 108, 248
Degradable biomaterials, 89, 106, 248
Design principles, 12
Desmin, 47
Desmosome junctions, 51
Differentiation, 30, 50, 69, 248, 254, 305
Direct cell-to-cell signaling, 49
Disciplines in TE, 26
Donor organs, 3, 4
Drive mechanism, 93
Ductile, 95
Dunnill, P, 7

Ectoderm, 162
Eight-step process, 9
Elastin, 48, 115, 165, 239
Electrical stimulation bioreactors, 221
Electrospinning, 198, 206, 209, 210
Embryogenesis, 54, 67, 69, 160, 270, 305
Encephalopathy, 297
Endocrine signaling, 49
Endoderm, 162
Endoplasmic reticulum, 46
Endoscopic balloon dilation, 245
Endothelial cells, 29, 54, 68, 103, 164
Endothelial progenitor cells, 166
Endotracheal intubation, 244
Engineering, 23
Enzymatic degradation, 99
Eukaryotic cells, 43
Extensometer, 93
Extracellular matrix, 47

Fibronectin, 48, 89, 105, 180, 302
Fibrosis, 101, 103
Folkman, J, 159
Foreign body reaction, 101
Functional assessment, 16
Functional electrical stimulation, 224
Functional performance, 17

Gap junctions, 50, 51
Gastric electrical stimulation, 223
Gauge length, 94
Gene therapy, 28, 180
Genetic manipulation, 10, 11, 15
Glysoaminoglycans, 48

INDEX

Goblet cells, 238
Golgi apparatus, 46
Granulation tissue, 102
Grips, 93
Growth of TE, 25

Hemangioblasts, 162
Hematopoietic stem cells, 9, 58, 60, 72, 162, 169, 256
Hemidemosome junctions, 51
Heparin, 48
Hepatectomy, 301
Hepatic artery, 297
Hepatic portal vein, 297
Hepatic stem cells, 306
Hepatocytes, 19, 25, 111, 296
High throughput, 21
Histological metrics, 17
Hooke's Law, 94
Human embryonic stem cells, 70, 169, 305
Hyaluronan, 48
Hydrogels, 16, 89, 108, 110, 139, 142, 212, 248, 272, 309
Hypoxia, 160

Idealized biomaterial, 116
Idealized bioreactors, 202
In vivo assessment, 10, 11, 15, 16, 272
In vivo vascularization, 172
Induced pluripotent stem cells, 71
Injury, 30, 55, 101, 167, 248, 267, 298
Inner membrane, 45
Inner nuclear membrane, 43
Input, 13
Integrins, 11, 47, 48, 87, 88, 104, 105, 115, 219, 307
Intercellular adhesion molecule-1, 165
Interdigital transducer, 214
Interior thyroid artery, 239
Interior vesical artery, 267
Intermediate filaments, 46
Intermembrane space, 45
Intraluminal pressure, 17
Intraocular lenses, 91
Intussusceptive angiogenesis, 161
Ischemia, 166

Junctional adhesion proteins, 50

Kidney transplantation, 3

Lamina propria, 239
Laminin, 48, 63, 107, 115, 134, 179, 302, 307
Langer, R, 18

Left ventricular assist devices, 21
Life sciences, 23
Linear Region, 94, 96
Liver development, 302
Liver regeneration, 301
Liver transplantation, 298, 299, 300
Living donor liver transplantation, 299
Load indicator, 93
Lower limb electrical stimulation, 224
Lumenization, 163
Lysosomes, 45

Magnetic nanoparticles, 146, 215
Mason, C, 7
Material deformation, 93
Material degradation, 97
Matrix, 45
Mechanical devices, 20
Mechanical testing system, 93
Medical sciences, 23
Membrane attack complex, 103
Mesenchymal stem cells, 9, 15, 41, 72, 307, 315
Mesoderm, 162
Mesodermal stem cells, 160, 162, 169, 174, 175
Metals, 107
Microcontact printing, 144
Microencapsulated cells, 6
Microfilaments, 46
Microfluidics, 133, 143, 144, 145
Microgravity, 21
Microtubules, 46
Mitochondria, 45
Monocyte chemoattractant protein-1, 165
Mucus, 238
Multipotent, 69
Myocardial infarction, 23, 57, 166
Myosin heavy chain, 17

Natural biomaterials, 106, 107
Nerem, R, 7
Neurogenic bladder dysfunction, 267
Neuromodulation, 268
Neuromuscular electrical stimulation, 223
Nondegradable biomaterials, 106
Nuclear envelope, 43, 44
Nuclear pore complex, 43
Nucleolus, 43
Nucleus, 43

Organ engineering, 28, 32, 254
Organ printing, 131, 132, 140, 141, 146, 150
Organ-on-a-chip, 132, 133, 145
Osteocytes, 72

Outer membrane, 45
Outer nuclear membrane, 43
Output, 13
Oxidative stress, 21

Pantip-2, 243
Paracrine signaling, 49
Performance metrics, 16
Perfusion bioreactors, 215
Perfusion seeding, 214
Phagocytosis, 46
Phagolysosomes, 46
Photoresponsive culture surface, 308
Physical degradation, 99
Plasma membrane, 47
Plastic deformation, 94
Platelet activation, 104
Poly(lactic-co-glycolic acid), 278
Polymeric scaffolds, 31, 89, 107, 111, 132, 138, 139, 248, 272, 304
Polymers, 19, 90, 107, 149, 209
Polymethyl methacrylate, 92
Post-intubation stenosis, 244
Post-tracheostomy stenosis, 244
Prokaryotic cells, 43
Protein therapy, 28, 29, 30

Rapid prototyping, 132, 133, 142
Regenerative medicine, 5, 6, 7, 32, 167
Reparative medicine, 32
Replica molding, 144
Resident stem cells, 58, 69, 72
Ribosomes, 45
Roboflasks, 208
Robotic protein printing, 307

Scaffold cellularization, 10, 11, 15, 16, 99, 139, 180, 185, 197, 205, 212, 257
Scientific Collaboration, 22
Scientific hurdles, 15
Screening, 21
Sefton, MV, 6
Selective layer sintering, 143
Self-organization, 108, 112, 132, 133, 134
Self-organization technology, 130
Self-renewal, 68
Seminal papers, 18
Sensor technology, 10, 11, 15, 200, 201, 203, 218
Sharma, CP, 87
Sir Nicholas Harold Lloyd Ridley, 91
Small intestinal submucosa, 275
Smart materials, 113
Soft lithography, 132, 143, 307
Solid free form fabrication, 142

Space radiation, 21
Spheroid culture, 313
Split liver transplantation, 300
Sprouting angiogenesis, 161
Stalk cells, 164
Stem cell transplantation, 56, 57
Stencils, 308
Stereolithography, 143
Stress-strain curve, 94
Stress-strain plot, 93
Stretch bioreactors, 219
Stretch-activated receptors, 219
Subclavian artery, 239
Submucosa, 239
Subpassaging, 62, 63
Superior vesical artery, 267
Surface acoustic waves, 214
Surgical bladder augmentation, 269
Synthetic biomaterials, 106, 107, 116

Taylor's cone, 209
Technological hurdles, 15
Tensile properties, 13, 90, 92, 93, 94, 95, 116, 138
Tensile strength, 95
Tensile test, 93
Therapeutic angiogenesis, 29, 166
Tight junctions, 50, 163
Tip cells, 164
Tissue biopsy, 62
Tissue engineered ventricles, 4
Tissue engineering, 5
Tissue Fabrication, 9
Totipotent, 43, 69, 70
Toughness, 95
Toxins, 21
Trachea, 238, 249, 250
Tracheal resection and reconstruction, 246
Tracheal stenosis, 243
Trachealess, 242
Tracheostomy, 244
Transcription, 66
Transcutaneous electrical stimulation, 223
Translation, 66
Tri-leaflet heart valves, 4
Tumor angiogenesis factor, 159
Twitch force, 17

Ultimate tensile strength, 94, 95
Unipotent, 69, 70, 72
Unspecialized cells, 68

Vacanti, JP, 6
Vagus nerve, 240

Vagus nerve stimulation, 223
Vascular cell adhesion molecule-1–VCAM-1, 165
Vascular endothelial growth factor, 161
Vascularization, 10, 11, 15, 30, 59, 158
Vasculogenesis, 161

Western blotting, 17
Wolffian duct, 270

Yield strength, 94, 95
Young's modulus, 94

IEEE Press Series in Biomedical Engineering

The focus of our series is to introduce current and emerging technologies to biomedical and electrical engineering practitioners, researchers, and students. This series seeks to foster interdisciplinary biomedical engineering education to satisfy the needs of the industrial and academic areas. This requires an innovative approach that overcomes the difficulties associated with the traditional textbooks and edited collections.

Series Editor: Metin Akay, University of Houston, Houston, Texas

1. *Time Frequency and Wavelets in Biomedical Signal Processing*
Metin Akay

2. *Neural Networks and Artificial Intelligence for Biomedical Engineering*
Donna L. Hudson, Maurice E. Cohen

3. *Physiological Control Systems: Analysis, Simulation, and Estimation*
Michael C. K. Khoo

4. *Principles of Magnetic Resonance Imaging: A Signal Processing Perspective*
Zhi-Pei Liang, Paul C. Lauterbur

5. *Nonlinear Biomedical Signal Processing, Volume 1, Fuzzy Logic, Neural Networks, and New Algorithms*
Metin Akay

6. *Fuzzy Control and Modeling: Analytical Foundations and Applications*
Hao Ying

7. *Nonlinear Biomedical Signal Processing, Volume 2, Dynamic Analysis and Modeling*
Metin Akay

8. *Biomedical Signal Analysis: A Case-Study Approach*
Rangaraj M. Rangayyan

9. *System Theory and Practical Applications of Biomedical Signals*
Gail D. Baura

10. *Introduction to Biomedical Imaging*
Andrew G. Webb

11. *Medical Image Analysis*
Atam P. Dhawan

12. *Identification of Nonlinear Physiological Systems*
David T. Westwick, Robert E. Kearney

13. *Electromyography: Physiology, Engineering, and Non-Invasive Applications*
Roberto Merletti, Philip Parker

14. *Nonlinear Dynamic Modeling of Physiological Systems*
Vasilis Z. Marmarelis

15. *Genomics and Proteomics Engineering in Medicine and Biology*
Metin Akay

16. *Handbook of Neural Engineering*
Edited by Metin Akay

17. *Medical Image Analysis, Second Edition*
Atam P. Dhawan

18. *Advanced Methods of Biomedical Signal Processing*
Edited by Sergio Cerutti, Carlo Marchesi

19. *Epistemology of the Cell: A Systems Perspective on Biological Knowledge*
Edward R. Dougherty, Michael L. Bittner

20. *Micro and Nanotechnologies for Engineering Stem Cells and Tissues*
Murugan Ramalingam, Esmaiel Jabbari, Seeram Ramakrishna,
Ali Khademhosseini

21. *Introduction to Neural Engineering for Motor Rehabilitation*
Dario Farina, Winnie Jensen, Metin Akay

22. *Introduction to Tissue Engineering: Applications and Challenges*
Ravi Birla